Generation: Distribution,
cations and Implications

Power Generation: Distribution, Applications and Implications

Edited by **Lauren Marini**

$\mathcal{CL}$LANRYE
INTERNATIONAL

New Jersey

Published by Clanrye International,
55 Van Reypen Street,
Jersey City, NJ 07306, USA
www.clanryeinternational.com

**Power Generation: Distribution, Applications
and Implications**
Edited by Lauren Marini

International Standard Book Number: 978-1-63240-413-8 (Hardback)

Printed in the United States of America.

Contents

Preface

The main aim of this book is to educate learners and enhance their research focus by presenting diverse topics covering this vast field. This is an advanced book which compiles significant studies by distinguished experts in the area of analysis. This book addresses successive solutions to the challenges arising in the area of application, along with it; the book provides scope for future developments.

The distribution, applications as well as implications of power generation are described in this book. The consumption of renewable energy sources such as wind energy or solar energy is currently of great interest. Nevertheless, since their availability is sporadic and unstable, this can lead to frequency variation, grid instability and total or partial loss of load power supply. This reason keeps them far from being appropriate sources to be directly connected to the main utility grid. Also, the availability of a static converter as output unit of the generating plants brings voltage and current harmonics into the electrical system which causes a negative result on system power quality. By integrating distributed power generation systems closed to the loads in the electric grid, we can get rid of the need to transfer energy over long distances through electric grid. This book discusses different power generation and distribution systems with an evaluation of some types of existing disturbances and a study of distinct industrial applications such as battery charges.

It was a great honour to edit this book, though there were challenges, as it involved a lot of communication and networking between me and the editorial team. However, the end result was this all-inclusive book covering diverse themes in the field.

Finally, it is important to acknowledge the efforts of the contributors for their excellent chapters, through which a wide variety of issues have been addressed. I would also like to thank my colleagues for their valuable feedback during the making of this book.

Editor

Part 1

Power Generation and Distribution Systems

Integration of Hybrid Distributed Generation Units in Power Grid

Marco Mauri[1], Luisa Frosio[1] and Gabriele Marchegiani[2]
[1]Politecnico di Milano,
[2]MCMEnergyLab s.r.l
Italy

1. Introduction

The use of renewable energy sources either as distributed generators in public AC networks or as isolated generating units supplying is one of the new trends in power-electronic technology. Renewable energy generators equipped with electronic converters can be attractive for several reasons, such as environmental benefits, economic convenience, social development.

The main environmental benefit obtained by using renewable sources instead of traditional sources, is the reduction in carbon emission. Many countries have adopted policies to promote renewable sources in order to respect the limits on carbon emission imposed by international agreements.

Moreover, renewable energies can be economically convenient in comparison with traditional sources, if the economic incentives for grid connected renewable sources are taken into account or in other particular situations to supply stand alone loads. In some cases, it can be more convenient to supply an isolated load with renewable local source instead of extending the public grid to the load or to supply it with diesel electric generators. In this case, in order to evaluate the economic benefits of renewable energy solution, It is necessary to take in account either the cost of the fuel and the cost of its transport to the load, that can be located in remote and hardly reachable areas.

In addition to the economic benefits, the use of distributed renewable generation units contributes to decentralise the electrical energy production, with a positive impact on the development of remote areas. The exploitation of local renewable sources supports local economies and lightens the energy supply dependency from fuels availability and prices fluctuations.

The integration in the electric grid of distributed power generation systems, located close to the loads, reduces the need to transfer energy over long distances through the electric grid. In this way several benefits are achieved, such as the reduction of bottle-neck points created by overcharged lines, the increase of global efficiency and the limitation of thermal stress on grid conductors. Renewable distributed generation units, if properly controlled and designed can improve the power flow management on the grid and reduce the probability of grid faults, so increasing the power quality of the energy supply.

It's important to evaluate also the possible drawbacks of the increasing number of renewable energy sources on the power-supply stability and quality, both in grid connected and stand-alone configurations, in order to prevent possible problems with a proper design and management of this generation units.

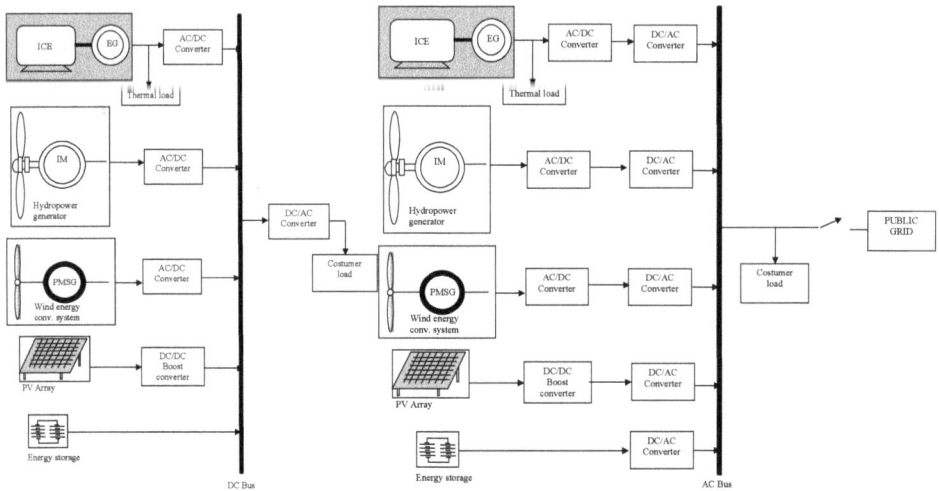

Fig. 1. Example of a HDGU structure with DC and AC bus

The negative influences on power quality of renewable energies derives mainly from two typical characteristics of renewable sources: their randomly varying availability and the presence of a static converter as output interface of the generating plants (with exception for hydroelectric).

The sudden variation of generated power from renewable sources can lead to frequency fluctuation and to grid instability, in grid connected systems, and to total or partial loss of loads power supply in stand-alone systems.

The output interface converter introduces in the electric system voltage and current harmonics that affect negatively the system power quality. On the other hand, if properly designed and controlled, the interface converter can efficiently support the grid in normal operations (reactive power and voltage control) and during grid faults. To reduce this influence, many actions can be taken, typical of the traditional power converters' design technique, such as: low emission topologies, high switching frequency, optimised control techniques and regulators and output passive filters (typically in LC configuration)

To mitigate the effect of renewable sources randomness on power quality different actions are possible: equipping the renewable generation unit with a storage system, integrating to the renewable source a non-renewable source (e.g. a diesel generation unit), integrating in the renewable source a storage system and a non-renewable programmable source.

What is interesting to talk about, as a specific aspect of renewable generation units, is the electric system that lays beyond the interface converter (renewable source, energy storage, power flow control, ecc…) and that allows to transform the random renewable sources into a regulated, flexible and grid-friendly (in grid connected mode) or load-friendly (in stand-alone mode) generation units.

First off all it will be analysed a first solution consists of a renewable source equipped with an energy storage system in stand-alone mode. Secondly a grid connected application of diffused hybrid generation units itwill be analysed.

In the following sections some guidelines of the design of renewable generation units will be analysed, focusing on the aspects that influences the power quality of the electrical system. Considering that the renewable sources are mainly wind and photovoltaic, the hybrid generation unit that will be considered, it is composed of a wind turbine, a photovoltaic generator and a battery bank. The hydroelectric source will not be considered, because of its atypical features (no need of static power converter, power availability often regulated by means of natural or artificial basins) that makes it similar to a traditional regulated thermic generation units.

2. Renewable generation units in stand alone mode

When it is necessary to supply electrical loads in stand-alone configuration there is no particular standard that define the power quality requirements. Anyway, the mains concepts of power quality regulation in grid connected systems can be used also for stand-alone systems and as general guidelines, a power quality level close to the one guaranteed for loads fed by the grid can be choose.

On the following sections, a list of the main power quality concepts is presented, focusing on how this concepts, originally dedicated to grid connected systems, can be adapted to stand-alone configurations, in order to obtain some guidelines to improve the performance of these systems.

For general purpose and to simplify the analysis, a stand-alone system, including some loads and one renewable hybrid generation unit equipped with a battery bank is considered.

These considerations can be generalised and adapted to a system of several renewable generation units parallel connected in order to supply the loads in stand-alone configuration. The main considerations regarding the battery bank can be adapted also to other energy storage systems, such as flyback wheels, hydraulic reservoirs, fuel cells, etc.

3. Hybrid renewable generation units in stand-alone mode

3.1 Requirements on voltage waveform (voltage amplitude in normal operating conditions)

In the considered renewable power units, the interface between power generators and loads is made by one ore more static converters. It is important that one of this static converter is a voltage controlled inverter (interface converter) that generates a voltage waveform corresponding to a certain reference given to its control system. By regulating the voltage reference signal, it's possible to obtain a sinusoidal voltage waveform with desired frequency and amplitude, within the limits of the power converter.

In order to maintain the nominal voltage level on the loads, it is important to include in the voltage regulator of the interface converter a compensation term that takes into account the voltage drops on the line impedance, by measuring the current drawn by the loads as it is shown in Fig.2.

Some consideration can be made about the Fig.2:

- The line connecting the power plants and the loads, in small application is a low voltage line, thus characterised by a mainly resistive impedance. The compensation term in the regulator should take into account the line impedance value and typology (reactive or resistive).

- Depending on the interface converter configuration and on the grid number of phases, the actuator that defines the switching command can be chosen according to different techniques: space vector, PWM, hysteresis, ecc...
- The dynamic of the interface converter regulator defines the ability of the converter to maintain a constant voltage on the loads even in case of sudden changes in the load power demand. To prevent the voltage amplitude to move from its nominal value, even in case of sudden changes in the loads, the interface converter regulator should be as fast as possible. On the other hand an excessive speed of the regulator can produce an instability behaviour of the system, for the presence of high frequency harmonic components in the measured voltage and currents, caused both by the loads and by the converter itself. To prevent this instability, that causes the generation of a highly distorted voltage waveform, the interface converter regulator should be slow enough to be immune from high frequency harmonics in the grid, but fast enough to control the voltage variation on the loads. The regulator indicated in the scheme can be different regulator type (P, PID, hysteresis, fuzzy,etc), but its parameters should be set in order to place the cut off frequency of the regulator at least one decade higher that the voltage frequency.

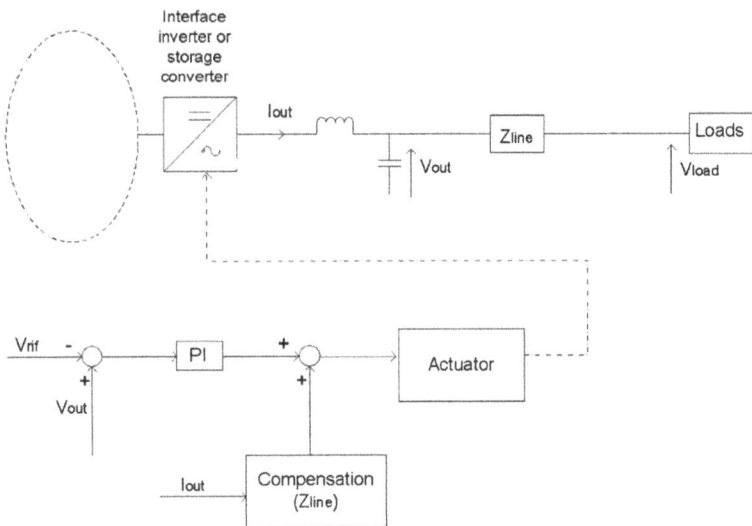

Fig. 2. Stand-alone voltage regulation schema

In case of more than one generation unit connected in parallel in an stand-alone configuration (islanded microgrid) a different control algorithm is needed for the interface converters. Several techniques have been studied to manage the parallel operation of stand-alone inverters and to assure a correct power sharing between the generation units. The more complex control techniques relay on a communication system between generation units. Other techniques can be implemented if no communication system is installed.

The accomplishment, in every node of the islanded microgrid, of the power quality requirements on the voltage value is a problem concerning the coordination between the regulation actions of each interface converter.

3.2 Requirements on voltage frequency

As the system voltage waveform is generated by the interface converter, its frequency can be set by acting on the frequency of the reference voltage (and current) in the voltage regulator (and, if present, in the current regulator).

As for the voltage regulation, in stand-alone microgrids several control techniques can be adopted to maintain the frequency of the supply in the range required for the power quality, with or without communication between paralleled generation units. The only difference from voltage regulation is that the control system of each generation units measure the same electrical frequency in every node of the system. To maintain the nominal frequency of the supply, the control actions of each interface converter must be coordinated through a signal coming from a higher level control system, if a communication system is present, or through a correct setting of the regulation parameters of each converter, if no communication is available.

3.3 Overvoltages and voltage dips

The electrical system of a micro hybrid plants is composed by three elements: the hybrid power plant, the loads, and the low voltage grid converter. Usually the loads are concentrated in a small area, not bigger than a medium rural village, and the hybrid power plant is located as near as possible to the loads, so the low voltage grid has usually a reduced extension and presents a very simple radial structure. If the interface converter and its regulator working correctly, overvoltage on the low voltage grid can only be generated by faults or environmental phenomena (lightening) on the grid. Due to the reduced extension of the grid, these events are very rare and can be reduced by carefully designing and installing the grid components.

The considerations about the overvoltages can be extended also for voltage dips (sags). The main cause of voltage dips on grid connected loads is the fast reclosing action of switches in order to eliminate transient faults. In stand alone systems no fast reclosing procedure is necessary due to the low extension of the grid, so, voltage dips are not a problem to solve in this systems.

3.4 Flickers

Flickers are fast variations of the voltage supplied to the loads. These voltage oscillations are generated by repetitive load connection and disconnection or by their discontinuous current absorption. Usually, the loads that origins flickers are big industrial loads, such as welders and arching furnaces. Stand alone hybrid power systems do not usually fed loads of this kind, anyway, if that may occur, the component who might prevent the flicker is, once again, the interface converter and its regulator.

Depending on its regulator speed and performances, the interface converter can be compensate the fast variations in the current absorbed by the loads and prevent flickers.

3.5 Harmonics

High frequency harmonic components in the electric system can affect the grid current and the grid voltage too. Due to the impedances of the system, current harmonic components can produce voltage harmonics, and vice versa. Anyway, the primary causes of voltage and current harmonics are quite different.

The harmonic components in grid current are produced by the loads equipped with electronic devices that absorb high frequency current components. These harmonic components can be reduced only by acting directly on the loads.

The voltage harmonic components are introduced in the system by the interface converter and are produced mainly by the switching of electronic components. Voltage harmonics can also be presents due to regulator malfunctioning or due to harmonic components at frequency lower than the cut off frequency of the power bus regulator. Generally, the power bus presents a parallel connection of the converters dedicated to the renewable generators, that inject power into the power bus, and of the converter dedicated to the storage system, that exchange power in both directions between the power bus and the storage system. To eliminate the grid voltage harmonics produced by the voltage harmonics on the power bus, it is necessary to guarantee the stability of the voltage level of the power bus. This can be achieved with a proper power flow control algorithm and by adding filters at the output of the other static converters connected to the power bus. Moreover, the dynamic behaviour of the storage system and its converter, can affect the harmonic content of the power bus voltage and, consequently, also of the grid voltage. The storage system converter regulator should maintain the energy balance between loads and sources on the common power bus. Any energy imbalance on the common bus causes fluctuations on the power bus voltage. More fast is the storage converter regulator and the more stable is the voltage on the power bus. A rapid compensation of energy unbalance on the power bus can be achieved imposing to the storage system fast and sharp charging and discharging operations, but these operations can lead to its premature ageing.

If some voltage harmonics still remain on the power bus and are transferred to the grid voltage, a solution to eliminate these can be the reducing of the cut off frequency of the inverter regulator, in order to make it able to compensate this lower harmonic components, too. Anyway, the cut off frequency of the interface inverter regulator should not be less at least one decade of the grid frequency.

3.6 Supply interruptions

In stand alone systems, the main issue regarding electric Power Quality is to guarantee the supply continuity. The main causes of power supply interruption in stand alone systems are due to fault in the renewable power plants. The electric grids that connect the renewable power plant to the loads is usually very short and simple so a fault event on the grid is very rare.

In stand alone systems, the supply interruptions caused by a lack of energy of the renewable power plants can be analysed using the concept of Loss of Power Supply or Loss of Load (LPS or LOL) and Loss of Power Supply Probability or Loss of Load Probability (LPSP or LOLP). The LPS and LPSP concepts are more or less equivalent to the LOL and the LOLP concepts, so only the LPS and LPSP indexes will be considered.

The LPS is the total energy required by the load that is not supplied during an interruption. The LPSP index depends on the duration of the interruptions and on the load power demand during interruptions. If a supply interruption occurs when no power is requested by the loads, it has no effect on the LPSP value.

The LPSP is a global index that defines the hybrid system availability during a particular period of time, usually one year; it takes into account the sum of the LPS and the energy demand during the year. As the renewable sources have a seasonally behaviour, the one year period is a good choice to evaluate the performances of hybrid systems.

The LPSP index expresses the probability to have a supply interruption on the stand alone loads, due to a lack of power of the renewable power plant. In an existing system, the LPSP can be defined as the sum of the energy not supplied to the loads during the supply interruptions occurred in a year, divided by the total energy required by the loads during the year as indicated in (1)

$$\text{LPSP} = \sum_{i=1}^{N} E_{1\_NS,i} / E_{1\_tot} \tag{1}$$

where N is the number of supply interruptions during one year, $E_{l\_NS,i}$ is the total energy required by the toad and not supplied during the interruption I and E_{l_tot} is the total energy required by the loads during the year. The LPSP can also be expressed in function of the power demand of the loads during the supply interruption and the duration of the supply interruptions as indicated in (2)

$$\text{LPSP} = \left(\sum_{i=1}^{N} \int_{\Delta_{SI,i}} P_{1\_NS,i}(t) \cdot dt \right) / E_{1\_tot} \tag{2}$$

where $P_{L\_NS,i}$ is the load power demand during the supply interruption I and $\Delta_{SI,i}$ is the duration of each supply interruption.

The expression (2) of the LPSP is more precise and it allows to take into account also the partial losses of supply. The partial losses of supply are produced by the intentional separation of the loads in case of lack of energy from the renewable power plants, or due to others critical conditions such as a low state of charge of the energy storage system. The separation of a part of the total loads is possible if, during the designing of the stand alone system, loads have been separated at least in two groups: privileged loads (PL) and non privileged loads (NPL). NPL loads are connected to the renewable generation unit through a controlled switch.

The separation of NPL is used to prevent the excessive discharging of the storage system and to extend as far as possible the supply of the PL. During partial LPS, the power not supplied to the loads is equal to the power demand of the non privileged loads. The LPSP can be expressed as indicated in (3).

$$\text{LPSP} = \frac{\overbrace{\sum_{i=1}^{N_T} \int_{\Delta_{SI,i}} P_{1\_NS,i}(t) \cdot dt}^{global\,LPS} + \overbrace{\sum_{j=1}^{N_P} \int_{\Delta PSI,j} P_{NPL\_NS,j}(t) \cdot dt}^{partial\,LPS}}{E_{1\_tot}} \tag{3}$$

This separation between global LPSs, that affects all the loads, and partial LPSs, that affects only the NPL, is useful if an economical evaluation of the LPSP must be carried out, e.g. to perform a tecno-economical optimum sizing of the renewable power plant.

A specific value of LPSP can be one of the requirements to be achieved through a proper design of the stand alone system, as a quality requirement of the supply service to the loads. The design value of LPSP can be calculated from the curve of the load demand during the year, the weather data (wind speed, solar radiation, temperature) in the system location, the size of the storage system and of the renewable generators of the hybrid power plant and the power management control strategies implemented in the renewable power plant.

From the size of the renewable generators and the weather data of the location it is possible to calculate the power available form the renewable energies, $P_{gen}(t)$.. The power demand curve of the two groups of loads ($P_{PL}(t)$ and $P_{NPL}(t)$) during the year can be known form enquiry on the territory or, more often, must be assumed by the designer knowing the type and number of loads that will be connected to the system. To define the working operations of the battery bank, its rated capacity, C_{nbat}, and its minimum admitted State Of Charge, SOC_m, should be defined. The value of SOC_m is usually indicated by the battery manufacturers, as the minimum rate of discharge of each discharging cycle that guarantees the required lifetime of the battery bank. In normal operation the battery maintains the energy balance between the sources and the loads, until its discharging rate doesn't exceed the minimum thresholds. When SOC_m is reached, the loads are disconnected and the renewable energy is used only to charge the storage system.

To evaluate the design LPSP, it is possible to divide the reference period T (one year) in time steps Δt usually equivalent to one hour: If the energy demand from the loads is higher than the sum of the energy available from the sources and the energy still cumulated in the storage system, there is a of loss of supply and the LPS(t) can be calculated as indicate in (4).

$$
\left.
\begin{aligned}
&\text{if} \;\; \overbrace{\left(P_{PL}(t)+P_{NPL}(t)\right)\cdot\Delta t}^{\text{Energy\_demand\_from\_loads}} > \overbrace{P_{gen}(t)\cdot\Delta t}^{\text{Energy\_from\_sources}} + \overbrace{\left(SOC(t)-SOC_m\right)\cdot C_{nbat}/100}^{\text{Energy\_available\_in\_storage}} \\
&\Rightarrow LPS(t)=\left(P_{PL}(t)+P_{NPL}(t)\right)\cdot\Delta t - P_{gen}(t)\cdot\Delta t - \left(SOC(t)-SOC_m\right)\cdot C_{nbat}/100 \\
&\text{if} \;\; \left(P_{PL}(t)+P_{NPL}(t)\right)\cdot\Delta t \le P_{gen}(t)\cdot\Delta t + \left(SOC(t)-SOC_m\right)\cdot C_{nbat}/100 \;\; \Rightarrow \;\; LPS(t)=0
\end{aligned}
\right\} \text{for } 100\% > SOC(t) > SOC_m \quad (4)
$$

When the battery SOC (t) reaches the threshold SOC_m the loads are disconnected, the battery is charged with the energy coming from the renewable sources and the loads are not supplied. The global supply interruption goes on until the battery is recharged at a certain level of charge, SOC_{reload} situated between the fully charge condition and the minimum threshold, SOC_m.

$$
LPS(t) = \left(P_{PL}(t) + P_{NPL}(t)\right)\cdot\Delta t \quad \text{for} \quad SOC_m < SOC(t) < SOC_{reload} \quad (5)
$$

More refined and complex strategies can be adopted to reduce as far as possible the LPSP on privileged loads, e.g. it is possible to disconnect the NPL when the battery SOC(t) exceed an imposed threshold, SOC_{NPL}, higher than SOC_m. In such way , the remaining energy stored in the battery is used to supply only the PL if the renewable energy is not enough and the design equations (4) became the following.

$$
\left.
\begin{aligned}
&\text{if} \;\; \left(P_{PL}(t)+P_{NPL}(t)\right)\cdot\Delta t > P_{gen}(t)\cdot\Delta t + \left(SOC(t)-SOC_{NPL}\right)\cdot C_{nbat}/100 \\
&\Rightarrow LPS(t)=\left(P_{PL}(t)+P_{NPL}(t)\right)\cdot\Delta t - P_{gen}(t)\cdot\Delta t - \left(SOC(t)-SOC_{NPL}\right)\cdot C_{nbat}/100 \\
&\text{if} \;\; \left(P_{PL}(t)+P_{NPL}(t)\right)\cdot\Delta t \le P_{gen}(t)\cdot\Delta t + \left(SOC(t)-SOC_{NPL}\right)\cdot C_{nbat}/100 \Rightarrow LPS(t)=0
\end{aligned}
\right\} \text{for } 100\% > SOC(t) > SOC_{NPL}
$$

$$
\left.
\begin{aligned}
&\text{if} \;\; P_{PL}(t)\cdot\Delta t > P_{gen}(t)\cdot\Delta t + \left(SOC(t)-SOC_m\right)\cdot C_{nbat}/100 \\
&\Rightarrow LPS(t)=P_{NPL}(t)\cdot\Delta t + P_{PL}(t)\cdot\Delta t - P_{gen}(t)\cdot\Delta t - \left(SOC(t)-SOC_m\right)\cdot C_{nbat}/100 \\
&\text{if} \;\; P_{PL}(t)\cdot\Delta t \le P_{gen}(t)\cdot\Delta t + \left(SOC(t)-SOC_m\right)\cdot C_{nbat}/100 \Rightarrow LPS(t)=P_{NPL}(t)
\end{aligned}
\right\} \text{for } SOC_{NPL} > SOC(t) > SOC_m
$$

$$
LPS(t)=\left(P_{PL}(t)+P_{NPL}(t)\right)\cdot\Delta t, \quad \text{for} \quad SOC_m < SOC(t) < SOC_{reload} \quad (6)
$$

Analysing the (6), it can be seen that the LPSP, in stand alone systems supplied by a generically hybrid renewable power plants with a storage systems, depends mainly on two design aspects: sizing of the renewable generators and of the storage system and to the control strategies adopted for the energy management.

A compromise between the power quality of the electrical supply and the costs of the system should be reached in the designing of renewable power plants in particular the use of a fuel generation unit to supply the load in case of a lack of energy from the renewable energy sources and the storage system can decrease the LPSP of the system up to zero.

4. Design aspects of HGDU in stand alone mode

As highlighted previously, the Power Quality in stand alone system supplied by a renewable hybrid generation unit depends mainly on the design of the generation units. Three main aspects of the design of renewable hybrid power plants for stand alone operation will be analysed, trying to point which are the possible choices that leads to a higher power quality of supply.

4.1 Sizing

The sizing of renewable generation units is the aspect that mostly affects the continuity of supply in stand alone systems. As it has been highlighted previously, the LPSP of a stand alone system depends on the energy available from the renewable generators and on the energy stored in the battery bank. This two variables are closely related to the sizing both of the renewable generators and of the storage systems, and depends on the control strategy used to manage the power flux in the renewable power plant also.

The optimal sizing of renewable hybrid power plants is a quite complex problem, because it concerns the optimisations of several variables.

In this section, some considerations about the optimum sizing methods are presented, in order to underline the main constraints affecting the sizing of the hybrid power plant and to highlight the drawbacks of a sizing choice in the continuity of the loads supply.

To design a hybrid renewable power plant equipped with storage system, the typology and the size of the renewable generators and of the storage system should be fixed.

4.1.1 Renewable generators and storage system

The choice of the renewable generators type (photovoltaic, wind turbine, hydroelectric, geothermal,...) depends on the renewable sources availability in the location where the system will be installed. The most widespread and easily exploitable renewable source is solar radiation, followed by the wind. Many studies shows that wind and solar radiation distributions have often a complementary behaviour, thus making convenient in many cases to combine wind and solar renewable generators in the hybrid power plants. In the following, this two main renewable sources will be considered, but the analysis can be easily extended also to other renewable generators.

The choice of the storage system should take into account, first of all, that storage systems can be classified into two groups: energy storage systems (batteries, hybrid compressed air systems,...) and power storage systems (super capacitors, flywheels,...). In stand alone systems the first need is to store exceeding energy produced by the renewable sources in order to deliver it to the loads when necessary. For this reason, the more suitable storage system for stand alone renewable energy power plants are batteries. In some cases high power storage systems (usually super capacitors) can be added in parallel to batteries, in order to reduce the electrical stress when high power peak are requested by the loads.

4.1.2 Sizing variables of the renewable generators and of the storage system

In the case of a hybrid wind/PV power plant, when the renewable generators are working in their maximum power points, the expressions of the power available from the two renewable source are:
power points, the expressions of the power available from the two renewable source are:

$$P_W(t) = \eta_W \cdot \frac{1}{2} \cdot \rho_a \cdot c_p \cdot v_w^3(t) \cdot A_w \qquad \text{for the wind turbine}$$

$$P_{PV}(t) = \eta_{PV} \cdot J_t(t) \cdot A_{PV} \qquad \text{for the photovoltaic generator}$$

where:
η_W and η_{PV} are the total efficiencies of the renewable generators,
ρ_a is the air density, $[kg/m^3]$
c_p is the power coefficient of the wind turbine, depending on its shape and typology,
$v_w(t)$ is the instantaneous wind speed at hub height, $[m/s]$
$J_t(t)$ is the instantaneous solar radiation on the tilted surface of the solar panels, $[W/m^2]$
A_w is the area swept by the wind turbine and A_{PV} is the total area covered by the PV panels, $[m^2]$
A_w is the area swept by the wind turbine and A_{PV} is the total area covered by the PV panels, $[m^2]$
The two design parameters on which depends the power available from the renewable generators are the two areas A_w and A_{PV}.
Some more detailed calculations can be done for the wind generator, considering that the power extracted from the wind turbine is typical limited by three speed thresholds, defining the control strategy of the wind generator controller. When the wind speed is lower than the cut-in threshold (v_{cut-in}) the wind turbine is blocked and no power is extract from the wind generator. When the wind speed is between the cut in threshold and the regulation threshold (v_{reg}) the wind generator is kept in the working point corresponding to the maximum power. When the wind speed is between the regulation threshold and the cut-off threshold ($v_{cut-off}$), the wind generator is regulated to supply a constant power, equal to its maximum rated power.
Taking into account the wind speed thresholds, the power available from the wind generator is:

$$P_W(t) = \begin{cases} 0 & \text{for} \quad v_w(t) < v_{cut-in} \\ \eta_W \cdot \frac{1}{2} \cdot \rho_a \cdot c_p \cdot v_w^3(t) \cdot A_w & \text{for} \quad v_{cut-in} < v_w(t) < v_{reg} \\ P_{n,max} & \text{for} \quad v_{reg} < v_w(t) < v_{cut-off} \\ 0 & \text{for} \quad v_w(t) > v_{cut-off} \end{cases} \qquad (7)$$

The maximum energy stored in the battery bank is the product of the battery nominal voltage (V_{nbat}) and rated capacity (C_{nbat}):

$$E_{nbat} = C_{nbat} \cdot V_{nbat} \quad [\text{Wh}] \tag{8}$$

To guarantee a certain expected lifetime of the battery, expressed in number of charging and discharging cycles, the batteries suppliers indicated a minimum state of charging that it not possible to exceed in the battery lifetime. The minimum state of charge admitted (SOC_m) depends on the desired expected lifetime of the battery, expressed in number of cycles. The battery lifetime is affected by many operating conditions, such as the ambient temperature, the amplitude of high frequency ripple in the battery voltage and current, the possibility to charge the battery following the optimum V-I curves, etc.

The available capacity of the battery bank C_{avbat} is lower than its nominal capacity and depends on the value of SOC_m:

$$C_{avbat} = C_{nbat} \cdot (100 - SOC_m) / 100 \tag{9}$$

In the same way, the available energy in the storage system, E_{avbat}, is defined as :

$$E_{avbat} = C_{avbat} \cdot V_{nbat} = E_{nbat} \cdot (100 - SOC_m) / 100 \tag{10}$$

The battery bank voltage depends on the battery converter configuration and on the voltage levels in the renewable power plant.

The number of desired charging/discharging cycles in the battery lifetime depends on economical evaluations that take into account the cost of battery replacements and the expected lifetime of the entire stand alone system. To guarantee the desired number of charging/discharging cycles at specific operating conditions of the battery it is necessary to define a correct value of SOC_m. The main design variable that should be defined by the designer of the renewable power plant is the desired available capacity of the battery, which defines the total energy that can be managed (stored or supplied) by the storage system without exceed the minimum state of charge SOC_m.

4.1.3 Sizing constrains

The two main constraints affecting the sizing of the renewable generators and of the storage system are the overall cost of the system and the continuity of supply to the loads.

The overall cost of the system is composed by two contributions: the initial capital investment and the maintenance cost. The initial cost is strictly related to the size of the renewable generators and of the battery bank. The main contribution to the maintenance cost is due to the battery replacements during the lifetime of the system, that is usually much longer than the battery lifetime. Globally, the battery cost has a great impact on the total cost of the system, because it affects both the initial investment and the maintenance cost.

The LPSP is a constraint in the design of renewable power plants that have a big influence on the overall cost. As it can be seen from the LPSP expression (), the availability of loads energy supply increases if both the renewable generators and the battery bank sizes are increased. In some cases, depending on the load power demand and the sources power availability, the LPSP can be increased only using bigger renewable generators. To reach higher values of LSPS especially in case of a low overlap between the source availability and the load profile, it is necessary to increase the battery capacity also in order to improve the supply continuity.

However, the renewable generators size increasing can allow to reach the desired level of LPSP with a less economical effort. A drawback of this design choice is the possibility to have some wasted energy in the system life cycle. The wasted energy is the amount of energy that could be converted from the renewable generators but can't be used to supply the loads nor to charge the batteries, because they are already completely charged. A waste of energy appears when the load demand and the energy availability form the sources are not overlapped and the storage system capacity is too small to treat the whole energy available from the sources and not directly supplied to the load. The wasted energy doesn't implies a waste of money, thus it can be a designer choice to oversize the renewable generators (or to undersize the battery bank) in order to achieve an acceptable LPSP with lower costs than if the same LPSP is achieved avoiding any waste of energy in system.

The size of the renewable generators and of the storage system is the result of a compromise between the total cost of the system and the continuity of supply to the loads, often reached through an iterative procedure.

Usually, no fixed value is set for the LPSP during the designing of the system. The LPSP target value is usually fixed by the designer taking into account the loads typology and profile and the renewable sources availability. In some cases the supply of some loads should be guaranteed in case of complete lack of the renewable sources for a certain period of time. In the telecommunication application, for example, the supply of each telecommunication station is usually guaranteed for three days with no contribution of the renewable sources. When this performances are required, the size of the minimum storage system is directly dependent on the load power and on the time period where the supply must be guaranteed. If it necessary to guarantee some days of supply only by batteries it is possible to have a very expensive storage systems, but also have a very high quality of supply. Considering that the stand-alone loads present a condition of very high quality supply, comparable to the one guaranteed form UPS in grid connected systems, the costs of renewable systems able to guarantee for a certain period the power supply with no renewable sources, are completely justified.

4.2 Plant configurations

In all hybrid power plants configurations of there is a **power bus**, that is the section of the circuit where all the sources, the interface inverter and the energy storage systems are connected in parallel. Maintaining the energy balance on the power bus of the hybrid power plants is the main condition to maintain the stability of the system.

There are many different power plant configurations, that can differ by the nature of the renewable sources, by the type of the hybrid storage system and by the kind of converters used. These hybrid power plants configuration are usually classified in function of the nature of the power bus (AC bus at frequency line, DC bus or high frequency AC bus).

The main choice to be made regarding the hybrid plant topology concerns, first of all, the type common power sharing bus (DC or AC bus) and, secondly, the typology of the static converters dedicated to the sources and to the battery bank

4.2.1 Common power bus

The two typical AC bus and DC bus configurations of a wind/PV hybrid power plant equipped with a battery bank are reported in figure 3 and 4.

To select the bus configuration of the hybrid power plant several aspects should be taken into account:

Type of loads and sources. If the loads operate at DC (computer servers, DC lamps, DC motors,...), the sources are only PV arrays and the storage system is made by electrochemical batteries, which produce power at DC, then the obvious choice is to use a DC bus, in order to avoid an interface DC/AC converter to feed the loads and to reduce the complexity of the sources and the battery converters.

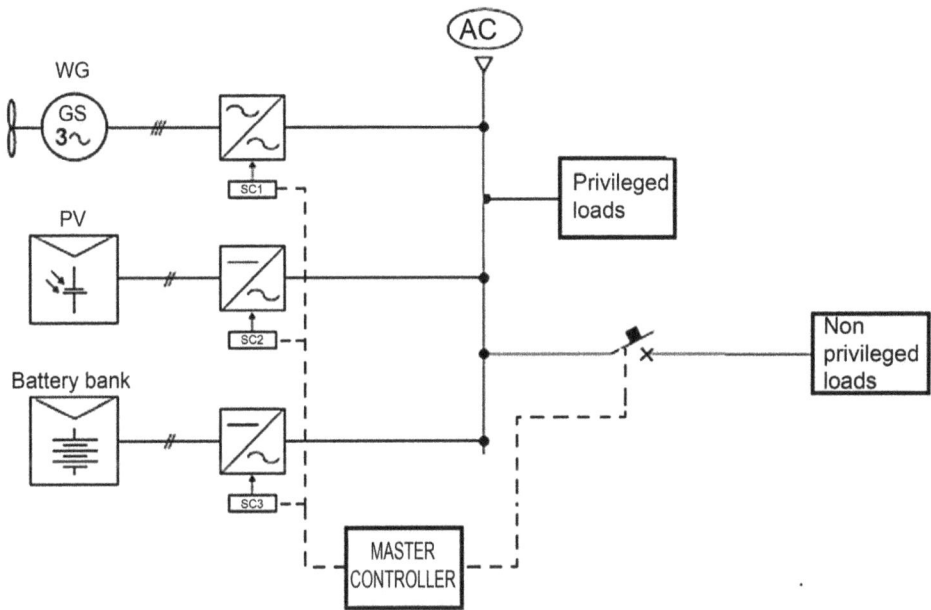

Fig. 3. AC bus configuration of hybrid wind/PV generation unit with battery bank, in islanded mode

If the loads are industrial machines operating at line frequency, the sources are AC generators (doubly fed or synchronous generators) and the storage system is made by electrochemical batteries, then an AC line may be a better system since the loads can be directly connected, thus eliminating the DC/AC interface converter.

No sources and very few loads require high frequency (mainly fluorescent lighting and compressors). For this reason the high frequency AC bus won't be analysed in the following, even if it leads to reduce the size of the passive components, in comparison with grid frequency AC bus.

Control system. Converters on a DC bus system can be simpler than those on a line-frequency AC bus system, and the control for the DC bus is simplified since frequency control is not necessary. In DC bus configurations only active power flow has to be regulated and it is controlled simply acting on the DC voltage level of the common DC bus. Moreover, the flow direction is closely related to current direction. Hence, active power control can be based only on current flow. In AC bus systems active and reactive power flow should be controlled, monitoring both the amplitude and the frequency of the AC bus voltage. Secondly, the phase shifting between voltage and currents should be monitored in order to control the power flow direction.

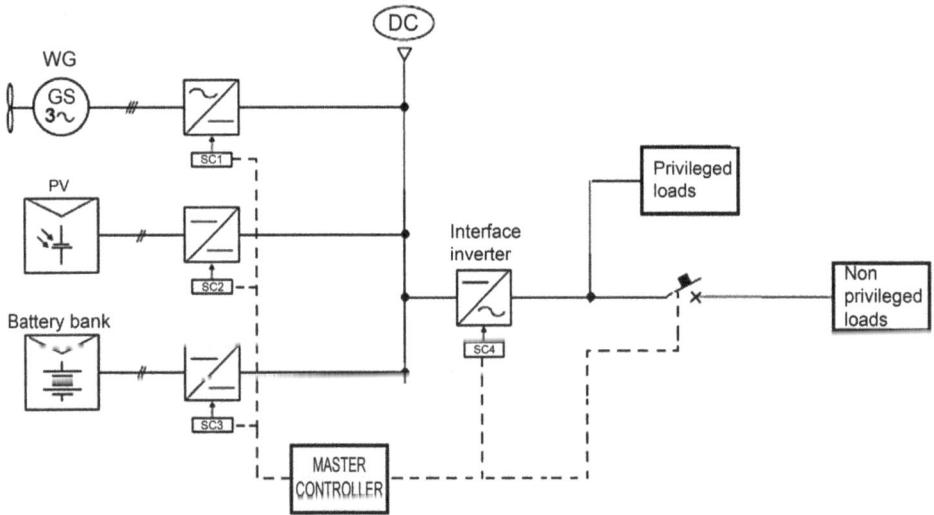

Fig. 4. DC bus configuration of hybrid wind/PV generation unit with battery bank, in islanded mode

Modularity. In low voltage AC bus hybrid systems, AC bus voltage amplitude and frequency are the same of standard low voltage AC public grid, in order to directly connect the AC loads to the AC bus. Any existing power source unit (PV module + DC/AC converters or wind turbine + doubly fed asynchronous generator, etc.) designed to be connected to the AC grid can be added in parallel on the AC bus of a hybrid system. In this way, the generating units can be acquired from different suppliers and additional generating units can be added if the load demand grows without big complications, paying only attention to not exceed the maximum power capability of wirings, and of the common structures of the hybrid systems.

In DC bus hybrid systems, the DC bus voltage level has not a standard value, as for AC bus systems, because it depends on the voltage level of the sources and of the configuration chosen for the source converters.

For example, In order to avoid the output transformer to adapt the voltage level of the DC/AC interface converter to the AC voltage (220 V) needed to feed standard loads, the DC bus voltage should be fixed around 400 V. In this case it may be necessary to add a DC/DC conversion stage between the renewable generation units (and the battery bank) and the DC bus, to adapt the output voltage level of the sources (and of the battery bank) to the required DC bus voltage level.

Another design choice could be to insert a transformer between the DC/AC interface converter and the loads, and to adapt the DC bus voltage level to the more convenient value for the renewable generation units.

Because the DC bus voltage does not present a standard voltage value only dedicated generation units can be connected in parallel to the DC bus. This condition limits the possibility to change suppliers of the system components and to connect additional generation units if required.

Efficiency. The global efficiency of the system depends on many aspects: renewable sources peak power related to the loads rated power, number and typology of converters, and power flow management. It's difficult to state which one of the two configurations is the more efficient in a general sense. Without taking into account a specific project, only some preliminary consideration can be made.

One of the main causes of power losses is the storage battery bank, due to the self-discharge rate and its charging and discharging efficiencies. So, the overall efficiency of an hybrid system is affected by the rate of active energy that has to be stored in the battery bank before being delivered to the loads. Moreover, the power flow management algorithm and the renewable source peak power related to the loads rated power have a great influence on the overall efficiency of an hybrid generation system.

In AC bus configurations transformers can be used to adapt generating units and storage system voltage level to the standard AC low voltage required to the load. In DC systems, additional conversion stages could be required to raise the output voltage of generation units and the storage system, or the system can be managed at low DC voltage, with no additional conversion stage but with an output transformer between the DC/AC interface converters and the loads. Both transformers efficiency and conversion stages efficiency depends on the voltage level of the system (higher voltages mean lower currents and consequently lower losses on conductors and lower switching losses) and they change according to the working condition of the system: usually transformers have the higher efficiency at rated power, while converters have a higher efficiency at low power levels.

Even if AC bus configuration has one converter less this does not implies a higher efficiency. It is necessary to analyse the structure of the converters to understand how many conversion stages are necessary to reach the loads voltage levels. In addition, it is necessary to consider that in the DC bus configuration, the DC/DC battery converter can be eliminated, if the voltage level of the battery bank is adequate to the DC bus voltage. This solution can increment the system efficiency, but can lead to a faster ageing of the batteries.

Reliability. In AC bus systems the critical component is the battery converter. Usually the battery converter is a VSI and it provides the voltage reference on the AC bus. The battery converter act as the grid in grid connected systems: it set the frequency and the amplitude of the AC bus voltage and it maintains the power balance between loads and sources. The power source converters are current controlled inverter, in order to inject on the AC bus the current corresponding to the maximum power available from the sources or to a reduced power level indicated by the power flow control system. If there is a fault on battery converter, the system collapses because there is no longer a voltage reference on the AC bus. Even if there is energy available from the sources, without the battery converter the load are not supplied. If there is a fault on one of the source converters the system can continue to supply the load with the energy coming from the other sources.

In DC bus systems there are two critical components: the interface converter and the battery converter. Clearly, if the interface converters breaks the loads can no longer be supplied. The battery converter maintains the power balance between the loads and the sources by regulating the DC bus voltage level around its rated value. If the battery converter shutdown the system can still work but only if the power available from the sources is higher than the power required form the loads and it is necessary to adopt a different power flow control strategy.

Overall costs. The overall cost of the system (Lyfe Cycle Cost) can be divided in two parts: initial investment costs and life operation costs. The initial investment depends mainly on the renewable generators peak power and on the battery bank nominal capacity. Also the converters number and design can affect the initial investment costs. Solution with less converters, such as the AC configuration or the DC configuration without battery converter, have smaller initial costs. The use of simpler converters topology with a small number of controlled switching components (usually IGBT) can reduce the initial cost of the system too.

The life operation cost is mainly affected by the number of replacement of the battery bank during the system life. The average life of PV panels is supposed to be at least of 20 years, the same as for the control and conversion equipment. In 20 years of operation the wind generator could require some maintenance operations, especially on the bearings and, if present, on the breaking system (air compressor, pincers, etc). Anyway, the expected life of an hybrid power plants can be considered of 20 years.

The battery expected lifetime is usually indicated by the manufacturers in number of cycle at a certain Depth of Discharge (DOD) and in standard operating conditions. The expected lifetime of batteries increases if the DOD of the cycle is reduced, thus it can often be convenient to oversize the battery bank nominal capacity in order to discharge it always at low DOD and to increase its expected lifetime. It must also be taken into account that if the batteries are working at ambient temperature higher than 20°C their expected life time decreases and approximately it goes halves every 10°C of temperature increment. Also the quality of the electrical voltage and current supplied to the batteries can affect its expected lifetime: a high frequency ripple in the battery current, coming from the battery converter or the others converters in the system can accelerate the battery ageing. If the batteries are charged following the optimal cycles indicated by the manufacturer (V-I curve) and are not periodically treated with a full charge/discharge cycle, their expected lifetime can be shorter than the one indicated by the manufacturer at rated condition.

As the battery lifetime depends on many conditions that can't be known in advanced, it can be seen that it's quite complicated to evaluate the exact number of batteries replacement that must be taken in place during the whole life of the hybrid system. A rough calculation can be done assuming that batteries will undergo a charge/discharge cycle each day with a maximum DOD of 30%. In this case one can expect from gel lead-acid batteries (the most performing technology for sealed lead acid batteries in stationary applications) a lifetime of 4000 cycles in standard conditions, that can be shortened to 2000 cycles considering the high operating temperatures, the high frequency ripple on the battery current and the non optimal charging and discharging operations. Considering one charging/discharging cycle each day, during the entire life of the hybrid system the batteries will need to be replaced at least 3 ÷ 4 times.

As the batteries have a very high cost and are difficult to transport, a very efficient way to reduce the cycle cost of the hybrid system is to adopt solutions that protect the batteries form excessive stress (thermal stress, cycling stress, electrical stress) in order to prevent them from premature ageing. It could be found that more complex battery converters, that can have a higher initial investment cost, are able to reduce the ripple on battery current, so reducing the number of battery replacements required and, consequently, reducing the life cycle costs of the system.

The number of charging/discharging cycles on the batteries can be reduced not only through a proper sizing of the renewable generators and of the storage system, but also by adopting appropriate power flow control algorithms. Regarding these problems there is not a big difference between the DC or the AC bus configuration. More differences can be found in considering the limitation of the high frequency ripple that can be achieved in the two different configurations, also taking into account the possibility to eliminate the battery converter on DC bus configuration.

5. Hybrid renewable generation units in grid connected mode

The aspects concerning the introduction of small renewable generation units dispersed on the territory into the low voltage (or medium voltage grid) is analysed in the literature under the general subject of "Dispersed Generation". The main problem concerning the dispersed generation is how to integrate a growing number of dispersed renewable generation units with the grid, without causing problems to the grid stability and to the grid regulation and protection system. The main solution to this problem is to adopt the Smart Grid configuration. Several definition exist in literature of the Smart Grid concept, depending on the aim of the author and on its point of view on the system. In general one Smart Grid can be considered as small low voltage (rarely medium voltage) grids including some dispersed generation units, some loads and at least a storage system, equipped with a communication system that are able to be connected to the low voltage (or medium voltage) grid or, possibly, to work in stand alone configuration. Whether they are integrated in a Smart Grid or connected to the low voltage grid, dispersed renewable generation units can support the Power Quality of the grid supply. To achieve this purpose, the main condition is to equip the renewable generation units with storage systems. Secondarily, to improve further the benefits of dispersed generation on power quality of supply, it can be needed to provide the renewable generation units with a communication system.

In the following sections, for generality purpose, hybrid generation units equipped with a battery storage system will be analysed. In particular the effects of this dispersed renewable generation units on the different power quality requirements will be examined and some design solutions allowing to improve the performances of this systems will be presented.

In the following the main concepts of Power Quality regulation in grid connected systems are analyzed, focusing on how small hybrid generation units, connected on the LV (or MV) grid, can affect them.

Most of the solutions reported below are currently not feasible for LV and MV public networks, that aren't designed to accept active loads, such as renewable generation units. Anyway, as many studies and test sites are being put in place and it seems to be a common will to modify the LV and MV networks structure and management in order to make them able to accept an increasing number of diffused generation, the following analysis will be carried on considering that the MV and the LV network are able to accept and to actively interact with an indefinite number of small hybrid generation units.

Currently, diffused generation units connected to MV and LV networks are required to maintain a "passive behavior", in the sense that they must inject in the grid only the active power available from the sources and separate themselves from the grid whenever a deviation from the nominal values of grid voltage or frequency occurs, that is whenever a problem on the grid appears. In this conditions, diffused generation units can't improve the existing power quality level, they can only avoid to worsen it, by controlling their injected current harmonic content and, eventually, by reducing their output power variations. To increment the Power Quality level of the LV and MV grids, diffused renewable generation units should be enabled by the standards and by appropriate changes on the grid structures, to have an active role. For example small generation units should be allowed to produce also reactive power and to support the grid in case of grid faults. To achieve this result many studies are being done on the network, to individuate possible problems correlated to the injection of active and reactive power in MV and LV nodes, and to solve them by an adjustment of the grid regulation (voltage and frequency regulations should be coordinated with the diffused generators) and protection system (the presence of diffused generators can cause the inversion of some current fluxes and may generate problems during the automatic reclosure operation of MV breakers). In the following the solutions to be adopted on MV and LV network to include diffused generators in their regulation structures will not be considered, but it will be assumed that this solutions will be implemented. The attention will be focused on the diffused generators side, trying to understand which are the principal features that must have diffused hybrid generation units in order to support and increase the Power Quality level of modified MV and LV networks. From these considerations some design constraints will follow, and they will be analyzed in the next section.

It must be noted that all the following considerations are referred only to small diffused hybrid generation units, to be connected on LV and MV network. The big renewable power plants connected to HV, usually big wind power plants, are not considered.

5.1 Requirements on voltage waveform

Usually the voltage regulation on public grids is made at the High Voltage and possibly even at the Medium Voltage level. In Italy, for example, the last step in voltage regulation is made in the primary substation (HV/MV substation) by setting the tap changers of the HV/MV transformers. Because of the voltage drops in the MV and LV lines, the voltage level on LV distribution network can vary from values close to the rated one near the secondary substations, to values close to the minimal admitted threshold at the end of long LV lines in high load conditions.

Usually, because of the predominantly inductive behaviour of the line impedances, the voltage and frequency regulations can be decoupled in such a way that grid nodes voltage levels are controlled by acting on the reactive power fluxes and the grid frequency is controlled by acting on the active power injections in the grid nodes.

Diffused renewable generation units connected to the low voltage network can cause unexpected local increments of the voltage level. The effect of the diffused generation units on the grid voltage depends on many aspects such as the generation unit size and position along the LV line and the power factor of the complex power injected in the grid. Nowadays in Italy there is no possibility to regulate the power factor of small renewable diffused generation connected to the LV grid, as it's stated that all the generation units connected to the LV grid must have an output power factor equal to 1 (only active power injection). Independently form the grid national rules, that will evolve together with the technical developments, the only way for diffused generation units to assist the voltage regulation in the LV network is to control the reactive power injected in the grid.

The hybrid power plants analyzed in this chapter can contribute to the improvement of voltage profiles on the LV and MV networks and can be actively involved in the grid voltage regulation only if they are able to regulate their reactive power injection in the grid.

Looking at the voltage regulation operated nowadays on the big power plants connected to the HV network, one can imagine that the voltage regulation on the MV and LV diffused generators could be organized in, at least, two steps: primary voltage regulation and secondary voltage regulation.

Each generator participating to the **primary voltage regulation** reacts to any node voltage variations with a change in its output reactive power. The relation between the output power variation and the voltage variation that causes it is called *statism*, and has a different value for each generator. A solution to apply the primary voltage regulation on the diffused generators could be to introduce a regulation statism also in each renewable generation unit, in such a way that its output reactive power changes depending on the grid voltage variations (droop control for microgrids). The droop control is implemented directly in the interface converter and the statism value can be calculated by the generation unit control system or can be imposed by an eternal higher level supervisor. The maximum current value of the interface converters limits the global complex power that the generation unit can inject in the grid. Thus, the reactive power set point should be coordinated with the active power set point resulting from the frequency regulation, not to exceed the current limit of the converter interface. If the reactive power statism is settled directly by the generation unit controller, it will be its duty to coordinate the active and reactive power set points. If the statism is imposed by an external supervisor, the hybrid power plant must send to it the information regarding its actual state (actual current, current limits of the converter, sensible component temperature,…), and the external supervisor will correct consequently the statism value of the power plant, taking into account the information received by all the diffused generators of a certain area. In this case a communication system is needed to implement the primary regulation, but the hybrid generation units of an area can be coordinated together and a more precise primary voltage regulation can be achieved.

The **secondary voltage regulation** is implemented by the external supervisor, after an action of the primary regulation, by sending to each power plant the variation in the reactive power production. The secondary voltage regulation has the aim to avoid the circulation of reactive power fluxes between generation units.

In this case a communication system is necessary between the central supervisor and the hybrid generation units of an area. The hybrid power plants send to the supervisor the information regarding their actual state and the external controller calculate consequently the reactive power injection variation of each generation unit.

The reactive power that can be injected or absorbed by a hybrid generation unit doesn't depend on the energy availability from the sources nor on the state of charge of the battery bank, as the energy available from the renewable generators and from the storage is active and not reactive energy. To supply a certain value of reactive power, the hybrid generation unit needs to absorb a minimal amount of active power sufficient to cover its global internal losses (losses in the converters, on the wirings, in the passive components, …) and the supply of the electronic control system (electronic boards, sensors, relays,…). This minimum active power can be provided by the renewable sources or by the battery bank or even, if the interface converter allows it, can be absorbed by the grid. In this third case, the hybrid generation units can provide the reactive power required by the grid even if there is not at all availability of active power neither from the renewable sources nor from the storage system.

If the minimum active power needed for the power plant operation is guaranteed, the reactive power injected (or absorbed) by the hybrid generation unit is limited only by the sizing of its components. The current circulating in each component should not over goes the maximal design value, thus the output active and reactive power of the hybrid system are limited in order to respect this condition.

In the design of the grid connected hybrid power plants it should be taken into account the maximum output current that can be supplied, considering both the active and the reactive contribute.

What explained about the voltage regulation concerns the hybrid generation units with a **DC bus configuration**. When this generation units are connected in parallel to the grid, they have only one interface point (interface converter) that can supply reactive power to the grid.

If the hybrid generation units have an **AC bus configuration**, all the source converters and the storage converters of all the power plants are connected in parallel on the grid, that can be seen as a common extension of all the AC buses. In this case all the grid connected converters can supply reactive power to the grid, within the limits of their maximum rated current. It's a design choice to decide which converters are enabled to participate to the voltage regulation, and if the converters will receive directly the reactive power set point form the external regulator or if the external regulator will send a global reactive power set point for the entire generation units and then the generation nit internal controller will split the reference reactive power value to each converter.

5.2 Requirements on voltage frequency

Unlike the voltage amplitude, the voltage frequency is a global variable, that is the same for all the grid nodes. As said before the voltage regulation is correlated to the active power, in particular, if the total active power injection in the grid nodes is equal to the total active power absorbed by the loads, the grid frequency remains constant. If the active power injected in the grid nodes is higher than the active power absorbed, the grid frequency increases and viceversa. Hybrid distributed generation units could contribute to the grid frequency regulation, as they are able to regulate the active power injected in the grid and they can also absorb active power and store it in the battery bank. Actually, the contribution of a single hybrid generation units is irrelevant among the total amount of active power exchanged in the grid nodes, thus a variation in the active power supplied by a single hybrid generation unit wouldn't have any effect on the grid frequency. Only if the number of diffused generators is relevant and if all the generation units are coordinated by a higher level supervisor, the diffused generation can have an impact on the grid frequency regulation. Each generation unit must be equipped with a communication system that enables it to receive from the grid operator the active power reference values, and it must guarantee a certain active power availability to follow the reference set point. Looking at the frequency regulation operated nowadays on the big power plants connected to the HV network, one can imagine that the frequency regulation on the MV and LV diffused generators could be organized in at least two steps: primary regulation and secondary regulation.

Each generator participating to the **primary frequency regulation** reacts to any frequency variations with a change in its output active power. The relation between the output power variation and the frequency variation that causes it is called *statism*, and has a different value for each generator. A solution to apply the primary frequency regulation on the diffused generators could be to introduce a regulation statism also in each renewable generation unit, in such a way that its output power changes depending on the grid frequency variations (droop control for microgrids). The droop control is implemented directly in the interface converter and the statism value can be calculated by the generation unit control system or can be imposed by an eternal higher level supervisor. It could be useful to adapt the statism value to the energy availability of the system: in case of a grid frequency reduction, an hybrid generation unit with a low energy availability will react with a lower increase of the injected power than a hybrid generation unit that have the battery fully charged. If the statism is set directly by the generation units controller, it will be its duty to calculate the energy availability of the system and adapt consequently the statism and no communication system is required in this case for the primary frequency regulation. If the statism is imposed by an external supervisor, the hybrid power plant must send to it the information regarding its actual state (battery state of charge, current availability of renewable sources, weather forecast,...), and the external supervisor will correct consequently the statism value of the power plant, taking into account the information received by all the diffused generators of a certain area. In this case a communication system is needed to implement the primary regulation, but the hybrid generation units of an area can be coordinated together and a better exploitation of the renewable sources can be performed.

The **secondary frequency regulation** is implemented by the external supervisor, who send to each power plant the variation in the active power production necessary to bring the grid frequency to the nominal value after an action of the primary regulation. In this case a communication system is necessary between the central supervisor and the hybrid generation units of an area. The hybrid power plants send to the supervisor the information regarding their actual state and the external controller calculate consequently the power injection variation of each generation unit.

It must be noted that unlike the traditional power plants, the hybrid generation units can also absorb active power from the grid. This working condition is possible if the interface converter is bidirectional and if the control system is enabled to manage this condition. One could think to exploit this characteristic, as it allow a larger range of active power regulation. For example it could useful to set two different value for the positive statism, that regulates the positive output power variations, and the negative statism, that regulates the variations in the active power absorbed by the generation unit. In this way generation units with a low battery state of charge can be controlled to react with a low injected power increment if the grid frequency decreases, and with a high absorbed power increment of the grid frequency raises. This kind of control can improve the battery management and the overall system coordination, but need a very strong coordination of the diffused generation unit, to avoid the condition where some generation units are discharging their batteries in the grid and some other generation units are absorbing active power to recharge their storage system. This exchange of active power between the storage systems is an unwanted condition because it leads to an great waste of energy caused by the charging and discharging efficiencies of the storage systems.

To guarantee the capability to participate to the primary and secondary frequency regulations, the generation units should keep a **primary and secondary active energy reserve**. This energy reserve can be stored in the battery bank of the hybrid generation units, by setting some threshold on the battery state of charge. The entity of the frequency regulation reserve must be fixed during the designing process together with the network operator, depending on the size of the power plant.

The value of the active energy reserve that must be guaranteed in the hybrid generation units in order to participate to the primary and secondary frequency regulation affects the size of the storage system and the size of the renewable generators. The storage system must be able to contain an energy at least equal to the sum of the two regulation reserves and the renewable generators must be able to supply enough active power for the ordinary operation and for recharge the storage system with a periodicity depending on the frequency of the frequency regulation operations.

What explained about the frequency regulation concerns the hybrid generation units with a **DC bus configuration**. When this generation units are connected in parallel to the grid, they have only one interface point (interface converter) and they can maintain the same internal power management algorithm used for the islanded operation. So, the interface converter follows the active power set points coming from the frequency regulation and the generation converters extract the maximum power available from the renewable sources (or limit this power to a set point value) and the battery converter maintains the energy balance on the DC.

If the hybrid generation units have an **AC bus configuration**, all the source converters and the storage converters of all the power plants are connected in parallel on the grid, that can be seen as a common extension of all the AC buses. In this case the renewable generator converters operate as current generators, injecting in the grid the maximum power available from the sources, or, if they can work in RPPT mode, limiting the injected active power to a certain set point given by the generation unit supervisor. The storage system converters perform the voltage and frequency regulation on the gird, as the interface converter does for generation units with DC bus configuration. The external regulator has the same functions than in the case of DC bus configuration plants. The coordination of the renewable generators and the battery bank of each power plant is demanded to the power plant general supervisor, which, for example, sends active power set points to each renewable source converters when the RPPT operation is needed. Moreover each power plant general supervisor should send to the external regulator information concerning the present availability of the renewable source and, possibly, the weather forecasts. Thus, even if in AC configuration all the converters of the power plants seems to be independently connected to the grid, the interface between each hybrid power plant and the external supervisor is made only by the power plant general controller, and not by each converter separately.

In the case of AC bus configuration the problem of active power circulation between storage systems is much more relevant than for Dc bus configuration. When all the storage systems are connected in parallel to the grid, they must exchange active power in both direction with the grid, which means that the battery can absorb power indistinctly form any other generator connected to the grid, even form the other storage system operating in discharging mode. The only way to avoid the exchange of active power between storage systems is that the external supervisor disables the charging of all the batteries when there is at least a storage system in discharging operation. This solution is, not feasible, or, at least very restrictive, thus, a rate of active power exchange between storage systems have to be accepted in parallel operation of AC bus generation units.

6. Design aspects of HGDU in grid connected mode

As highlighted previously, the Power Quality in grid connected systems can be supported and improved by an active participation of the diffused renewable generation units. On the following three main aspects of the design of renewable hybrid power plants for grid connected operation will be analyzed, trying to point which are the possible choices that leads to a higher power quality of supply.

6.1 Sizing

The sizing of renewable generation units is the aspect that mostly affects their capability to participate to the frequency regulation (primary and secondary regulation reserve) and the voltage regulation (high current injection in case of voltage dips), to supply grid services (peak shaving and load leveling) and to guarantee the preferred loads continuity of supply during gird black outs. In grid connected systems the sizing of hybrid diffused generation units is based on technical-economic optimization. Unlike islanded systems, in grid connected generation units it's possible to evaluate the price of the active power (and possibly also reactive power) produced, depending on the kind of services the generation unit offers. One can imagine that in the future the produced energy will be differently paid, in function of its contribution to the grid power quality: for example energy produced in peak shaving operation should be rewarded at a higher price than energy produced in low load hors. Moreover, also grid services not directly correlated to the production of active power, such as the valley filling and the reactive injection to regulate the grid voltage, should be paid, in agreement with the network operator. The initial investment cost should be compared to the revenues coming from the active energy sale and from the supplying of all grid services.

Sizing constraints

As said before, the sizing of gird connected hybrid generation units must be done comparing the overall cost of the hybrid system with the revenues coming from the sale active energy and of grid services to the network operator.

As for the islanded systems, the overall cost of the system is made up of two contributions: the initial capital investment and the maintenance cost. The initial cost is strictly related to the size of the renewable generators and of the battery bank. The main contribution to the maintenance cost is due to the battery replacements during the lifetime of the system, that is usually much longer than the battery lifetime. Globally, the battery cost, that depends on the battery nominal capacity, has a great impact on the total cost of the system, as it affects both the initial investment and the maintenance cost.

Together with the network operator its should be decided the active energy price in normal conditions (normal supply to the grid) and in case of grid services (active energy supplied during for peak shaving, for load leveling, to guarantee de supply of privileged loads during grid break down, ...). Then, the hybrid system designer should decide which services provide to the grid and the system sizing is calculated consequently.

A possible strategy for the hybrid system sizing, considering the different grid services to supply is presented in the following.

The grid services requiring an injection (or absorption) of active power affect the renewable generators and battery bank sizing. To the design of the hybrid system it's necessary to get an estimated curve of the active power required to the generation unit by the grid, in order to comply the grid services performances established at the beginning of the project. The estimation of this curve is the most difficult part of the project and is a necessary step, as it allows to correctly size the storage system and the renewable generators. For the peak shaving, load leveling and frequency regulation services, it can be planned, together with the grid operator, the expected curve of active power required to the generation unit. This curve can be considered as the equivalent of the load demand profile for the islanded systems, as it represents the minimum active power supply that should be guaranteed by the generation unit. An index can be calculated for the grid connected systems, analogue to the LPSP for the islanded systems, in order to define the capability of the hybrid generation unit to satisfy the grid services guaranteed at the beginning of the project. In the following this index will be referred as LGSP (Loss of Grid Service Probability). As for the islanded systems, the LGSP can be calculated as:

$$LGSP = \left(\sum_{i=1}^{N} \int_{\Delta_{SI,i}} P_{gs\_NS,i}(t) \cdot dt \right) / E_{gs\_tot} \tag{11}$$

where N is the number of failed service supply to the grid in one year;
Egs_tot is the total active energy required to fulfill the grid services guaranteed to the grid operator;
Pgs_NS,i is the power not supplied during the failed ith service supply period
SI,i is the duration of each service supply interruption.
The design value of LGSP is chosen by the hybrid generation unit designer, on the basis of the economical compromise between the overall cost of the system, the revenue from the grid service supplying, and the penalties to be paid in case of loss of grid service supply. It should be noted that in grid connected systems, the over sizing of the hybrid generation unit is not critical, as, except for special conditions of very low load demand, the excessive energy produced by the generation units can always be sold to the grid, even if its price can decrease when the energy need of the grid is low. Consequently, a reasonable design choice could be to fix the LGSP desired value to zero. From this condition the minimal size of the renewable generators and of the storage system can be derived.
The second step is to verify the need of adding a high power storage system to supply high active power peaks to the grid, for example in case of peak shaving service.
If the renewable generation unit must guarantee the continuity of supply to some privileged loads, it should be verified if the energy capability of the hybrid system is enough to guarantee the loads supply during the maximum expected duration of grid interruptions. If the energy capability of the system is not enough, the renewable generators and in particular the storage system sizes should be increased.
It should be noted that the concept of wasted energy changes from islanded systems to grid connected systems. In grid connected systems the produced energy can almost always be sold to the grid, and so the rate of wasted energy can be considered always zero. A real waste of energy can occur when the hybrid system must provide a valley filling service. In this case, when the load is low the hybrid generation units must absorb power from the grid and store it into the battery bank, and the renewable generators must be shut down. During the valley filling period, al the energy available from the renewable sources is not converted and is wasted energy. It must be said that usually the load demand is very low during the night, that is when the photovoltaic source is not available, so the wasted energy in grid connected system is very low and the economical drawbacks of a system over sizing are less relevant than in the case of islanded systems. It must be noted that if the hybrid generation units must provide a valley filling service, the storage system must be properly managed, in order to be able to store the energy absorbed by the grid during the valley filling operation. If the battery are fully charged when the valley filling is required, the

energy absorbed from then grid will be dissipated in the breaking resistance of the hybrid power plant and is, actually, wasted energy. If no breaking resistance is included in the hybrid generation unit, the valley filling service can be provided only if the battery state of charge is low enough to enable the storage of all the active energy that is required to be absorbed from the grid.

The considerations written above refers to the energy sizing of the renewable generation unit. The power (or thermal) sizing is calculated considering the maximum active power peak and the maximum reactive power peak that the hybrid generation unit will be required to supply to the grid, and their contemporary probability. The maximum active power peak is presumably due to the peak shaving operation, while the maximum reactive power peak is usually caused by the voltage regulation in case of a significant voltage amplitude reduction (voltage dip).

6.2 Plant configurations

There are many different power plants configuration, that can differ by the nature of the renewable sources, by the type of the hybrid storage system and by the kind of converters used. The different hybrid power plants configuration are usually classed in function of the nature of the power bus (AC bus at frequency line, DC bus or high frequency AC bus).

The main choice to be made regarding the hybrid plant topology, as for islanded systems, concern first of all the type common power sharing bus (DC or AC bus) and, secondly the typology of the static converters dedicated to the sources and to the battery bank.

Some of the aspects considered for the islanded system regarding the DC bus or the Ac bus consideration are still valid also for grid connected systems, such as the consideration regarding the modularity and the overall costs. Some other considerations, listed below, should be modified and adapted to the grid connected hybrid power plants.

Kind of loads and sources. Grid connected generation units must supply grid frequency AC voltage (or current) both in grid connected and in transient islanded operation in case of grid break down. If the sources are AC generators (doubly fed or synchronous generators) and the storage system is made by electrochemical batteries, then a line frequency microgrid may be a better system since the grid can be directly connected to the AC bus, thus eliminating the DC/AC interface converter.

Control system. For the internal control system of each hybrid power plant, the same consideration made for the islanded systems can be referred also for the grid connected configurations. Converters on a DC bus system tend to be simpler than those on a AC bus system, and the control for a DC bus system is simplified since frequency control and reactive power internal fluxes control are not a consideration. In AC bus systems both active and reactive power internal flow should be controlled, monitoring both the amplitude and the frequency of the AC bus voltage. Moreover, the phase shifting between voltage and currents should be monitored in order to control the power flow direction.

Some more considerations should be done regarding the external control system needed to coordinate the operations of several hybrid generation units grouped in a so called "smart grid" and connected in parallel to the grid. As explained previously, this coordination control is necessary in order to enable the diffused generation units to participate to the grid voltage and frequency regulation and to provide grid services.

For **DC bus configuration** systems connected in parallel with the grid, the only interface toward the grid and the other generation units is always the interface converter. The hybrid generation units receives active (P) and reactive (Q) power set points from the external smart grid regulator or from the internal droop control implemented for the voltage and frequency regulations. The interface converters implements the P and Q set points and the generation unit internal energy balance is maintained by the storage converter, by controlling the DC bus voltage. The internal control of the hybrid power plant is not affected by the external system, but it goes on working with the same logic both in grid connected operation and in back up islanded operation. Moreover, when the generation unit is working in back up islanded operation it makes no difference for the energy balance regulator if the generation unit is in single islanded configuration or in parallel with other generation units forming an islanded microgrid configuration. Thanks to the interface converter and its regulator, there is a complete decoupling between the internal hybrid power plants system and the external system that is the grid, in grid connected operation, or other generation units and the privileged loads, in back up islanded operation.

In **AC bus systems**, all the generation units converters are connected in parallel to the grid. The hybrid generation units receives P and Q set points by the external smart grid regulator or by the internal droop control implemented for the voltage and frequency regulations. The renewable generator converters, in normal operating conditions, inject all the active power available into the grid. The battery converter injects the reactive power required by the Q set point and injects or absorb the active power equivalent to the difference between the P set point and the total power produced by the renewable generators. When the AC bus system passes in the back up islanded operation only the storage converter control changes its operation, while the renewable converters are not sensitive to the change in the external system configuration, as for the DC bus configuration.

The main difference between the DC bus and the AC bus configuration concerning the control system is related to the active power exchange between storage systems of different generation units connected in parallel to the grid or forming an islanded smart grid during the back up operation. If the hybrid generation units are not required to provide the valley filling service (that is to absorb active power when the load demand is low) for the DC bus systems the circulation of active power between storage systems can be avoided by using unidirectional interface converter. In such a way the active power flux is always directed toward the grid and the battery bank of each generation unit can be recharged only by the renewable generators of the power plants. If the valley filling operation is required, the interface converters must be bidirectional and the only way to avoid the exchange of active power between storage systems is that the external smart grid controller blocks the charging of any storage system when there is at least one storage system in discharging operation. Usually, all the generation units of a smart grid are coordinated to provide the same grid service, as their single active or reactive contribution would not be perceived by the grid, given the small size of the diffused generation units among the grid inertia. It can be considered that all the hybrid generation units connected to a smart grid, and referring to a single external controller, will be operating in valley filling mode at the same time, thus all the storage systems will be in charging operation at the same time and they will not exchange power. In all other working conditions different form the valley filling, the interface converter will be controlled to allow an active power flux only directed toward the grid, in order to prevent the active power exchange between the battery banks.

In the AC bus configuration all the battery converters are in parallel on the grid and they are necessarily bidirectional, to allow the charge and discharge of the batteries, even if the valley filling service is not required. To avoid the exchange of active power between storage systems, the external smart grid controller should prevent the charging of any storage system when there is at least one storage system in discharging operation. This control technique is very binding and requires a continuous monitoring of the storage system operation in every working condition of the generation units, and not only during the valley filling operation as for DC bus systems. The only alternative several for AC bus parallel connected generation units is to accept a rate of active power exchange between storage systems.

From what explained above, It can be stated that the DC configuration is preferable regarding the external coordination of several hybrid generation units connected in parallel, as the interface converter ensures a complete decoupling between the internal system of the generation unit and the external system (grids and other generation units connected in parallel).

Efficiency. As for islanded systems, the global efficiency of the grid connected generation units depends on many aspects: renewable source peak power related to the loads rated power, number and typology of converters, and power flow management. The considerations made for the islanded systems can be referred also for grid connected generations units, by adding the following comments.

One of the main cause of power losses in the system is the storage battery bank, thus, the overall efficiency of an hybrid system is affected by the rate of active energy that has to be stored in the battery bank before being delivered to the grid. The rate of the stored energy can be reduced by optimizing the internal power flow management algorithm and by the choice of the grid services that the renewable generation units provides.

In AC bus configurations transformers can be used to adapt generating units and storage system voltage level to the LV or MV grid voltage. In DC systems, additional conversion stages could be required to raise the output voltage of generation units and the storage system, or the system can be managed at low DC voltage, with no additional conversion stage but with an output transformer between the DC/AC interface converters and the grid. The efficiencies of the additional conversion stages and of the transformer should be evaluated for different operating conditions of the generation units.

Generally, for medium size generation units connected to the MV grid, the transformer is necessary in both the configurations, to rise the power plant output voltage to the MV grid level.

Reliability. In AC bus systems the critical component is the battery converter. Usually the battery converter implements the active and reactive power regulation to provide the grid services (peak shaving, load leveling,…) and to participate to the voltage and frequency regulations.

If the battery converter breaks down, the renewable generator converters can go on feeding active power into the grid, but the active power can no longer be regulated by the storage system, as require provide grid services and to participate to the power and voltage regulation.

In DC bus systems there are two critical components: the interface converter and the battery converter.

If the interface converters breaks the hybrid generation units is separated from the grid and the loads and can no longer supply neither reactive nor active power.

The battery converter maintains the power balance between the active power injected into the grid and the active power generated by the renewable sources If the battery converter breaks down, the renewable generator converters can go on feeding active power into the DC bus, but the active power balance is no longer maintained on the DC bus. The generation unit can go on supplying into the grid the active power available from the renewable sources only if the interface converter control changes its operation. If the battery converter doesn't work anymore, the interface converter must maintain the energy balance on the DC bus, by regulating the output active power in order to maintain the DC bus voltage to its nominal value. It's not very reasonable to contemplate this working condition, as the system is not providing grid services and needs to be repaired. Thus, it can be stated that the DC bus generation units can't inject any power into the grid when the battery converter breaks down, even if the interface converter is still working.

7. References

Bollen, M.H.J. (2003). What is power quality. *Electric Power System Research,* Vol.66, No.1, (July 2003), pp. 5-14

De Brabandere K. (2007). Control of Microgrids, *Proceedings of Power Engineering Society General Meeting,* pp. 1-7, Tampa, Florida, USA, June 24-27, 2007

Diaf S. et. Others (2007). A methodology for optimal sizing of autonomus hybrid PV/wind system. *Energy Policy,* Vol.35, No.11, (November 2007), pp. 5708-5718,

Elster S. (2006). Re Hybrid systems: Coupling of Renewable energy sources on AC and DC side of the inverter. *ReFOCUS Journal,* Vol.7, No.5, (September-October 2006), pp. 46-48

Kolhe, M. (2009). Techno-Economic Optimum Sizing of Stand Alone Solar Photovoltaic System. *IEEE Transaction on Energy Conversion,* Vol.24, No.2, (June 2009), pp. 511-519

Modeling of Photovoltaic Grid Connected Inverters Based on Nonlinear System Identification for Power Quality Analysis

Nopporn Patcharaprakiti[1,2], Krissanapong Kirtikara[1,2],
Khanchai Tunlasakun[1], Juttrit Thongpron[1,2], Dheerayut Chenvidhya[1],
Anawach Sangswang[1], Veerapol Monyakul[1] and Ballang Muenpinij[1]
[1]King Mongkut's University of Technology Thonburi, Bangkok,
[2]Rajamangala University of Technology Lanna, Chiang Mai
Thailand

1. Introduction

Photovoltaic systems are attractive renewable energy sources for Thailand because of high daily solar irradiation, about 18 MJ/m^2/day. Furthermore, renewable energy is boosted by the government incentive on adders on electricity from renewable energy like solar PV, wind and biomass, introduced in the second half of 2000s. For PV systems, domestic rooftop PV units, commercial rooftop PV units and ground-based PV plants are appealing. Applications of electricity supply from PV plants that have been filed total more than 1000 MW. With the adder incentive, more households will be attracted to produce electricity with a small generating capacity of less than 10 kW, termed a very small power producer (VSPP). A possibility of expanding domestic roof-top grid-connected units draw our attention to study single phase PV-grid connected systems. Increased PV penetration can have significant [1-2] and detrimental impacts on the power quality (PQ) of the distribution networks [3-5]. Fluctuation of weather condition, variations of loads and grids, connecting PV-based inverters to the power system, requires power quality control to meet standards of electrical utilities. PV can reduce or improve power quality levels [6-9]. Different aspects should be taken into account. In particular, large current variations during PV connection or disconnection can lead to significant voltage transients [10]. Cyclic variations of PV power output can cause voltage fluctuations [11]. Changes of PV active and reactive power and the presence of large numbers of single phase domestic generators can lead to long-duration voltage variations and unbalances [12]. The increasing values of fault currents modify the voltage sag characteristics. Finally, the waveform distortion levels are influenced in different ways according to types of PV connections to the grid, i.e. direct connection or by power electronic interfaces. PV can improve power quality levels, mainly as a consequence of increase of short circuit power and of advanced controls of PWM converters and custom devices. [13]

Grid-connected inverter technology is one of the key technologies for reliable and safety grid interconnection operation of PV systems [14-15]. An inverter being a power conditioner of a PV system consists of power electronic switching components, complex control systems [16]. In addition, their operations depends on several factors such as input weather condition, switching algorithm and maximum power point tracking (MPPT) algorithm implemented in grid-connected inverters, giving rise to a variety of nonlinear behaviors and uncertainties [17]. Operating conditions of PV based inverters can be considered as steady state condition [18], transient condition [19-20], and fault condition such islanding [21-22]. In practical operations, inverters constantly change their operating conditions due to variation of irradiances, temperatures, load or grid impedance variations. In most cases, behavior of inverters is mainly considered in a steady state condition with slowly changing grid, load and weather conditions. However, in many instances conditions suddenly change, e.g. sudden changes of input weather, cloud or shading effects, loads and grid changes from faults occurring in near PV sites [23]. In these conditions, PV based inverters operate in transient conditions. Their average power increases or decreases upon the disturbances to PV systems [24]. In order to understand the behavior of PV based inverters, modeling and simulation of PV based inverter systems is the one of essential tools for analysis, operation and impacts of inverters on the power systems [25].

There are two major approaches for modeling power electronics based systems, i.e. analytical and experimental approaches. The analytical methods to study steady state, transient models and islanding conditions of PV based inverter systems, such as state space averaging method [26], graphical techniques [27-28] and computation programming [29]. In using these analytic methods, one needs to know information of system. However, PV based inverters are usually commercial products having proprietary information; system operators do not know the necessary information of products to parameterize the models. In order to build models for nonlinear devices without prior information, system identification methods are exposed [33-34]. In the method reported in this paper, specific information of inverter is not required in modeling. Instead, it uses only measured input and output waveforms.

Many recent research focuses on identification modeling and control for nonlinear systems [35-37]. One of the effective identification methods is block oriented nonlinear system identification. In the block oriented models, a system consists of numbers of linear and nonlinear blocks. The blocks are connected in various cascading and parallel combinations representing the systems. Many identification methods of well known nonlinear block oriented models have been reported in the literature [38-39]. They are, for example, a NARX model [40], a Hammerstein model [41], a Wiener model [42], a Wiener-Hammerstein model and a Hammerstein-Wiener model [43]. Advantages of a Hammerstein model and a Wiener model enables combination of both models to represent a system, sensors and actuators in to one model. The Hammerstein-Wiener model is recognized as being the most effective for modeling complex nonlinear systems such PV based inverters [44].

In this paper, real operating conditions weather input variation, i.e. load variations and grid variations, of PV- based inverters are considered. Then two different experiments, steady state and transient condition, are designed and carried out. Input-output data such as currents and voltages on both dc and ac sides of a PV grid-connected systems are recorded. The measured data are used to determine the model parameters by a Hammerstein-Wiener nonlinear model system identification process. In the Section II, PV system characteristics are introduced. The I-V characteristic, an equivalent model, effects of radiation and temperature on voltage and current of PV are described. In the Section III, system identification methods, particularly a Hammerstein-Wiener Model is explained. In the following section, the experimental design and implementation to model the system is illustrated. After that, the obtained model from prior sections is analyzed in terms of control theories. In the last section, the power quality analysis is discussed. The output prediction is performed to obtain electrical outputs of the model and its electrical power. The power quality nature is analyzed for comparison with outputs of model. Subsequently, voltage and current outputs from model are analyzed by mathematical tools such the Fast Fourier Transform-FFT, the Wavelet method in order to investigate the power quality in any operating situations.

2. PV grid connected system (PVGCS) operation

In this section, PV grid connected structures and components are introduced. Structures of PBGCS consist of solar array, power conditioners, control systems, filtering, synchronization, protection units, and loads, shown in Fig. 1.

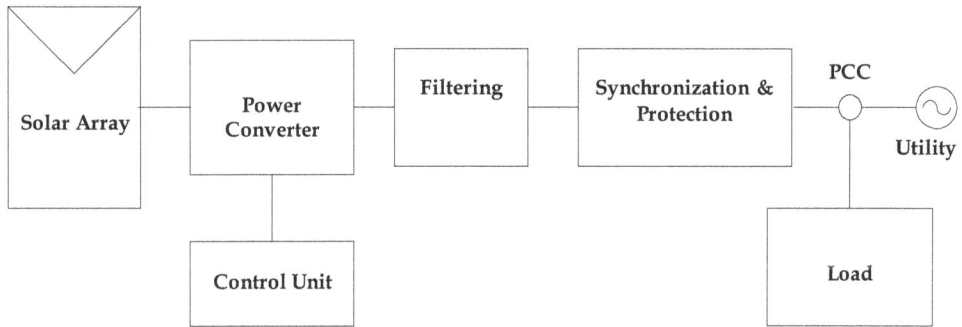

Fig. 1. Block diagram of a PV grid connected system

2.1 Solar array
Environmental inputs affecting solar array/cell outputs are temperature (T) and irradiance (G), fluctuating with weather conditions. When the ambient temperature increases, the array short circuit current slight increases with a significant voltage decrease. Temperature and I-V characteristics are related, characterized by array/cell temperature coefficients. Effects of irradiance, radiant solar energy flux density in W/m², apart from solar radiation at sea level, are determined by incident angles and array/cell envelops. Typical characteristics of relationship between environmental inputs (irradiance and temperature) and electrical parameters (current and voltage of array/cells) are shown in Fig. 2 [45]. In our experimental designs, operating conditions of PV systems under test is designed and based on typical operating conditions.

2.2 Operating conditions of a PV grid connected system
A PV system, generating power and transmitting it into the utility, can be categorized in three cases, i.e. a steady state condition, a transient condition and a fault condition like islanding. Three factors affecting the operation of inverters are input weather conditions, local loads and utility grid variations.

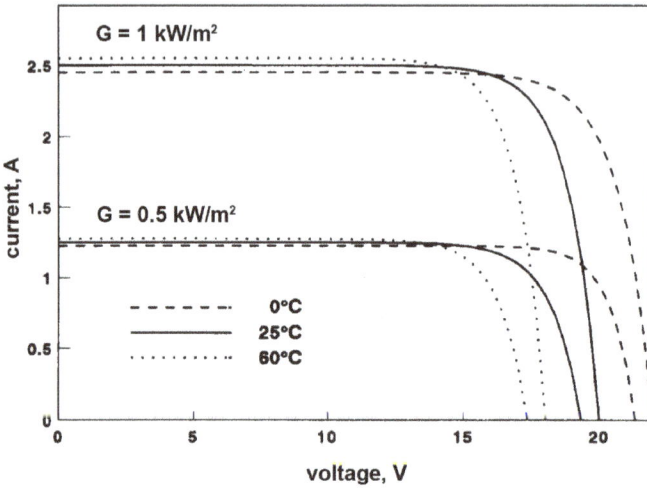

Fig. 2. Temperature and irradiance effects on I-V characteristics of PV arrays/cells [46]

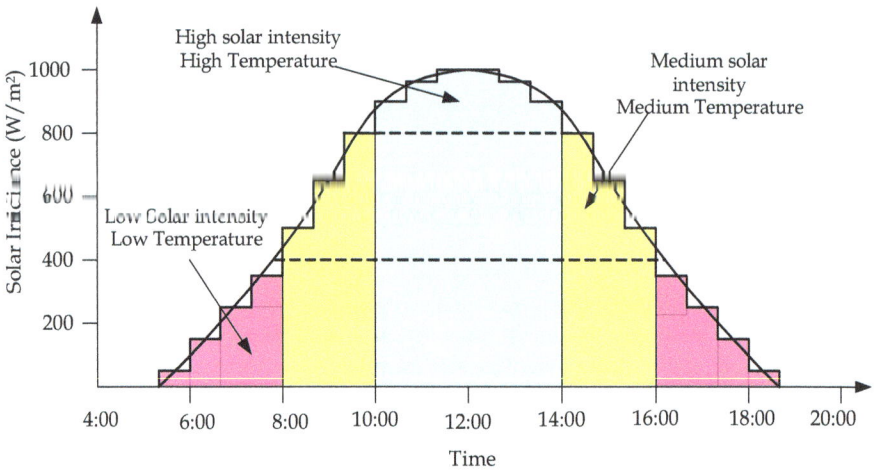

Fig. 3. Variations of solar irradiance and temperature throughout a day conditioning PVGCS operation

Firstly, under a steady state condition, input, load and utility under consideration are treated as being constant with slightly change weather condition. Installed capacities of PV systems in a steady state are low, medium and high capacity. According to the weather conditions throughout a day as shown in Fig. 3 [47-48], a low radiation about 0-400 W/m² is common in an early morning (6:00 AM-9:00 AM) and early evening (16:00 PM-19:00 PM), medium radiation of 400-800 W/m² in late morning (9:00 AM-11:00 PM) and early afternoon (14:00 PM-16:00 PM) and high radiation of 800-1000 W/m² around noon (11.00

AM - 14:OO PM). Loads fluctuate upon activities of customer groups, for example, a peak load   for industrial zones occurs in afternoon (13:00 - 17:00 PM) and a peak load for residential zones   occurs in evening(18:00 - 21:00 PM). Variations from steady state conditions impact power quality such as overvoltage, over-current, harmonics, and so on. In case of transients, there are variations in inputs, loads and utility. Weather variations such solar irradiance and temperature exhibit significant changes. Unexpected accidents happen. Local loads may sudden change due to activities of customers in each time. A utility has some faults in nearby locations which impact utility parameters such grid impedance. These conditions lead to short duration power quality problems with such spikes, voltage sag, voltage swell. In some extreme cases, abnormal conditions, such as very low solar irradiance or abnormal conditions such islanding, the gird-connected PV systems may collapse. The PV systems are black out and cut out of the utility grid. Such can affect power quality, stability and reliability of power systems.

2.3 Power converter
There are several topologies for converting a DC to DC voltage with desired values, for example, Push-Pull, Flyback, Forward, Half Bridge and Full Bridge [49]. The choice for a specific application is often based on many considerations such as size, weight of switching converter, generation interference and economic evaluation [50-51]. Inverters can be classified into two types, i.e. voltage source inverter (VSI) if an input voltage remains constant and a current source inverter (CSI) if input current remains constant [52-53]. The CSI is mostly used in large motor applications, whereas the VSI is adopted for and alone systems. The CSI is a dual of a VSI. A control technique for voltage source inverters consists of two types, a voltage control inverter, shown in Fig. 4(a) and a current control inverter, Fig. 4(b) [54].

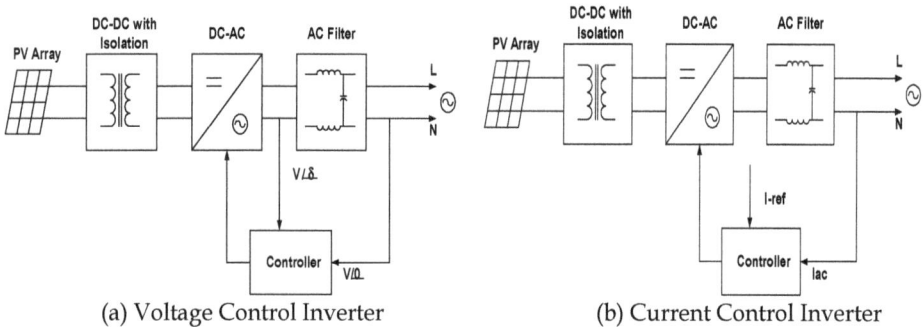

(a) Voltage Control Inverter   (b) Current Control Inverter

Fig. 4. Control techniques for an inverter

3. System Identification

System identification is the process for modeling dynamical systems by measuring the input/output from system. In this section, the principle of system identification is described. The classification is introduced and particularly a Hammerstein-Wiener model is explained. Finally, a MIMO (multi input multi output model with equation and characteristic is illustrated.

3.1 Principle of system identification

A dynamical system can be classified in terms of known structures and parameters of the system, shown in Fig.5, and classified as a "White Box" if all structures and parameters are known, a "Grey Box" , if some structures and parameters known and a "Black Box" if none are known [55].

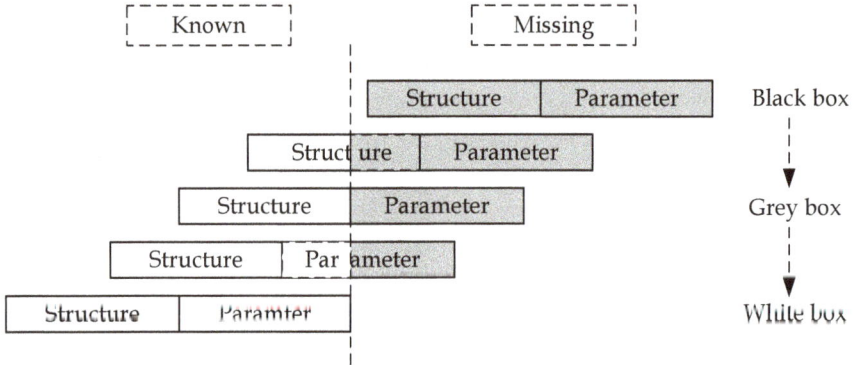

Fig. 5. Dynamical system classifications by structures and parameters

Steps in system identification can be described as the following process, shown in Fig. 6.

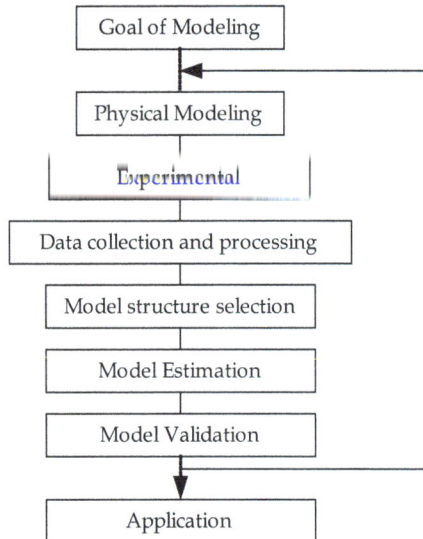

Fig. 6. System identification processes

Each step can be described as follows

3.1.1 Goal of modeling

The primary goal of modeling is to predict behaviors of inverters for PV systems or to simulate their outputs and related values. The other important goal is to acquire mathematical and physical characteristics and details of systems for the purposes of controlling, maintenance and trouble shooting of systems, and planning of managing the power system.

3.1.2 Physical modeling

Photovoltaic inverters, particularly commercial products, compose of two parts, i.e. a power circuit and a control circuit. Power electronic components convert, transfer and control power from input to output. The control system, switching topologies of power electronics are done by complex digital controls.

3.1.3 Model structure selection

Model structure selection is the stage to classify the system and choose the method of system identification. The system identification can be classified to yield a nonparametric model and a parametric model, shown in Fig 7. A nonparametric model can be obtained from various methods, e.g. Covariance function, Correlation analysis. Empirical Transfer Function Estimate and Periodogram, Impulse response, Spectral analysis, and Step response.

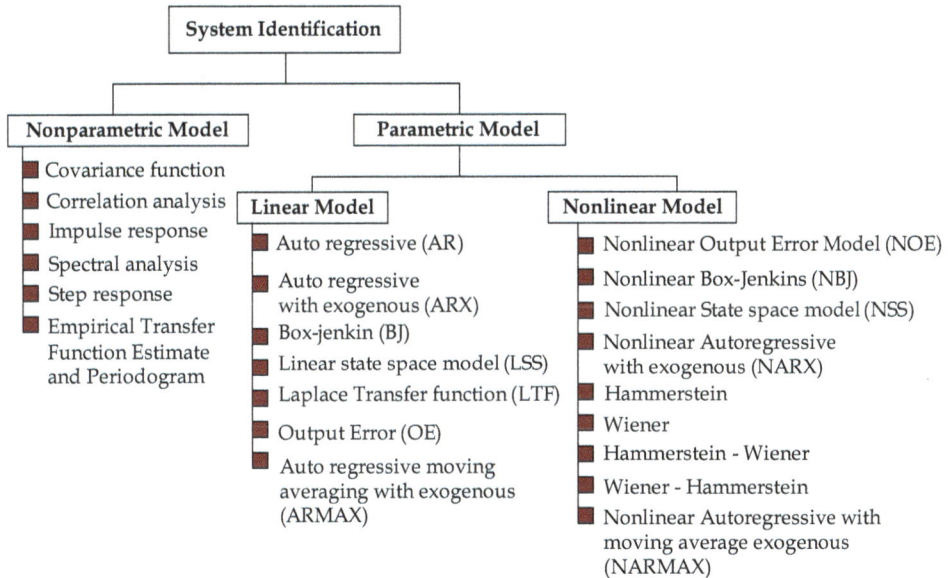

Fig. 7. Classification of system identification

Parametric models can be divided to two groups: linear parametric models and nonlinear parametric models. Examples of linear parametric models are Auto Regressive (AR), Auto Regressive Moving Average (ARMA), and Auto Regressive with Exogenous (ARX), Box-Jenkins, Output Error, Finite Impulse Response (FIR), Finite Step Response (FSR), Laplace Transfer Function (LTF) and Linear State Space (LSS). Examples of nonlinear parametric models are Nonlinear Finite Impulse Response (NFIR), Nonlinear Auto-Regressive with Exogenous (NARX), Nonlinear Output Error (NOE), and Nonlinear Auto-Regressive with

Moving Average Exogenous (NARMAX), Nonlinear Box-Jenkins (NBJ), Nonlinear State Space, Hammerstein model, Wiener Model, Hammerstein-Wiener model and Wiener-Hammerstein model [56]. In practice, all systems are nonlinear; their outputs are a nonlinear function of the input variables. A linear model is often sufficient to accurately describe the system dynamics as long as it operates in linear range Otherwise a nonlinear is more appropriate. A nonlinear model is often associated with phenomena such as chaos, bifurcation and irreversibility. A common approach to nonlinear problems solution is linearization, but this can be problematic if one is trying to study aspects such as irreversibility, which are strongly tied to nonlinearity. Inverters of PV systems can be identified based on nonlinear parametric models using various system identification methods.

3.1.4 Experimental design
The experimental design is an important stage in achieving goals of modeling. Number parameters such as sampling rates, types and amount of data should be concerned. Grid connected inverters have four important input/output parameters, i.e. DC voltage (Vdc), DC current voltage (Idc), AC voltage (Vac) and AC current (Iac). In experiments, these data are measured, collected and send to a system identification process. Finally, a model of a PV inverter is obtained, shown in Fig. 8.

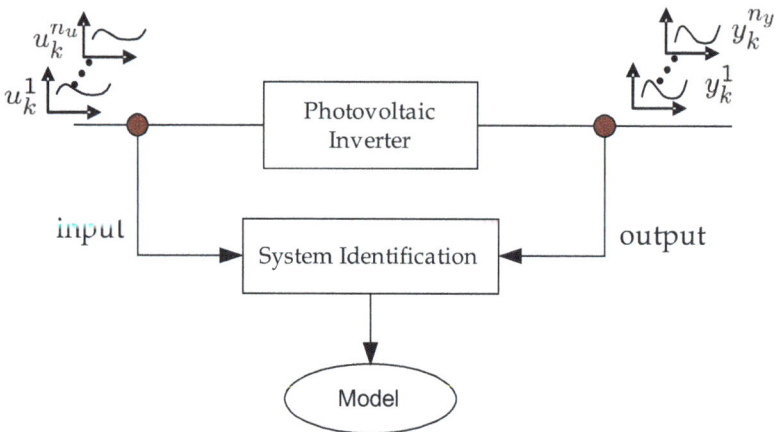

Fig. 8. Experimental design of a photovoltaic inverter modeling using system identification

3.1.5 Model estimation
Data from the system are divided into two groups, i.e., data for estimation (estimate data) and data for validation (validate data). Estimate data are used in the system identification and validate data are used to check and improve the modeling to yield higher accuracy.

3.1.6 Model validation
Model validation is done by comparing experimental data or validates data and modeling data. Errors can then be calculated. In this paper, parameters of system identification are optimized to yield a high accuracy modeling by programming softwares.

3.2 Hammerstein-Wiener (HW) nonlinear model

In this section, a combination of the Wiener model and the Hammerstein model called the Hammerstein-Wiener model is introduced, shown in Fig. 9. In the Wiener model, the front part being a dynamic linear block, representing the system, is cascaded with a static nonlinear block, being a sensor. In the Hammerstein model, the front block is a static nonlinear actuator, in cascading with a dynamic linear block, being the system. This model enables combination of a system, sensors and actuators in one model. The described dynamic system incorporates a static nonlinear input block, a linear output-error model and an output static nonlinear block.

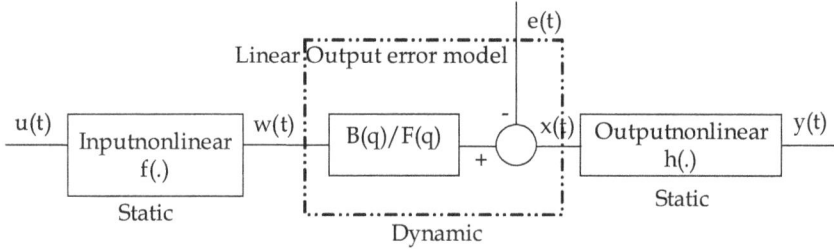

Fig. 9. Structure of Hammerstein-Weiner Model

General equations describing the Hammerstein-Wiener structure are written as the Equation (1)

$$
\left.
\begin{aligned}
w(t) &= f(u(t)) \\
x(t) &= \sum_i^{n_u} \frac{B_i(q)}{F_i(q)} w(t - n_k) \\
y(t) &= h(x(t))
\end{aligned}
\right\}
\tag{1}
$$

Which u(t) and y(t) are the inputs and outputs for the system. Where w(t) and x(t) are internal variables that define the input and output of the linear block.

3.2.1 Linear subsystem

The linear block is similar to an output error polynomial model, whose structure is shown in the Equation (2). The number of coefficients in the numerator polynomials B(q) is equal to the number of zeros plus 1, b_n is the number of zeros. The number of coefficients in denominator polynomials F(q) is equal to the number of poles, f_n is the number of poles. The polynomials B and F contain the time-shift operator q, essentially the z-transform which can be expanded as in the Equation (3). n_u is the total number of inputs. n_k is the delay from an input to an output in terms of the number of samples. The order of the model is the sum of b_n and f_n. This should be minimum for the best model.

$$
x(t) = \sum_i^{n_u} \frac{B_i(q)}{F_i(q)} w(t - n_k)
\tag{2}
$$

$$
\begin{aligned}
B(q) &= b_1 + b_2 q^{-1} + \ldots\ldots + b_n q^{-b_n + 1} \\
F(q) &= 1 + f_1 q^{-1} + \ldots\ldots + f_n q^{-f_n}
\end{aligned}
\tag{3}
$$

3.2.2 Nonlinear subsystem

The Hammerstein-Wiener Model composes of input and output nonlinear blocks which contain nonlinear functions f(•) and h(•) that corresponding to the input and output nonlinearities. The both nonlinear blocks are implemented using nonlinearity estimators. Inside nonlinear blocks, simple nonlinear estimators such deadzone or saturation functions are contained.

i.   **The dead zone (DZ) function** generates zero output within a specified region, called its dead zone or zero interval which shown in Fig. 10. The lower and upper limits of the dead zone are specified as the start of dead zone and the end of dead zone parameters. Deadzone can define a nonlinear function y = f(x), where f is a function of x, It composes of three intervals as following in the equation (4)

$$\left.\begin{array}{ll} x \le a & f(x) = x - a \\ a < x < b & f(x) = 0 \\ x \ge b & f(x) = x - b \end{array}\right\} \tag{4}$$

when x has a value between a and b, when an output of the function equal to $F(x) = 0$, this zone is called as a "zero interval".

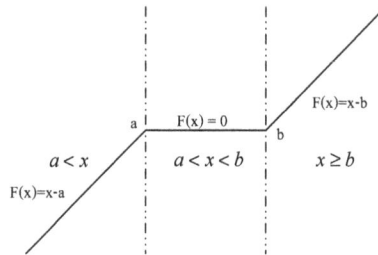

Fig. 10. Deadzone function

ii.   **Saturation (ST) function** can define a nonlinear function y = f(x), where f is a function of x. It composes of three interval as the following characteristics in the equation (5) and Fig. 11. The function is determined between a and b values. This interval is known as a "linear interval"

$$\left.\begin{array}{ll} x > a & f(x) = a \\ a < x < b & f(x) = x \\ x \le b & f(x) = b \end{array}\right\} \tag{5}$$

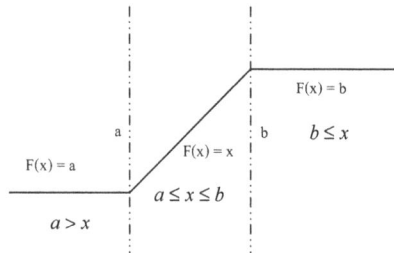

Fig. 11. Saturation function

iii. **Piecewise linear (PW) function** is defined as a nonlinear function, y=f(x) where f is a piecewise-linear (affine) function of x and there are n breakpoints (x_k, y_k) which k=1,...,n. $y\_k = f(x\_k)$. f is linearly interpolated between the breakpoints. y and x are scalars.

vi. **Sigmoid network (SN) activation function** Both sigmoid and wavelet network estimators which use the neural networks composing an input layer, an output layer and a hidden layer using wavelet and sigmoid activation functions as shown in Fig.12

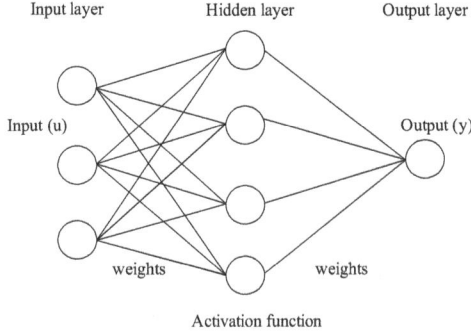

Fig. 12. Structure of nonlinear estimators

A sigmoid network nonlinear estimator combines the radial basis neural network function using a sigmoid as the activation function. This estimator is based on the following expansion:

$$y(u) = (u - r)PL + \sum_{i}^{n} a_i f((u - r)Qb_i - c_i) + d \qquad (6)$$

when u is input and y is output. r is the the regressor. Q is a nonlinear subspace and P a linear subspace. L is a linear coefficient. d is an output offset. b is a dilation coefficient., c a translation coefficient and a an output coefficient. f is the sigmoid function, given by the following equation (7)

$$f(z) = \frac{1}{e^{-z} + 1} \qquad (7)$$

v. **Wavelet Network (WN) activation function.** The term wavenet is used to describe wavelet networks. A wavenet estimator is a nonlinear function by combination of a wavelet theory and neural networks. Wavelet networks are feed-forward neural networks using wavelet as an activation function, based on the following expansion in the equation (8)

$$y = (u - r)PL + \sum_{i}^{n} as_i * f(bs(u - r)Q + cs) + \sum_{i}^{n} aw_i * g(bw_i(u - r)Q + cw_i) + d \qquad (8)$$

Which u and y are input and output functions. Q and P are a nonlinear subspace and a linear subspace. L is a linear coefficient. d is output offset. as and aw are a scaling coefficient and a wavelet coefficient. bs and bw are a scaling dilation coefficient and a wavelet dilation coefficient. cs and cw are scaling translation and wavelet translation coefficients. The scaling function f (.) and the wavelet function g(.) are both radial functions, and can be written as the equation (9)

$$f(u) = \exp(-0.5 * u' * u)$$
$$g(u) = (\dim(u) - u' * u) * \exp(-0.5 * u' * u) \tag{9}$$

In a system identification process, the wavelet coefficient (a), the dilation coefficient (b) and the translation coefficient (c) are optimized during model learning steps to obtain the best performance model.

3.3 MIMO Hammerstein-Wiener system identification

The voltage and current are two basic signals considered as input/output of PV grid connected systems. The measured electrical input and output waveforms of a system are collected and transmitted to the system identification process. In Fig. 13 show a PV based inverter system which are considered as SISO (single input-single output) or MIMO (multi input-multi output), depending on the relation of input-output under study [57]. In this paper, the MIMO nonlinear model of power inverters of PV systems is emphasized because this model gives us both voltage and current output prediction simultaneously.

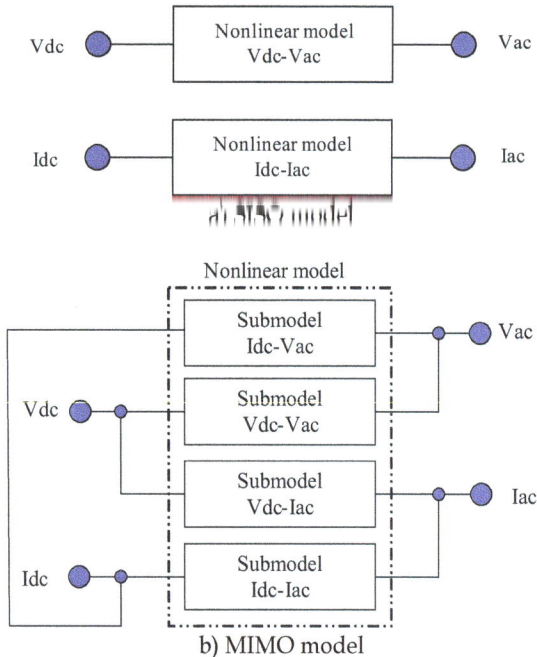

b) MIMO model

Fig. 13. Block diagram of nonlinear SISO and MIMO inverter model

For one SISO model, there is only one corresponding set of nonlinear estimators for input and output, and one set of linear parameters, i.e. pole b_n, zero f_n and delay n_k, as written in the equation (9). For SIMO, MISO and MIMO models, there would be more than one set of nonlinear estimators and linear parameters. The relationships between input-output of the MIMO model have been written in the equation (10) whereas v_{dc} is DC voltage, i_{dc} DC current, v_{ac} AC voltage, i_{ac} AC current. q is shift operator as equivalent to z transform. $f(\bullet)$ and $h(\bullet)$ are input and output nonlinear estimators. In this case a deadzone and saturation are selected into the model. In the MIMO model the relation between output and input has four relations as follows (i) DC voltage (v_{dc}) – AC voltage (v_{ac}), (ii) DC voltage (v_{dc}) – AC current (i_{ac}), (iii) DC current (i_{dc}) – AC voltage (v_{ac}) and (iv) DC current(v_{dc})–AC voltage (v_{ac}).

$$\left. \begin{aligned} v_{ac}(t) &= \frac{B(q)}{F(q)} f(v_{dc}(t-n_k)) + e(t) \\ i_{ac}(t) &= \frac{B(q)}{F(q)} f(i_{dc}(t-n_k)) + e(t) \end{aligned} \right\} \tag{10}$$

$$\left. \begin{aligned} V_{ac}(t) &= h\left(\frac{B_1(q)}{F_1(q)} f(v_{dc}(t-n_{k1})) + e(t)\right) \otimes h\left(\frac{B_2(q)}{F_2(q)} f(i_{dc}(t-n_{k2})) + e(t)\right) \\ I_{ac}(t) &= h\left(\frac{B_3(q)}{F_3(q)} f(v_{dc}(t-n_{k3})) + e(t)\right) \otimes h\left(\frac{B_4(q)}{F_4(q)} f(i_{dc}(t-n_{k4})) + e(t)\right) \end{aligned} \right\} \tag{11}$$

$$\begin{aligned} B_i(q) &= b_1 + b_2 + \ldots + b_{n_{bi}} q^{-n_{bi}+1} \\ F_i(q) &= f_1 + f_2 + \ldots + f_{n_{fi}} q^{-n_{fi}+1} \end{aligned} \tag{12}$$

Where n_{bi}, n_{fi} and n_{ki} are pole, zero and delay of linear model. Where as number of subscript i are 1,2,3 and 4 which stand for relation between DC voltage-AC voltage, DC current-AC voltage, DC voltage-AC current and DC current-AC current respectively. The output voltage and output current are key components for expanding to the other electrical values of a system such power, harmonic, power factor, etc. The linear parameters, zeros, poles and delays are used to represent properties and relation between the system input and output. There are two important steps to identify a MIMO system. The first step is to obtain experimental data from the MIMO system. According to different types of experimental data, the second step is to select corresponding identification methods and mathematical models to estimate model coefficients from the experimental data. The model is validated until obtaining a suitable model to represent the system. The obtained model provides properties of systems. State-space equations, polynomial equations as well as transfer functions are used to describe linear systems. Nonlinear systems can be describes by the above linear equations, but linearization of the nonlinear systems has to be carried out. Nonlinear estimators explain nonlinear behaviors of nonlinear system. Linear and nonlinear graphical tools are used to describe behaviors of systems regarding controllability, stability and so on.

4. Experimental

In this work, we model one type of a commercial grid connected single phase inverters, rating at 5,000 W. The experimental system setup composes of the inverter, a variable DC power supply (representing DC output from a PV array), real and complex loads, a digital power meter, a digital oscilloscope, , a AC power system and a computer, shown schematically in Fig 14. The system is connected directly to the domestic electrical system (low voltage). As we consider only domestic loads, we need not isolate our test system from the utility (high voltage) by any transformer. For system identification processes, waveforms are collected by an oscilloscope and transmitted to a computer for batch processing of voltage and current waveforms.

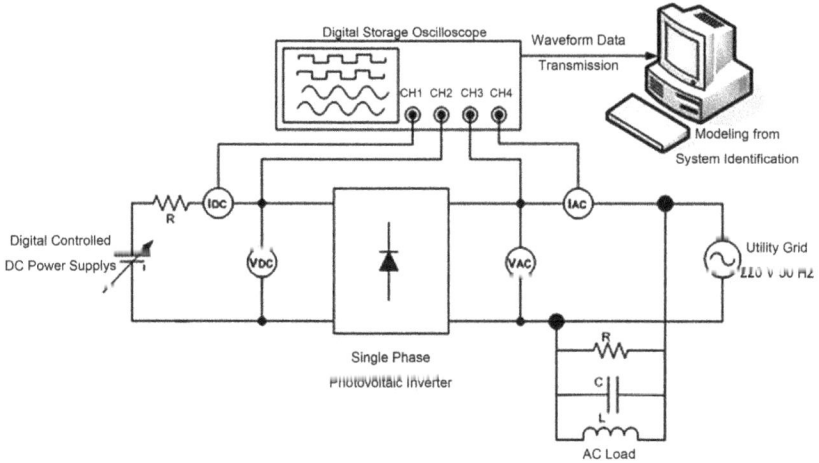

Fig. 14. Experimental setup

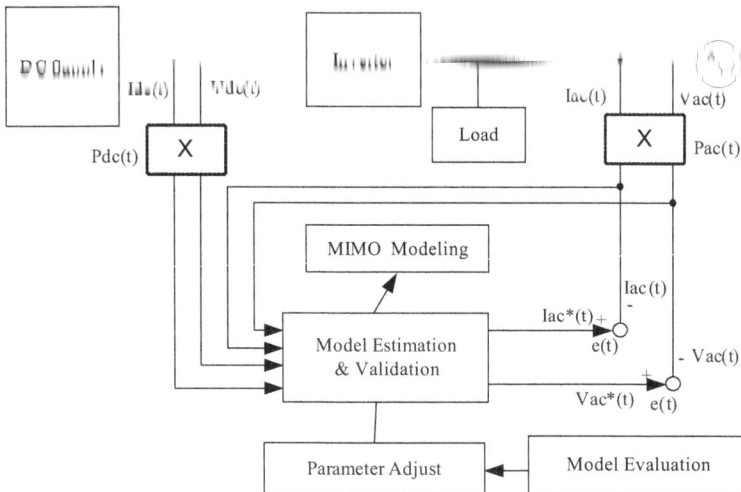

Fig. 15. An inverter modeling using system identification process

Major steps in experimentation, analysis and system identifications are composed of Testing scenarios of six steady state conditions and two transient conditions are carried out on the inverter, from collected data from experiments, voltage and current waveform data are divided in two groups to estimate models and to validate models previously mentioned.

The system identification scheme is shown in Fig.15. Good accuracy of models are achieved by selecting model structures and adjusting the model order of linear terms and nonlinear estimators of nonlinear systems. Finally, output voltage and current waveforms for any type of loads and operating conditions are then constructed from the models. This allows us to study power quality as required.

4.1 Steady state conditions

To emulate working conditions of PVGCS systems under environment changes (irradiance and temperature) affecting voltage and current inputs of inverters, six conditions of DC voltage variations and DC current variations. The six conditions are listed as Table 1. They are 3 conditions of a fixed DC current with DC low, medium and high voltage, i.e. , FCLV (Fixed Current Low Voltage), FCMV (Fixed Current Medium Voltage) and FCHV (Fixed Current High Voltage) which shown in Fig. 16. The other three corresponding conditions are a DC fixed voltage with DC low, medium and high current, i.e., FVLC (Fixed Voltage Low Current), FVMC (Fixed Voltage Medium Current), and FVHC (Fixed Voltage High Current) as shown in Fig.17.

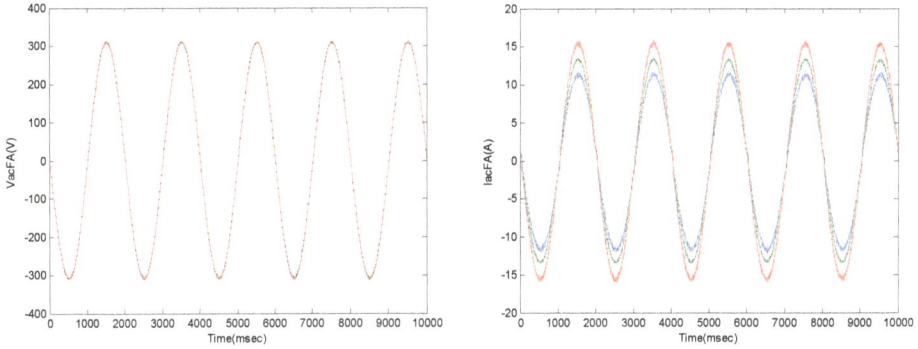

Fig. 16. AC voltage and current waveforms corresponding to FCLV, FCMV and FCHV conditions

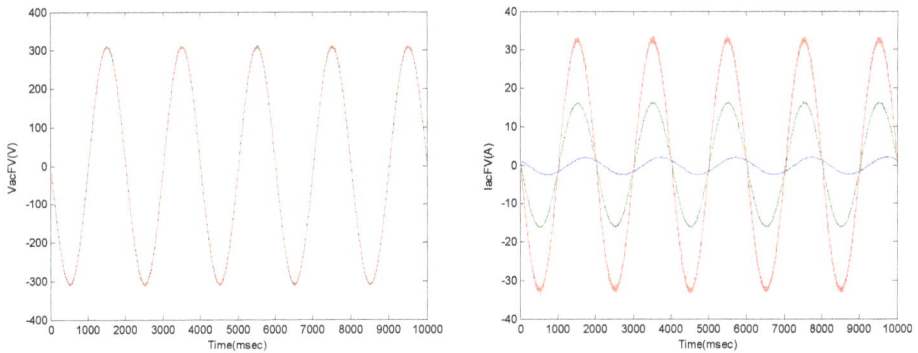

Fig. 17. AC voltage and current waveforms corresponding to FVLC, FVMC and FVHC conditions

No.	Case	Idc (A)	Vdc (V)	Pdc (W)	Iac (A)	Vac (A)	Pac (VA)
1	FCLV	12	210	2520	11	220	2420
2	FCMV	12	240	2880	13	220	2860
3	FCHV	12	280	3360	15	220	3300
4	FVLC	2	235	470	2	220	440
5	FVMC	10	240	2,400	10	220	2,200
6	FVHC	21	245	5,145	23	220	5,060

Table 1. DC and AC parameters of an inverter under changing operating conditions

4.2 Transient conditions

Transient conditions are studied under two cases which composed of step up power transient and step down power transient. The step up condition is done by increasing power output from 440 to1,540 W, and the step down condition from 1,540 to 440 W, shown in Table 2. Power waveform data of the two conditions are divided in two groups, the first group is used to estimate model, the second group to validate model. Examples of captured voltage and current waveforms under the step-up power transient condition (440 W or 2 A) to 1540 W or 7A) and the step-down power transient condition (1540 W or 7A) to 440 W (2A) are shown in Fig. 18 and 19, respectively.

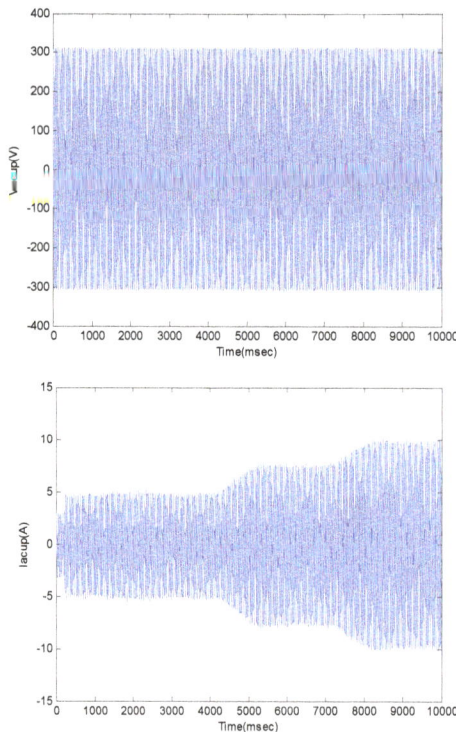

Fig. 18. AC voltage and current waveforms under the step up transient condition

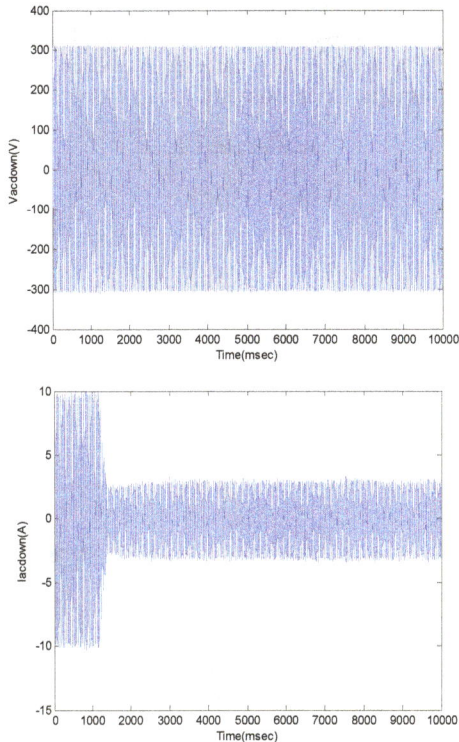

Fig. 19. AC voltage and current waveforms under the step down transient condition

Electrical parameters	Voltage, current and power for transient step down conditions		Voltage, current and power for transient step up conditions	
AC output voltage (V)	220	220	220	220
AC output current (A)	7	2	2	7
AC output power (W)	1540	440	440	1540

Table 2. Inverter operations under step up/down conditions

5. Results and discussion

In the next step, data waveforms are divided into the "estimate data set" and the "validate data set". Examples are shown in Fig. 20, whereby the first part of the AC and DC voltage waveforms are used as the estimate data set and the second part the validate data set. The system identification process is executed according to mentioned descriptions on the Hammerstein-Wiener modeling.

The validation of models is taken by considering (i) model order by adjusting the number of poles plus zeros. The system must have the lowest-order model that adequately captures the system dynamics.(ii) the best fit, comparing between modeling and experimental outputs, (iii) FPE and AIC, both of these values need be lowest for high accuracy of modeling (iv)

Nonlinear behavior characteristics. For example, linear interval of saturation, zero interval of dead-zone, wavenet, sigmoid network requiring the simplest and less complex function to explain the system. Model properties, estimators, percentage of accuracy, final Prediction Error-FPE and Akaikae Information Criterion-AIC are as follows [58]:

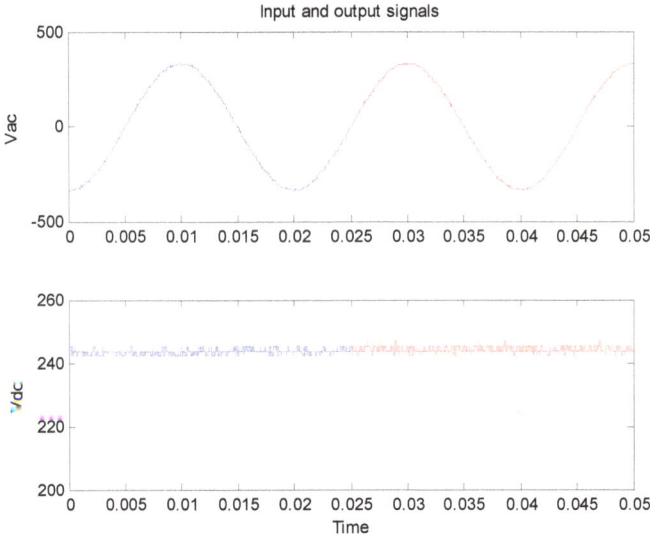

Fig. 20. Data divided into Estimated and validated data

Criteria for Model selection

The percentage of the best fit accuracy in operation (FY) is obtained from comparison between experimental waveform and simulation modeling waveform

$$Best\ fit = 100 * (1 - norm(y^* - y) / norm(y - \bar{y})) \tag{13}$$

where y* is the simulated output, y is the measured output and $\bar{y}$ is the mean of output. FPE is the Akaike Final Prediction Error for the estimated model, of which the error calculation is defined as equation (14)

$$FPE = V\left(\frac{1 + d/N}{1 - d/N}\right) \tag{14}$$

where V is the loss function, d is the number of estimated parameters, N is the number of estimation data. The loss function V is defined in Equation (15) where θ_N represents the estimated parameters.

$$V = \det\left(\frac{1}{N}\sum_1^N \varepsilon(t,\theta_N)\big(\varepsilon(t,\theta_N)\big)^T\right) \tag{15}$$

The Final Prediction Error (FPE) provides a measure of a model quality by simulating situations where the model is tested on a different data set. The Akaike Information

Criterion (AIC), as shown in equation (16), is used to calculate a comparison of models with different structures.

$$AIC = \log V + \frac{2d}{N} \tag{16}$$

Waveforms of input and output from the experimental setup consist of DC voltage, DC output current, AC voltage and AC output current. Model properties, estimators, percentage of accuracy, Final Prediction Error - FPE and Akaikae Information Criterion - AIC of the model are shown in Table 3. Examples of voltage and current output waveforms of multi input-multi output (MIMO) model in steady state condition (FVMC) having accuracy 97.03% and 91.7 % are shown in Fig 21.

Type	I/P	O/P	Linear model parameters [nb₁ nb₂ nb₃ nb₄] poles [nf₁ nf₂ nf₃ nf₄] zeros [nk₁ nk₂ nk₃ nk₄] delays	% fit Voltage Current	FPE	AIC
Steady state conditions						
FCLV	DZ	DZ	[4 4 3 5]; [5 5 3 6]; [3 4 4 2]	87.3 85.7	3,080.90	10.9
FCMV	PW	PW	[5 2 4 4]; [4 2 3 4]; [2 2 4 3];	84.5 86.4	729.03	6.59
FCHV	ST	ST	[2 2 3 4]; [1 2 1 2]; [2 1 3 2];	89.5 88.7	26.27	3.26
FVLC	SN	SN	[3 6 3 2]; [8 5 4 3]; [2 4 3 5];	56.8 60.5	0.07	2.57
FVMC	WN	WN	[3 4 2 5]; [4 2 3 4]; [2 3 2 4];	97.03 91.7	254.45	7.89
FVHC	WN	WN	[1 4 3 5]; [5 2 3 5]; [1 3 2 4];	88 94	3,079.8	10.33
Transient conditions						
Step Up	DZ	DZ	[3 4 2 4]; [4 5 4 3]; [2 3 5 5]; [4 5 2 2];	91.75 87.20	3,230	7.40
Step Down	PW	PW	[3 5 5 3]; [3 5 4 3]; [3 5 5 4]; [4 4 4 1];	85.99 85.12	3,233	10.0

Table 3. Results of a PV inverter modeling using a Hammerstein-Wiener model

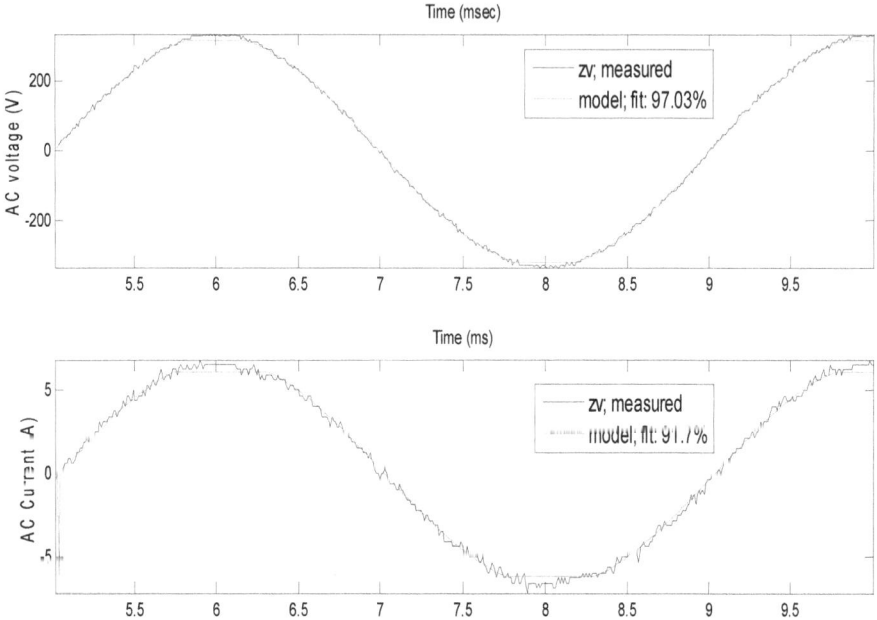

Fig. 21. Comparison of AC voltage and current output waveforms of a steady state FVMC MIMO model

In Table 3 n_{bi}, n_{fi} and n_{ki} are poles, zeros and delays of a linear model. The subscript (1, 2, 3 and 4) stands for relations between DC voltage-AC voltage, DC current-AC voltage, DC voltage-AC current and DC current-AC current respectively. Therefore, the linear parameters of the model are $[n_{b1}, n_{b2}, n_{b3}, n_{b4}]$, $[n_{f1}, n_{f2}, n_{f3}, n_{f4}]$, $[n_{k1}, n_{k2}, n_{k3}, n_{k4}]$.

The first value of percentages of fit in each type, shown in the Table 3, is the accuracy of the voltage output, the second the current output from the model. From the results, nonlinear estimators can describe the photovoltaic grid connected system. The estimators are good in terms of accuracy, with a low order model or a low FPE and AIC. Under most of testing conditions, high accuracy of more than 85% is achieved, except the case of FVLC. This is because of under such an operating condition, the inverter has very small current, and it is operating under highly nonlinear behavior. Then complex of nonlinear function and parameter adjusted is need for achieve the high accuracy and low order of model. After obtaining the appropriate model, the PVGCS system can be analyzed by nonlinear and linear analyses. Nonlinear parts are analyzed from the properties of nonlinear function such as dead-zone interval, saturation interval, piecewise range, Sigmoid and Wavelet properties. Nonlinear properties are also considered, e.g. stability and irreversibility In order to use linear analysis, Linearization of a nonlinear model is required for linear control design and analysis, with acceptable representation of the input/output behaviors. After linearizing the model, we can use control system theory to design a controller and perform linear analysis. The linearized command for computing a first-order Taylor series approximation for a system requires specification of an operating point. Subsequently, mathematical representation can be obtained, for example, a discrete time invariant state space model, a transfer function and graphical tools.

6. Applications: Power quality problem analysis

A power quality analysis from the model follows the Standard IEEE 1159 Recommend Practice for Monitoring Electric Power Quality [59]. In this Standard, the definition of power quality problem is defined. In summary, a procedure of this Standard when applied to operating systems can be divided into 3 stages (i) Measurement Transducer, (ii) Measurement Unit and (iii) Evaluation Unit. In comparing operating systems and modeling, modeling is more advantageous because of its predictive power, requiring no actual monitoring. Based on proposed modeling, the measurement part is replaced by model prediction outputs, electrical values such as RMS and peak values, frequency and power are calculated, rather than measured. The actual evaluation is replaced by power quality analysis. The concept representation is shown in Fig.22.

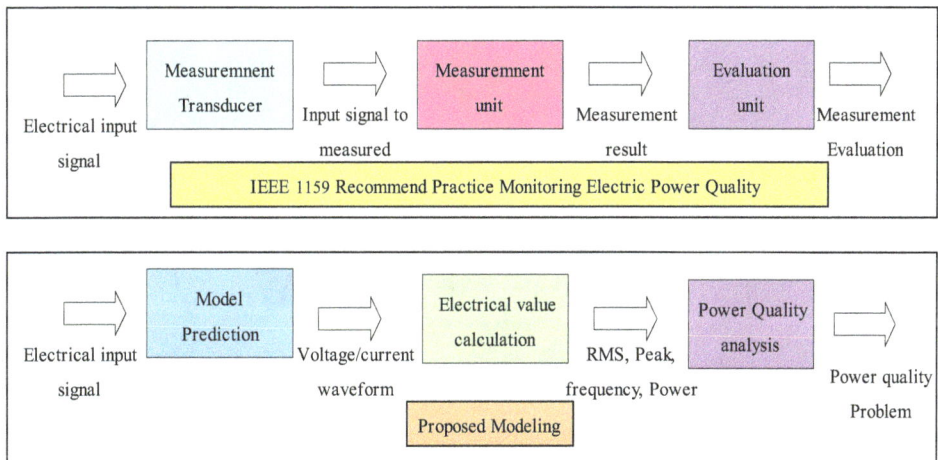

Fig. 22. Diagram of power quality analysis from IEEE 1159 and application to modeling

6.1 Model output prediction
In this stage, the model output prediction is demonstrated. From the 8 operation conditions selected   in experimental, we choose two representative case. One is the steady state Fix Voltage High Current (FVHC) condition, the other the transient step down condition.   To illustrate model predictive power, Fig.23 shows an actual and   predictive output current waveforms of the transient step down condition. We see good agreement between experimental results and modeling results.

6.2 Electrical parameter calculation
In this stage, output waveforms are used to calculate RMS, peak and per unit (p.u.) values, period, frequency, phase angle, power factor, complex power (real, reactive and apparent power) Total Harmonic Distortion - THD.

6.2.1 Root mean square
RMS values of voltage and current can be calculated from the following equations:

$$V_{rms} = \sqrt{\frac{1}{T}\int_0^T v^2(t)dt}$$

(17)

$$I_{rms} = \sqrt{\frac{1}{T}\int_0^T v^2(t)dt}$$

$$V_m = \sqrt{2}V_{rms}$$

(18)

$$I_m = \sqrt{2}I_{rms}$$

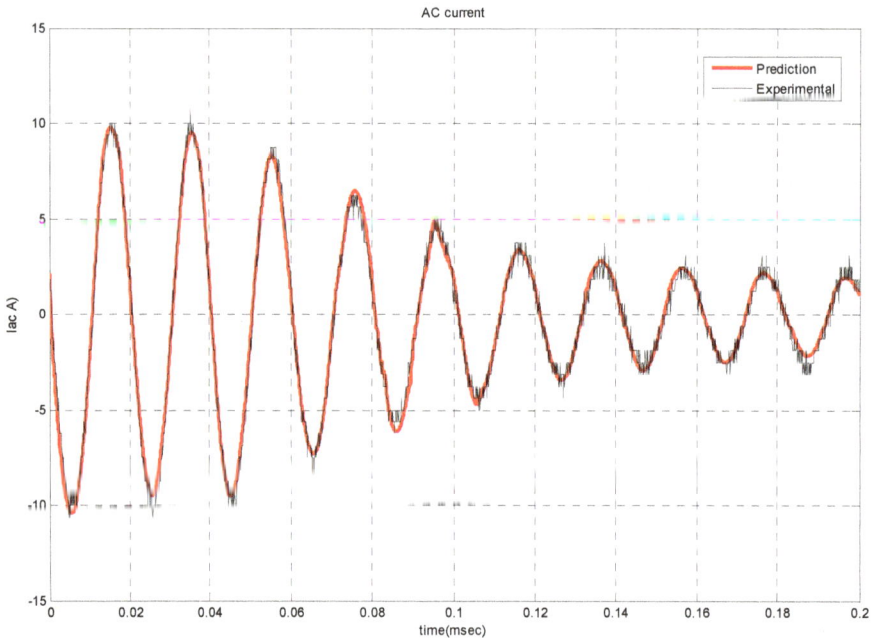

Fig. 23. Prediction and experiment results of AC output current under a transient step down condition

6.2.2 Period, frequency and phase angle
We calculate a phase shift between voltage and current from the equation (19), and the frequency (f) from equation (20).

$$\phi = \frac{\Delta t(ms)\cdot 360°}{T\ ms}$$

(19)

$$f = \frac{1}{T}$$

(20)

Δt is time lagging or leading between voltage and current (ms), T is the waveform period.

6.2.3 Power factor, apparent power, active power and reactive power

The power factor, the apparent power S (VA), the active power P (W), and the reactive power Q (Var) are related through the equations

$$PF = \cos\phi = \frac{P(W)}{S(VA)} \tag{21}$$

$$S = VI' \tag{22}$$

$$P = S\cos\phi \tag{23}$$

$$Q = S\sin\phi \tag{24}$$

6.2.4 Harmonic calculation

Total harmonic distortion of voltage (THD$_v$) and current (THD$_i$) can be calculated by the Equations 25 and 26, respectively.

$$\%THD_i = \frac{\sqrt{\sum_{h=2}^{\infty} I_{h(rms)}^2}}{I_{1(rms)}} x\ 100\% \tag{25}$$

$$\%THD_v = \frac{\sqrt{\sum_{h=2}^{\infty} V_{h(rms)}^2}}{V_{1(rms)}} x\ 100\% \tag{26}$$

Where Vh (rms) is RMS value of h th voltage harmonic , Ih (rms) RMS value of h th current harmonic,   V1 (rms) RMS value of fundamental voltage and I1 (rms)   RMS value of fundamental current

Parameter	Steady state FVHC condition			Transient step down condition		
	Experimental	Modeling	% Error	Experimental	Modeling	% Error
Vrms (V)	218.31	218.04	0.12	217.64	218.20	-0.26
Irms (A)	23.10	23.21	-0.48	4.47	4.45	0.45
Frequency (Hz)	50	50	0.00	50.00	50.00	0.00
Power Factor	0.99	0.99	0.00	0.99	0.99	0.00
THDv (%)	1.15	1.2	-4.35	1.18	1.24	-5.08
THDi (%)	3.25	3.12	4.00	3.53	3.68	-4.25
S (VA)	5044.38	5060.7	-0.32	972.85	970.99	0.19
P (W)	4993.94	5010.1	-0.32	963.12	961.28	0.19
Q (Var)	711.59	713.85	-0.32	137.24	136.97	0.19
V p.u.	0.99	0.99	0.00	0.98	0.99	-1.02

Table 4. Comparison of measured and modeled electrical parameters of the FVHC condition and the transient step down condition

We next demonstrate accuracy and precision of power quality prediction from modeling. Table 4 shows the comparisons. Two representative cases mentioned above are given, i.e. the steady state Fix Voltage High Current (FVHC) condition, and the transient step down condition. Comparison of THDs is shown in Fig. 23. Agreements between experiments and modeling results are good.

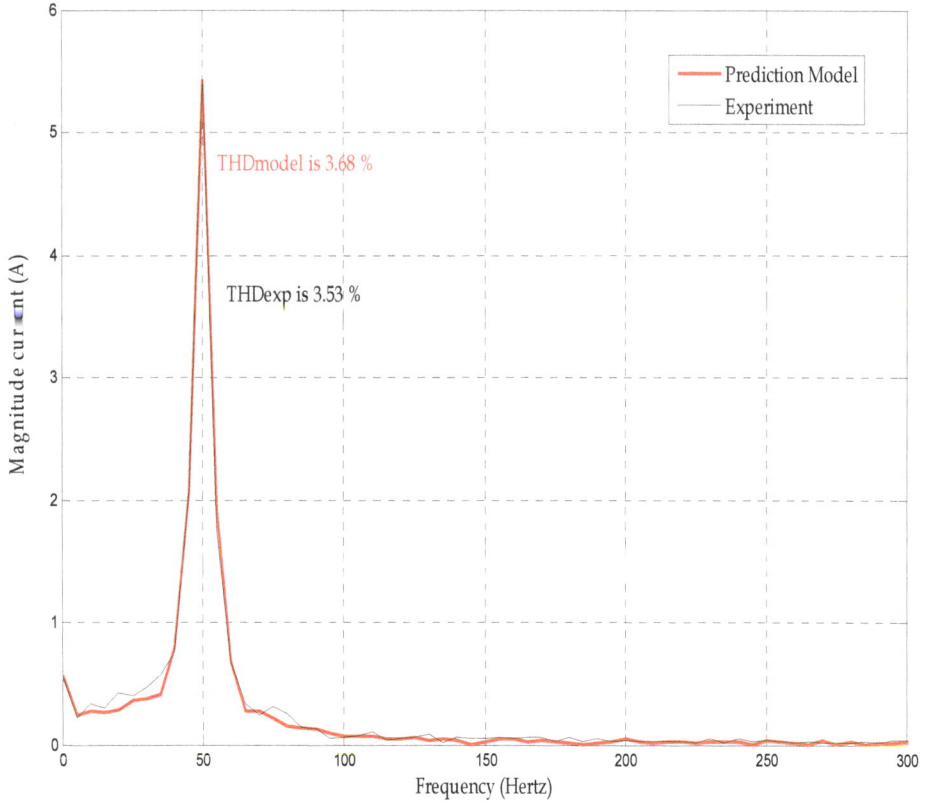

Fig. 24. Comparison of measured and modeled THD of AC current of the transient step down condition

6.3 Power quality problem analysis
The power quality phenomena are classified in terms of typical duration, typical voltage magnitude and typical spectral content. They can be broken down into 7 groups on transient, short duration voltage, long duration voltage, voltage unbalance, waveform distortion, voltage fluctuation or flicker, frequency variation. Comparisons of the Standard values and modeled outputs of the FVHC and the transient step down conditions are shown in Table 5. The results show that under both the steady state and the transient cases, good power quality is achieved from the PVGCS.

Type	Typical Duration	Typical Voltage Magnitude	Typical Spectral Content	Steady State FVHC	Transient Step down	Result
1.Transient						
- Impulsive	5 ns – 0.1ms					
- Oscillation						
- low frequency	0.3-50 ms	0.4 pu.	< 5 kHz	0.99 pu.	0.99 pu.	Pass
- medium frequency	5-20 ms	0-8 pu.	5-500 kHz			
- high frequency	0-5 ms	0.4 pu.	0.5–5 MHz			
2.Short Duration						
- voltage sag	10 ms-1 min	0.1-0.9 pu.	-	0.99 pu.	0.99 pu.	Pass
- voltage swell	10 ms-1 min	1.1-1.8 pu.				
3. Long Duration						
- overvoltage (OV)	> 1 min	> 1.1 pu.		0.99 pu.	0.99 pu.	Pass
- undervoltage (UV)	> 1 min	< 0.9 pu.	-			
- voltage Interruption	> 1 min	0 pu.				
4. Voltage Unbalance	Steady state	0.5-2%	-	-	-	-
5.Waveform distortion						
- Harmonic voltage	Steady state	< 5% THD	0-100th	1.20 %	1.24 %	Pass
- Harmonic Current	Steady state	< 20% THD	0-100th	3.25 %	3.68 %	Pass
- Interharmonic	Steady state	0-2%	0-6 kHz	-	-	-
- DC offset	Steady state	0-0.1%	< 200 kHz	-	-	-
- Notching	Steady state	-	-	-	-	-
- Noise	Steady state	0-1%	Broad band	-	-	-
6.Voltage fluctuation						
- Flicker	Intermittent	0.1-7%	< 25 Hz	0.01%	0.01%	Pass
7.Frequency variation			± 3 Hz			
- Overfrequency	< 10 s	-	> 53 Hz	50	50	Pass
- Underfrequency			< 47 Hz			

Table 5. Comparison modeling output with Categories and Typical Characteristics of power system electromagnetic phenomena

7. Conclusions

In this paper, a PVGCS system is modeled by the Hammerstein-Wiener nonlinear system identification method. Two main steps to obtain models from a system identification process are implemented. The first step is to set up experiments to obtain waveforms of DC inverter voltage/current, AC inverter voltage/current, point of common coupling (PCC) voltage, and grid and load current. Experiments are conducted under steady state and transient conditions for commercial rooftop inverters with rating of few kW, covering resistive and complex loads. In the steady state experiment, six conditions are carried out. In the transient case, two conditions of operating conditions are conducted. The second stage is to derive system models from system identification software. Collected waveforms are transmitted into a computer for data processing. Waveforms data are divided in two groups. One group is used to estimate models whereas the other group to validate models. The developed programming determines various model waveforms and search for model waveforms of maximum accuracy compared with actual waveforms. This is achieved through selecting model structures and adjusting the model order of the linear terms and nonlinear estimators of nonlinear terms. The criteria for selection of a suitable model are the "Best Fits" as defined by the software, and a model order which should be minimum.

After obtaining appropriate models, analysis and prediction of power quality are carried out. Modeled output waveforms relating to power quality analysis are determined from different scenarios. For example, irradiances and ambient temperature affecting DC PV outputs and nature of complex local load can be varied. From the model output waveforms, determination is made on power quality aspects such as voltage level, total harmonic distortion, complex power, power factor, power penetration and frequency deviation. Finally, power quality problems are classified.

Such modelling techniques can be used for system planning, prevention of system failures and improvement of power quality of roof-top grid connected systems. Furthermore, they are not limited to PVGCS but also applicable to other distributed energy generators connected to grids.

8. Acknowledgements

The authors would like to express their appreciations to the technical staff of the CES Solar Cells Standards and Testing Center (CSSC) of King Mongkut's University of Technology Thonburi (KMUTT) for their assistance and valuable discussions. One of the authors, N. Patcharaprakiti receives a scholarship from Rajamangala University of Technology Lanna (RMUTL), a research grant from the Energy Policy and Planning Office (EPPO) and Office of Higher education commission, Ministry of Education, Thailand for enabling him to pursue his research of interests. He is appreciative of the scholarship and research grant supports.

9. References

[1] Bollen M. H. J. and Hager M., Power quality: integrations between distributed energy resources, the grid, and other customers. Electrical Power Quality and Utilization Magazine, vol. 1, no. 1, pp. 51–61, 2005.

[2] Vu Van T., Belmans R., Distributed generation overview: current status and challenges. Inter-national Review of Electrical Engineering (IREE), vol. 1, no. 1, pp. 178–189, 2006.

[3] Pedro González, "Impact of Grid connected Photovoltaic System in the Power Quality of a Network", Power Electrical and Electronic Systems (PE&ES), School of Industrial Engineering, University of Extremadura.

[4] Barker P. P., De Mello R. W., Determining the impact of distributed generation on power systems: Part 1 – Radial distribution systems. PES Summer Meeting, IEEE, Vol. 3, pp. 1645–1656, 2000.

[5] Vu Van T., Impact of distributed generation on power system operation and control. PhD Thesis, Katholieke Universiteit Leuven, 2006.

[6] Muh. Imran Hamid and Makbul Anwari, "Single phase Photovoltaic Inverter Operation Characteristic in Distributed Generation System", Distributed Generation, Intech book, 2010

[7] Vu Van T., Driesen J., Belmans R., Interconnection of distributed generators and their influences on power system. International Energy Journal, vol. 6, no. 1, part 3, pp. 127–140, 2005.

[8] A. Moreno-Munoz, J.J.G. de-la-Rosa, M.A. Lopez-Rodriguez, J.M. Flores-Arias, F.J. Bellido-Outerino, M. Ruiz-de-Adana, "Improvement of power quality using distributed generation", Electrical Power and Energy Systems 32 (2010) 1069–1076

[9] P.R.Khatri, V.S.Jape, N.M.Lokhande, B.S.Motling, "Improving Power Quality by Distributed Generation" Power Engineering Conference, 2005. IPEC 2005. The 7th International.

[10] Vu Van T., Driesen J., Belmans R., Power quality and voltage stability of distribution system with distributed energy resources. International Journal of Distributed Energy Resources, vol. 1, no. 3, pp. 227–240, 2005.

[11] Woyte A., Vu Van T., Belmans R., Nijs J., Voltage fluctuations on distribution level introduced by photovoltaic systems. IEEE Transactions on Energy Conversion, vol. 21, no. 1, pp. 202–209, 2006.

[12] Thongpron, K.Kirtara, Effects of low radiation on the power quality of a distributed PV-grid connected system, Solar Energy Materials and Solar Cells Solar Energy Materials and Solar Cells, Vol. 90, No. 15. (22 September 2006), pp. 2501-2508.

[13] S.K. Khadem, M.Basu and M.F.Conlon, "Power quality in Grid connected Renewable Energy Systems : Role of Custom Power Devices", International conference on Renewable Energies and Power quality (ICREPQ'10), Granada, Spain, 23rd to 25th March, 2010

[14] Mohamed A. Eltawil Zhengming Zhao, "Grid-connected photovoltaic power systems: Technical and potential problems—A review" Renewable and Sustainable Energy Reviews Volume 14, Issue 1, January 2010, Pages 112-129

[15] Soeren Baekhoej Kjaer, et al., "A Review of Single-Phase Grid-Connected inverters for Photovoltaic Modules, EEE Transactions and industry applications, Transactions on Industry Applications, Vol. 41, No. 5, September 2005

[16] F. Blaabjerg, Z. Chen and S. B. Kjaer, "Power Electronics as Efficient Interface in Dispersed Power Generation Systems." IEEE Trans. on Power Electronics 2004; vol.19 no. 5. Pp. 1184-1194. 2000.

[17] Chi, Kong Tse, "Complex behavior of switching power converters", Boca Raton : CRC Press, c2004

[18] Giraud, F., Steady-state performance of a grid-connected rooftop hybrid wind-photovoltaic power system with battery storage, IEEE Transactions on Energy Conversion, 2001.

[19] G.Saccomando, J.Svensson, "Transient Operation of Grid-connected Voltage Source Converter Under Unbalanced Voltage Conditions", IEEE Industry Applications Conference, 2001.

[20] Li Wang and Ying-Hao Lin, "Small-Signal Stability and Transient Analysis of an Autonomous PV System", Transmission and Distribution Conference and Exposition, 2008.

[21] Zheng Shi-cheng, Liu Xiao-li, Ge Lu-sheng; , "Study on Photovoltaic Generation System and Its Islanding Effect," Industrial Electronics and Applications, 2007. ICIEA 2007. 2nd IEEE Conference on , 23-25 May 2007, pp.2328-2332

[22] Pu kar Mahat, et al, "Review of Islanding Detection Methods for Distributed Generation", DRPT 6-9 April 2008, Nanjing, China.

[23] Yamashita, H. et al. "A novel simulation technique of the PV generation system using real weather conditions Proceedings of the Power Conversion Conference, 2002 PCC Osaka 2002.

[24] Golovanov, N, "Steady state disturbance analysis in PV systems", IEEE Power Engineering Society General Meeting, 2004.

[25] D. Maksimovic , et al, "Modeling and simulation of power electronic converters," Proc. IEEE, vol. 89, pp. 898, Jun. 2001

[26] R. D. Middlebrook and S. Cuk "A general unified approach to modeling switching converter power stages," Int.J. Electron., vol. 42, pp. 521, Jun, 1977,

[27] Marisol Delgado and Hebertt Sira-RamírezA bond graph approach to the modeling and simulation of switch regulated DC-to-DC power supplies, Simulation Practice and Theory Volume 6, Issue 7, 15 November 1998, Pages 631-646.

[28] R.E. Araujo, et al, "Modelling and simulation of power electronic systems using a bond formalism", Proceeding of the 10th Mediterranean Conference on control and Automation – MED2002, Lisbon, Portugal, July 9-12 2002.

[29] H. I. Cho, "A Steady-State Model of the Photovoltaic System in EMTP", The International Conference on Power Systems Transients (IPST2009), Kyoto, Japan June 3-6, 2009

[30] Y. Jung, J. Sol, G. Yu, J. Choj, "Modelling and analysis of active islanding detection methods for photovoltaic power conditioning systems," Electrical and Computer Engineering, 2004. Canadian Conference on , vol.2, 2 5 May 2004, pp. 979 202

[31] Seul-Ki Kim, "Modeling and simulation of a grid connected PV generation system for electromagnetic transient analysis", Solar Energy, Volume 83, Issue 5, May 2009, Pages 664-678.

[32] Onbilgin, Guven, et al, "Modeling of power electronics circuits using wavelet theory", Sampling Theory in Signal and Image Processing, September, 2007.

[33] George A. Bekey, "System identification- an introduction and a survey, Simulation, Transaction of the society for modeling and simulation international", October 1970 vol. 15 no. 4, 151-166.

[34] J.sjoberg, et al., "Nonlinear black-box Modeling in system identification: a unified overview, Automatica Journal of IFAC, Vol 31, Issue 12, December 1995.

[35] K. T. Chau and C. C. Chan "Nonlinear identification of power electronics systems," Proc. Int. Conf. Power Electronics and D20ve Systems, vol. 1, 1995, p. 329.

[36] J. Y. Choi , et al. "System identification of power converters based on a black-box approach," IEEE Trans. Circuits Syst. I, Fundam. Theory Appl., vol. 45, pp. 1148, Nov. 1998.

[37] F.O. Resende and J.A.Pecas Lopes, "Development of Dynamic Equivalents for Microgrids using system identification theory", IEEE of Power Technology, Lausanne, pp. 1033-1038, 1-5 July 2007.

[38] N.Patcharaprakiti, et al, "Modeling of Single Phase Inverter of Photovoltaic System Using System Identification", Computer and Network Technology (ICCNT), 2010 , April 2010, Page(s): 462 – 466.

[39] Hatanaka, et al. , "Block oriented nonlinear model identification by evolutionary computation approach", Proceedings of IEEE Conference on Control Applications, 2003, Vol 1, pp 43- 48, June 2003.

[40] Li, G. Identification of a Class of Nonlinear Autoregressive Models With Exogenous Inputs Based on Kernel Machines, IEEE Transactions on Signal Processing, Vol 59, Issue 5, May 2011.

[41] T. Wigren, "User choices and model validation in system identification using nonlinear Wiener models", Proc13:th IFAC Symposium on System Identification, Rotterdam, The Netherlands, pp. 863-868, August 27-29, 2003.

[42] Alonge, et al, "Nonlinear Modeling of DC/DC Converters Using the Hammerstein's Approach" IEEE Transactions on Power Electronics, July 2007, Volume: 22, Issue: 4,pp. 1210-1221.

[43] Guo. F. and Bretthauer, G. "Identification of cascade Wiener and Hammerstein systems", In Proceeding. of ISATED Conference on Applied Simulation and Modeling, Marbella, Spain, September, 2003.

[44] N.Patcharaprakiti and et al., "Modeling of single phase inverter of photovoltaic system using Hammerstein–Wiener nonlinear system identification", Current Applied Physics 10 (2010) S532–S536,

[45] A Guide To Photovoltaic Panels, PV Panels and Manufactures' Data, January 2009.

[46] Tomas Markvart, "Solar electricity", Wiley, 2000

[47] D.R.Myers, "Solar Radiation Modeling and Measurements for Renewable Energy Applications: Data and Model Quality", International Expert Conference on Mathematical Modeling of Solar Radiation and Daylight—Challenges for the 21st, Century, Edinburgh, Scotland

[48] R.H.B. Exell, "The fluctuation of solar radiation in Thailand", Solar Energy, Volume 18, Issue 6, 1976, Pages 549-554.

[49] M. Calais, J. Myrzik, T. Spooner, V.G. Agelidis, "Inverters for Single-phase Grid Connected Photovoltaic Systems - An Overview," Proc. IEEE PESC'02, vol. 2. 2002. Pp 1995-2000.

[50] Photong, C. and et al, "Evaluation of single-stage power converte topologies for grid-connected Photovoltaic", IEEE International Conference on Industrial Technology (ICIT), 2010.

[51] Y. Xue, L.Chang, S. B. Kj r, J. Bordonau, and T. Shimizu, "Topologies of Single- Phase Inverters for Small Distributed Power Generators: An Overview," IEEE Trans. on Power Electronics, vol. 19, no. 5, pp. 1305-1314, 2004.

[52] S.H. Ko, S.R. Lee, and H. Dehbonei, "Application of Voltage- and Current- Controlled Voltage Source Inverters for Distributed Generation System," IEEE Trans. Energy Conversion, vol. 21, no.3, pp. 782-792, 2006.

[53] Yunus H.I. and et al., Comparison of VSI and CSI topologies for single-phase active power filters, Power Electronics Specialists Conference, 1996. PESC '96 Record., 27th Annual IEEE

[54] N. Chayawatto, K. Kirtikara, V. Monyakul, C Jivacate, D Chenvidhya, "DC–AC switching converter mode lings of a PV grid-connected system under islanding phenomena", Renewable Energy 34 (2009) 2536–2544.

[55] Lennart Ljung, "System identification : theory for the user", Upper Saddle River, NJ, Prentice Hall PTR, 1999.

[56] Nelles, Oliver, Nonlinear system identification : from classical approaches to neural networks and fuzzy models, 2001

[57] N.Patcharaprakiti and et al., "Nonlinear System Identification of power inverter for grid connected Photovoltaic System based on MIMO black box modeling", GMSTECH 2010 : International Conference for a Sustainable Greater Mekong Subregion, 26-27 August 2010, Bangkok, Thailand.

[58] Lennart Ljung, System Identification Toolbox User Guide, 2009

[59] IEEE 1159, 2009 Recommended practice for monitoring electric power quality.

3

Power Quality Improvement by Using Synchronous Virtual Grid Flux Oriented Control of Grid Side Converter

Vasanth Reddy Bathula and Chitti Babu B.

MIC College of Technology, NIT Rourkela

India

1. Introduction

The conventional energy sources are limited and have pollution to environment as more attention and interest have been paid on the utilization of renewable energy sources such as wind energy, fuel cells and solar energy. Distributed power generation system (DPGS) is alternative source of energy to meet rapidly increase energy consumption. These DPGS are not suitable to be connected   directly to the main utility grid. Rapid development of power electronic devices and technology, double sided converters are used to interface between DPGS and utility grid as they match the characteristics of the DPGS and the requirements of the grid connections. Power electronics improves the performance of DGPS and increase the power system control capabilities, power quality issues, system stability [1].

Rapidly increase in number of DGPS's leads to complexity in control while integration to grid. As a result requirements of grid connected converters become stricter and stricter to meet very high power quality standards like unity power factor, less harmonic distortion, voltage and frequency control, active and reactive power control, fast response during transients and dynamics in the grid. Hence the control strategies applied to DGPS become of high interest and need to further investigated and developed [3].

In this chapter, a virtual grid flux oriented vector control [2] (outer loop controller) and three different types of current controllers such as hysteresis current controller, current regulated delta modulator, modified ramp type current controller (inner current loop) techniques are proposed, with main focus on DC link voltage control, harmonic distortion, constant switching frequency, unity power factor operation of inverter. Vector control of grid connected inverter is similar to vector control of electric machine. Vector control uses decoupling control of active and reactive power. The control system for the vector control of grid connected converter consists of two control loops. The inner control loop controls the active and reactive grid current components. The outer control loop determines the active current reference by controlling the direct voltage. A cascaded control system, such as vector control is a form of state feedback. One important advantage of state feedback is that the inner control loop can be made very fast. For vector control, current control is the inner control loop. The fast inner current control nearly eliminates the influence from parameter variations, cross coupling, disturbances and minor non-linearity in the control process. Vector control uses PI-controllers in order to improve dynamic response and to reduce the cross coupling between active and reactive powers. Hysteresis current controller is used as inner current control loop to provide switching pulses to inverter, it has good dynamic response it is more suitable as inner current controller where we need fast acting inner current loop, but drawback of this hysteresis current controller is variation of switching frequency with parameters of grid voltage, filter inductor and output current is having lower order harmonics. In current regulated delta modulator switching frequency of inverter is limited by using latch circuit, but in this case also switching frequency is not maintaining constant during fundamental period. In Modified ramp type current controller switching frequency is limited, also maintain switching frequency constant. This controller ramp signal is generated at particular frequency to maintain switching frequency constant [4],[5],[6].

2. Configuration of DPGS and its control

The configuration of Distributed Power Generation System depends on input power sources (wind, solar etc) and different hardware configurations are possible. The basic structure of DPGS is shown in fig.1. The system consists of renewable energy sources, two back-to-back converters with conventional pulse width modulation techniques, grid filter, transformer and utility grid [1].

The input-side converter, controlled by an input side controller, normally ensures that the maximum power is extracted from the input power source and transmits the information about available power to the grid-side controller.

The main objective of the grid-side controller is to interact with the utility grid. The grid-side controller controls active power sent to the grid, control of reactive power transferred between the DPGS and the grid, control of the DC-link voltage, control of power quality and grid synchronization. Gird filter and transformer eliminates harmonics is inverter output voltage and ensures proper synchronization of inverter with grid.

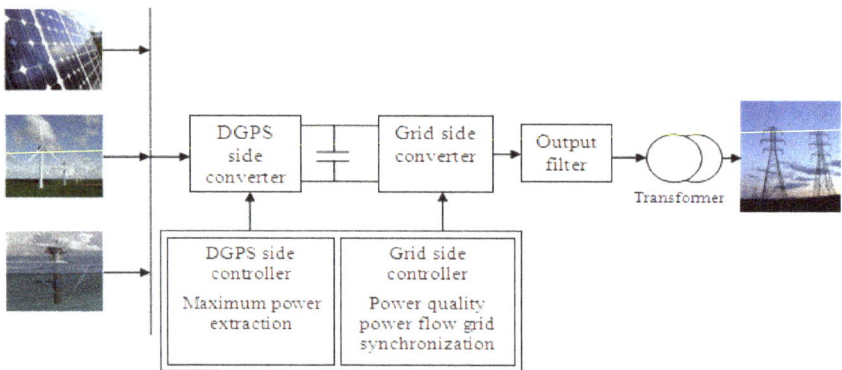

Fig. 1. General structure of distributed power generating system

2.1 Grid connected system requirements
The fundamental requirements of interfacing with the grid are as follows, the voltage magnitude and phase must equal to that required for the desired magnitude and direction

of the power flow. The voltage is controlled by the transformer turn ratio and/or the rectifier inverter firing angle in a closed-loop control system. The frequency must be exactly equal to that of the grid, or else the system will not work. To meet the exacting frequency requirement, the only effective means is to use the utility frequency as a reference for the inverter switching frequency.

Earlier, control and stabilization of the electricity system as taken care only by large power system like thermal, nuclear etc. due to large penetration of DPGS the grid operators requires strict interconnection called grid code compliance. Grid interconnection requirements vary from country to country. Countries like India where the wind energy systems increasing rapidly, a wind farm has to be able to contribute to control task on the same level as conventional power plants, constrained only by limitation of existing wind conditions. In general the requirements are intended to ensure that the DPGS have the control and dynamic properties needed for operation of the power system with respect to both short-term and long-term security of supply, voltage quality and power system stability. In this paper most significant requirement is power quality. The power quality measurement is mainly the harmonic distortion and unity power factor [2].

3. Virtual grid flux oriented control

Virtual grid flux vector control of grid connected Pulse Width Modulated (PWM) converter has many similarities with vector control of an electric machine. In fact grid is modeled as a synchronous machine with constant frequency and constant magnetization[2]. A virtual grid flux can be introduced in order to fully acknowledge the similarities between an electric machine and grid. In space vector theory, the virtual grid flux becomes a space vector that defines the rotating grid flux oriented reference frame, see in fig.2. The grid flux vector is aligned along d-axis in the reference frame, and grid voltage vector is aligned with q-axis. Finding the position of grid flux vector is equivalent to finding the position of the grid voltage vector. An accurate field orientation can be expected since the grid flux can be measured. The grid currents are controlled in a rotating two-axis grid flux orientated reference frame. In this reference frame, the real part of the current corresponds to reactive power while the imaginary part of the current corresponds to active power. The reactive and active power can therefore be controlled independently since the current components are orthogonal. Accurate field orientation for a grid connected converter becomes simple since the grid flux position can be derived from the measurable grid voltages. The grid flux position is given by

$$\cos(\theta_g) = \frac{e_{g\beta}}{|e_g|}, \ \sin(\theta_g) = -\frac{e_{g\alpha}}{|e_g|} \tag{1}$$

3.1 Action of Phase lock loop (PLL)

The implementation of the grid voltage orientation requires the accurate and robust acquisition of the phase angle of the grid voltage fundamental wave, considering strong distortions due to converter mains pollution or other harmonic sources. Usually this is accomplished by means of a phase lock loop (PLL). PLL determines the position of the virtual grid flux vector and provides angle (θ_g) which is used to generate unit vectors $\cos(\theta_g)$, $\sin(\theta_g)$ for converting stationary two phase quantities in stationary reference frame into rotating two phase quantities in virtual grid flux oriented reference frame. PLL is ensures the phase angle between grid voltages and currents is zero. That means PLL provides displacement power factor as unity.

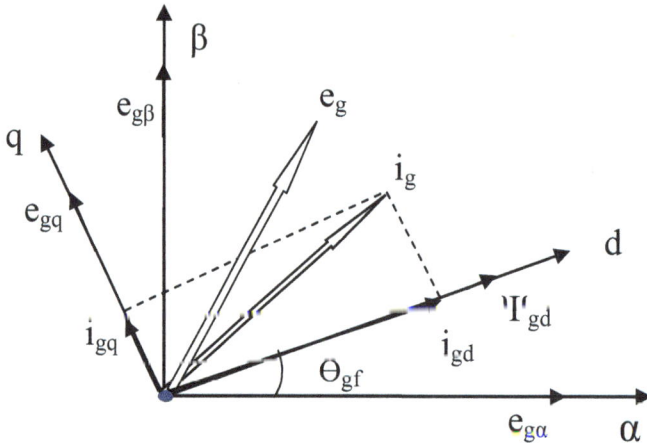

Fig. 2. Virtual grid flux oriented reference frame

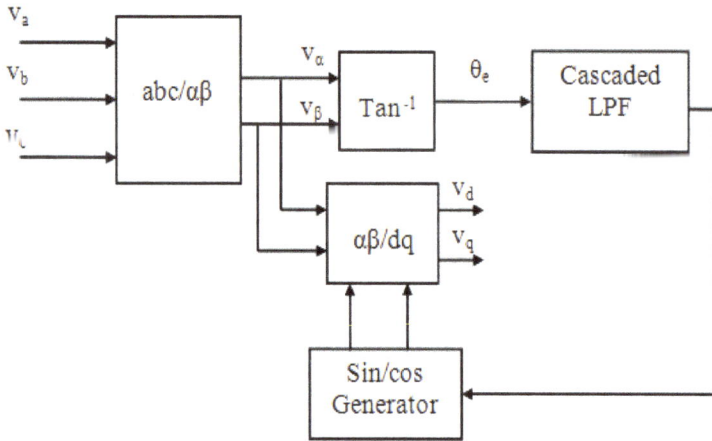

Fig. 3. Instantaneous PLL circuit

3.2 Control scheme for grid connected VSI

The block diagram of purposed system is shown in fig .4. The control system of vector controlled grid connected converter here consisting two control loops. The inner control loop having novel hysteresis current controller which controls the active and reactive grid current components. The active current component is generated by an outer direct voltage control loop and the reactive current reference can be set to zero for a unity power factor. The grid currents are controlled in a rotating two-axis grid flux orientated reference frame.

In this reference frame, the real part of the current corresponds to reactive power while the imaginary part of the current corresponds to active power. The reactive and active power can therefore be controlled independently since the current components are orthogonal [2].

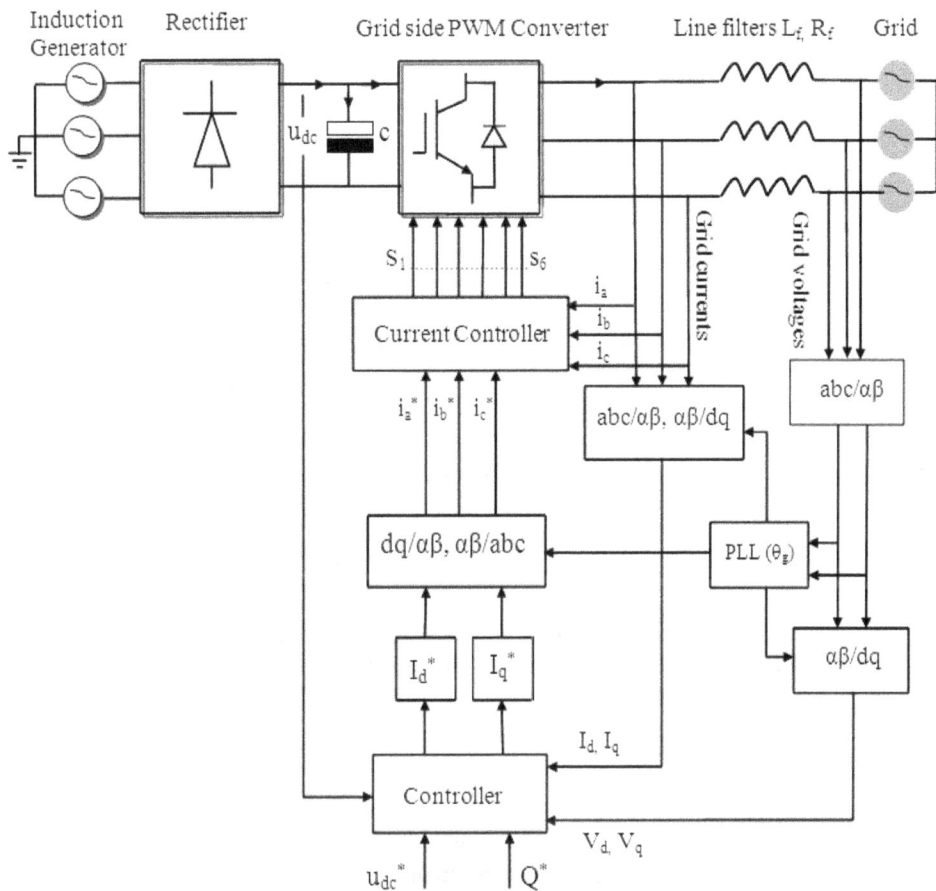

Fig. 4. Block diagram of virtual grid flux oriented control of grid connected VSI

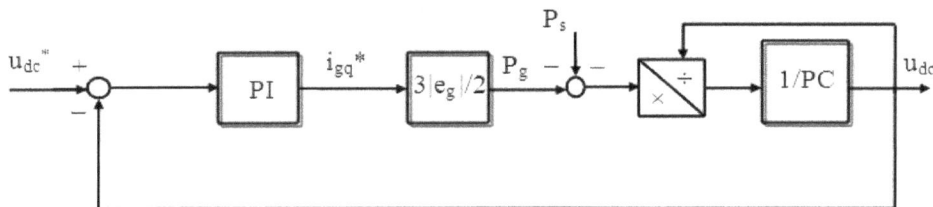

Fig. 5. Block diagram of closed loop control of dc link voltage

3.2.1 DC voltage controller (outer loop)

The following derivation of direct voltage controller assumes instantaneous impressed grid currents and perfect grid flux orientation. The instantaneous power flowing into grid can be written as

$$S_g = P_g + jQ_g = \frac{3}{2}e_g i_g^* = \frac{3}{2}(|e_g|i_q + j|e_g|i_d)$$ (2)

$$S_g = \frac{3}{2}(|e_g|i_q + j|e_g|i_d)$$ (3)

The active power is real part of equation.3.

$$P_g = \frac{3}{2}|e_g|i_{gq}$$ (4)

When neglecting capacitor leakage, the direct voltage link power is given by

$$P_{DC} = u_{DC}i_{DC} = u_{DC}C\frac{du}{dt}$$ (5)

Assuming the converter losses are neglected, the power balance in the direct voltage link is given by

$$u_{DC}C\frac{du_{DC}}{dt} = -P_s - P_g = -P_s - \frac{3}{2}|e_g|i_{gq}$$ (6)

Where P_s is the distributed energy system power is assumed to be independent of the DC voltage. A transfer function of between direct voltage and active grid current Ig is obtained as

$$u_{DC} \approx -\frac{3|e_g|}{2pCu_{DC}}i_{gq}$$ (7)

The transfer function is non-linear. It is acceptable to substitute the direct voltage with the reference set value since the objective is to maintain a constant direct voltage. The assumption gives linearized transfer function.

$$u_{DC} \approx -\frac{3|e_g|}{2pCu_{DC}^*}i_{gq}$$ (8)

Applying internal model control gives the direct voltage link controller as

$$F = \frac{\alpha}{p}G^{-1} = -\alpha\frac{2Cu_{dc}^*}{3|e_g|}$$ (9)

From eq.8, a P-controller is obtained for regulating the direct voltage. The P- controller is optimal for an integrator process in the sense that the P- controller eliminates the remaining error for steps in the reference value. However, there will be a remaining error for steps in the reference value. However, there will be a remaining error when the gird is loaded and active power flows between the direct voltage link and the grid. The remaining error can be eliminated by adding an integrator to the direct voltage link controller. The following is often adapted for selecting the controller integration time in traditional PI-controller design.

$$T_i = \frac{10}{\omega_c} \approx \frac{10}{\alpha} \tag{10}$$

The active reference current of the grid connected converter can be written as

$$i_{gq}^* = K_p(1 + \frac{1}{T_i p})(u_{dc}^* - u_{dc}) \tag{11}$$

$$k_p = -\alpha \frac{2Cu_{dc}^*}{3|e_g|} \tag{12}$$

Negative proportional gain is because the distributed energy source references are used for grid. A block diagram that represents the direct voltage control is shown in Fig. 5. Note that closed-loop bandwidth of the current control is assumed to be much faster than the closed-loop bandwidth of the direct voltage link.

3.2.2 Open loop reactive power control (outer loop)
The reactive power flowing into grid is controlled by the reactive current component. Simplest form of controlling reactive power is through open loop control. Taking imaginary part of eq.3 reactive reference current as

$$i_{gd}^* = \frac{2}{3e_{gq}} Q_g^* \tag{13}$$

The i_{gq}^* and i_{gd}^* current references are converted into three phase current references i_a^*, i_b^*, i_c^* which are given to current controller.

4. Current control approach to VSI

4.1 Objectives
The current control methods play an important role in power electronic systems, mainly in current controlled PWM voltage source inverters which are widely used in ac motor drives, active filters, and high power factor, uninterruptable power supply (UPS) systems, and continuous ac power supplies [5]. The performance of converter system is largely dependence on type of current control strategy. Therefore current controlled PWM voltage source inverters are one of the main subjects in modern power electronics. Compared to conventional open loop PWM voltage source inverter, the current controlled PWM voltage source inverters have fallowing advantages:
1. Control of instantaneous current waveform and high accuracy.
2. Peak current protection.
3. Overload rejection.
4. Extremely good dynamics.

5.   compensation of effects due to load parameter variations(resistance and reactance).
6.   Compensation of semiconductor voltage drop and dead times of converter.
7.   Compensation of DC link and AC side voltages.

4.1.1 Basic scheme of current controlled PWM

The main task of current controller in PWM VSI is to force the load current according to the reference current trajectory [6] by comparing to command phase currents with measured instantaneous values of phase currents, the current controller generates the switching states for the converter power devices in such a way that error current should be minimized. Hence current control implements two tasks, error compensation (decrease in current error) and modulation (determination of switching states).

4.1.2 Basic requirement and performance criteria

Basic requirements of current controller are as fallows [16]
1.   No phase and amplitude errors (ideal tracking) over a wide output frequency range.
2.   To provide high dynamic response of system.
3.   Limited or constant switching frequency of converter to ensure safety operation of converter semiconductor power devices
4.   Low harmonic content
5.   Good dc-link voltage utilization.

Some of the requirements like fast response and low harmonic distortion contradict each other.

4.2 Different current control techniques
4.2.1 Hysteresis current control technique

Hysteresis current controller is an instantaneous feedback system which detects the current error and produces driver commands for switches when error exceeds specified band.

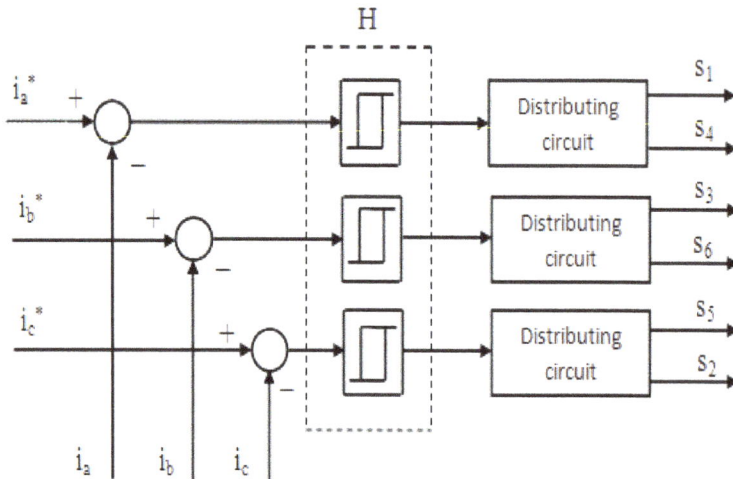

Fig. 6. Hysteresis current control scheme

The purpose of the current controller is to control the load current by forcing it to follow a reference one. This is achieved by the switching action of the inverter to keep the current within the hysteresis band. Simplified diagram of a typical three-phase hysteresis current controller is shown in fig. 6, The load currents are sensed and compared with the respective command currents using three independent hysteresis comparators having a hysteresis band H. the output signals of the comparators are used to active the inverter power switches. Based on the band, there are two types of current controllers, namely, the fixed band hysteresis current controller and sinusoidal band hysteresis current controller [5].

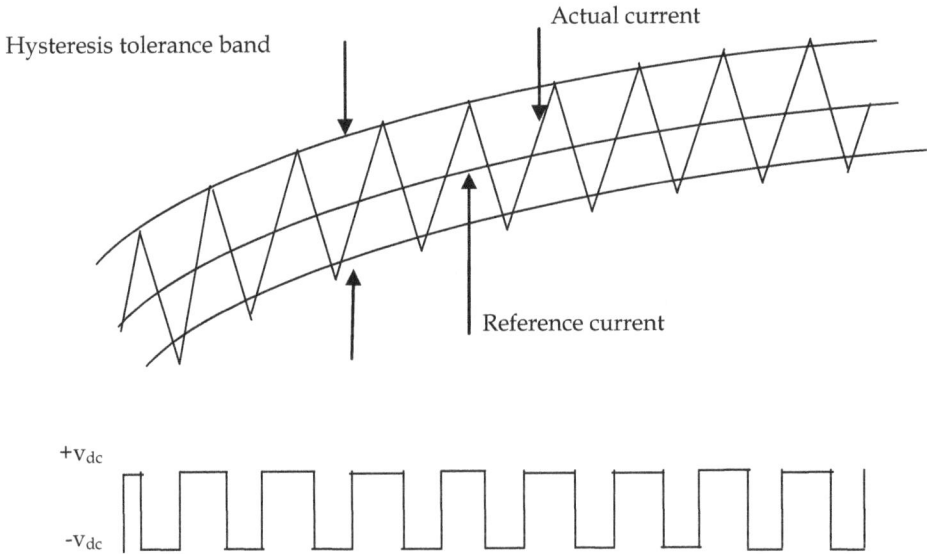

Fig. 7. Hysteresis current control operation

4.2.2 Current regulated delta modulator

Current regulated delta modulator operation is same as that of hysteresis current controller. Main advantage of delta modulator is switching frequency variation is very less and we can limit switching frequency to desirable frequency. Circuit diagram of current regulated delta modulator is shown in fig. 8. This consists of comparators and latch to limit switching frequency [7]. Actual grid currents compared with reference currents provides error currents, this error currents are flowing into comparator. These comparators are acting like hysteresis limiters which limit error current between two bands and generate pulses. These pulses are given to latching circuit as binary values 0 and 1. Latching circuit is operated when clock signal is enabling the latch. Switching frequency of inverter is decided clock frequency of latch. Operation of latch is that it holds the input until the clock signal enables the latch. When clock signal applied latch will enable and it will give output to gate drive circuit, which drives the inverter in such a way that error current should be minimized and grid current most follow the reference current value.

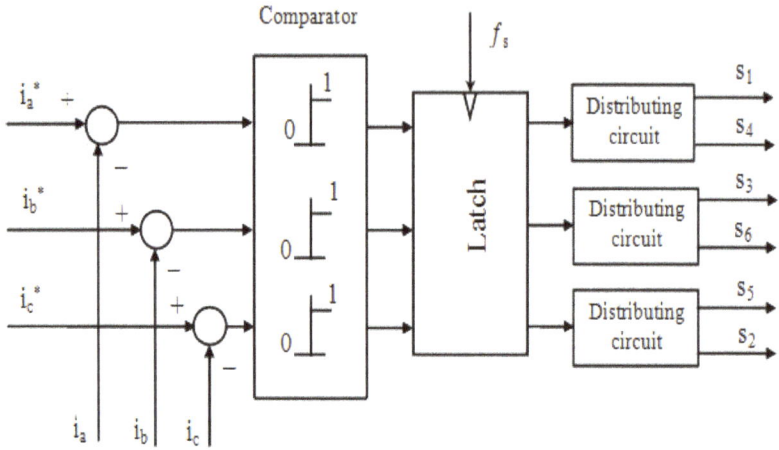

Fig. 8, Current regulated delta modulator scheme

The delta modulation offers an opportunity of on-line harmonic minimization of pulse width modulated inverter without conventional optimization processes, like selective harmonic elimination or harmonic weighting techniques.

4.2.3 Modified ramp type current controller
A conventional ramp-comparator controller is also shown in fig. 9, the phase shifters are bypassed. The actual values of the three-phase load currents are measured and compared to the reference currents.

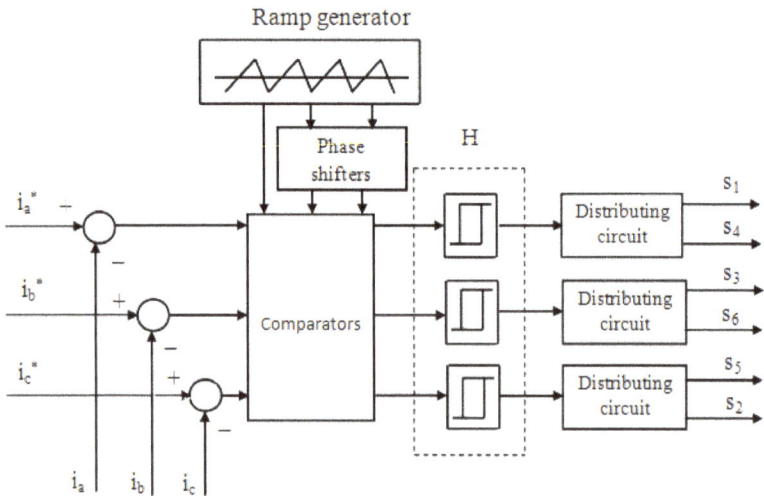

Fig. 9. Modified ramp type current control scheme

The generated error signals are compared to a triangular waveform of fixed frequency and amplitude. If the current error signal is positive and larger than the triangular wave, the switches are activated to apply +VB to the load. However, if current error signal is positive and smaller then the triangular wave, the switches are activated to apply –VB to the load. Some hysteresis has been added to the controller, in order to prevent multiple crossings of the error signals with the triangular wave. The ramp-comparator controller is a modulation system. The frequency of the triangular wave is the carrier frequency, while the error current signal is a modulated waveform.

Since this controller uses a fixed-frequency triangular wave, it has the effect of maintaining a constant switching frequency of the inverter. This is the main advantage of this controller. However, it has some disadvantages, as the output current has amplitude and phase errors. This results in a transmission delay in the system. Moreover, a zero vector is applied to the load. This means load is disconnected several instants over the fundamental period of the output voltage. In order to overcome the above difficulties, a new ramp- comparator controller is proposed. Fig.9 shows the schematic diagram of this controller, in which the phase shifters are included. The actual values of the three phase load currents are measured and compared to the three 120^0 phase-shifted triangular waveforms having the same fixed frequency and amplitude. The performance of this scheme is considered identical to those for three independent single-phase controllers. It is to be noted that there is no interaction between the operations of the three phases. As a result, the zero voltage vectors will be eliminated for balanced operation. This does not necessarily lead to the possibility of creating the positive and/or negative sequence sets due to the controller alone. The zero voltage vectors eliminate the necessity of neutral connections for some applications and, in such cases; no harmonic neutral current can flow in the load. Thus, there is no problem of incorporating the proposed controller in industrial motor drives. For applications in UPS's and active filters, the higher harmonics lead to more losses. However, the shifting of current harmonics to higher orders does not usually create noticeable problems in motor drives, as the machine inductance filters out the higher harmonics and limits the associated detrimental effects[5].

5. Simulation results of grid connected VSI

Simulation of virtual grid flux oriented control of grid connected VSI was done by using MATLAB/Simulink and results for different types of current controllers has shown below

5.1 Simulation results of hysteresis current controller

For the simulation of virtual grid flux oriented control of grid connected inverter the following are the set values

DC link reference voltage (u_{dc}^*)   = 2200V
DC link capacitance (C)   = 600UF
Reference Reactive power (Q^*)   = 0 VAr
Hysteresis band width (H)   = 20 Amp
Ramp generator frequency   = 2 KHz
Proportional Gain (K_p)   = 0.01
Integral Gain (K_i)   = 60

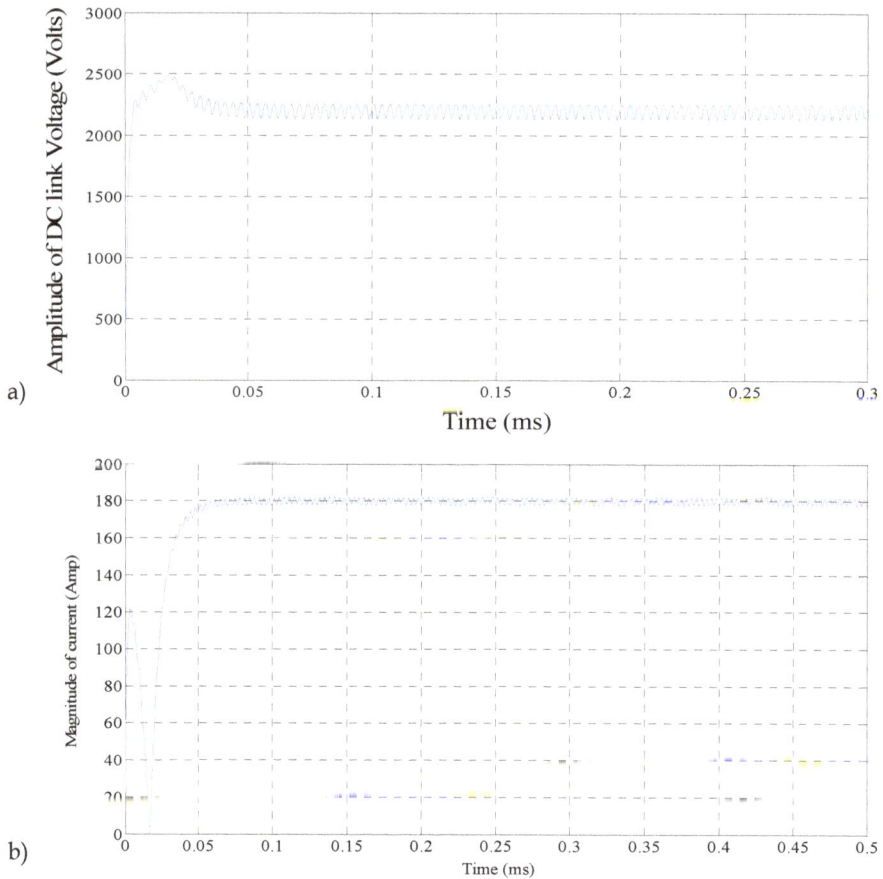

a)

b)

Fig. 10. (a) DC link voltage (b) Active component of grid current i_q^*

Above mentioned figures shows the waveforms of DC link voltage in fig. 10 (a) and active component of grid current reference I_q^* in fig. 10(b). The DC link voltage is maintained at its reference set value of 2.2KV which is shown is fig. 10(a). From fig. 10(b) active component of grid current i_q^* in synchronous reference frame is obtained from the PI controller using error of DC link voltage by controlling DC link voltage to its reference value. This grid current reached to steady state within 0.05sec this shows fastness of inner current control loop. the Form figures we can see that current reaches to zero value which is due to DC link voltage is more than the set reference value, then error is going to be negative which makes the current to decrease. Then PI controller will act on DC link voltage of maintained at its reference value then error is zero. Hysteresis current controller controls the current is flowing into the grid to maintain set value of DC link voltage.

Above mentioned figures shows the waveforms of reactive component of grid current in fig. 11(a) and three-phase reference current waveforms fig. 11(b). Reactive component of grid current is zero because of reactive power flowing to grid is taken as zero to maintain unity power factor at grid. Three-phase references current are obtained from the active and reactive grid current components after dq to ABC transformations.

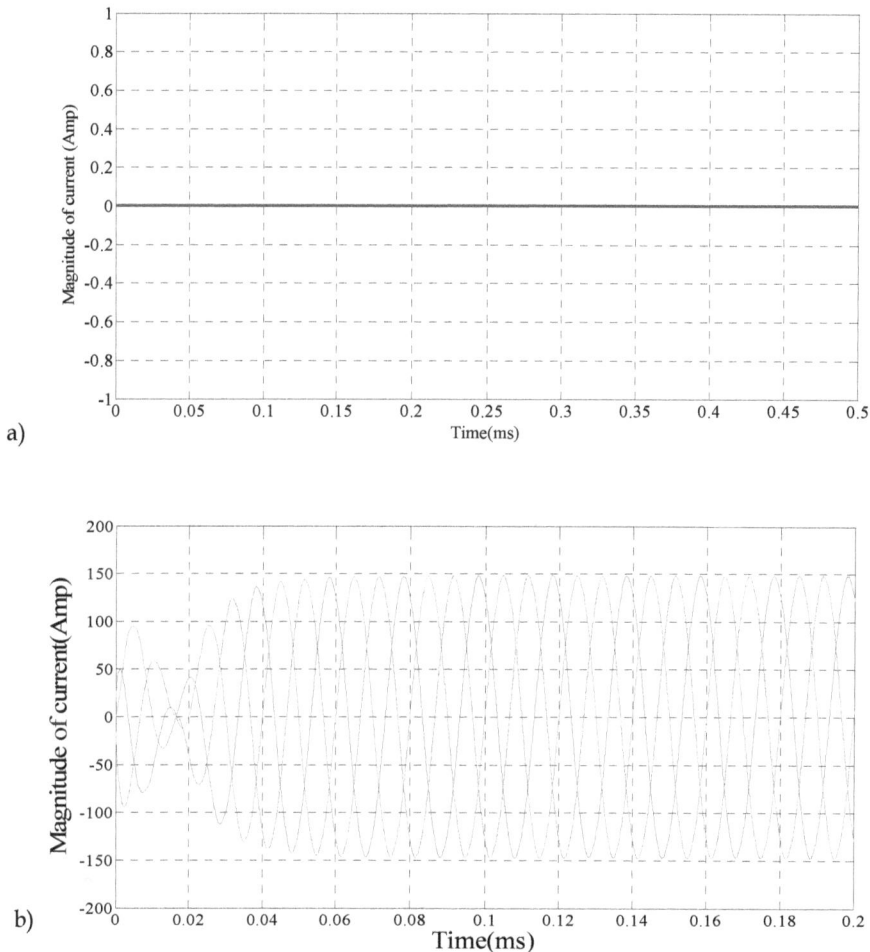

a)

b)

Fig. 11. (a) Reactive component of grid current (b) Three phase reference current waveform

Above mentioned figures shows the waveforms of three-phase grid current in fig. 12(a) and harmonic spectrum of grid current in fig. 12(b). the three phase grid current flows the actual reference current which we can seen from figures 12(b), 12(a).THD spectrum of grid current is having percentage of THD is around 4.41 and having fundamental component of 144.4 amps. And it having few lower order harmonics which are due to ripple in DC link voltage at six times of supply frequency. This is because of reference current I_{gq}^* is generated from PI-controller through control of DC-link voltage if there is oscillation in DC link voltage which will inject into the reference current, their by load current is affected by this oscillations. Ripple in DC link voltage can be eliminated by using resonant DC link Inductor which smoothness the current flowing in DC link capacitor their by ripple magnitude is decreased.

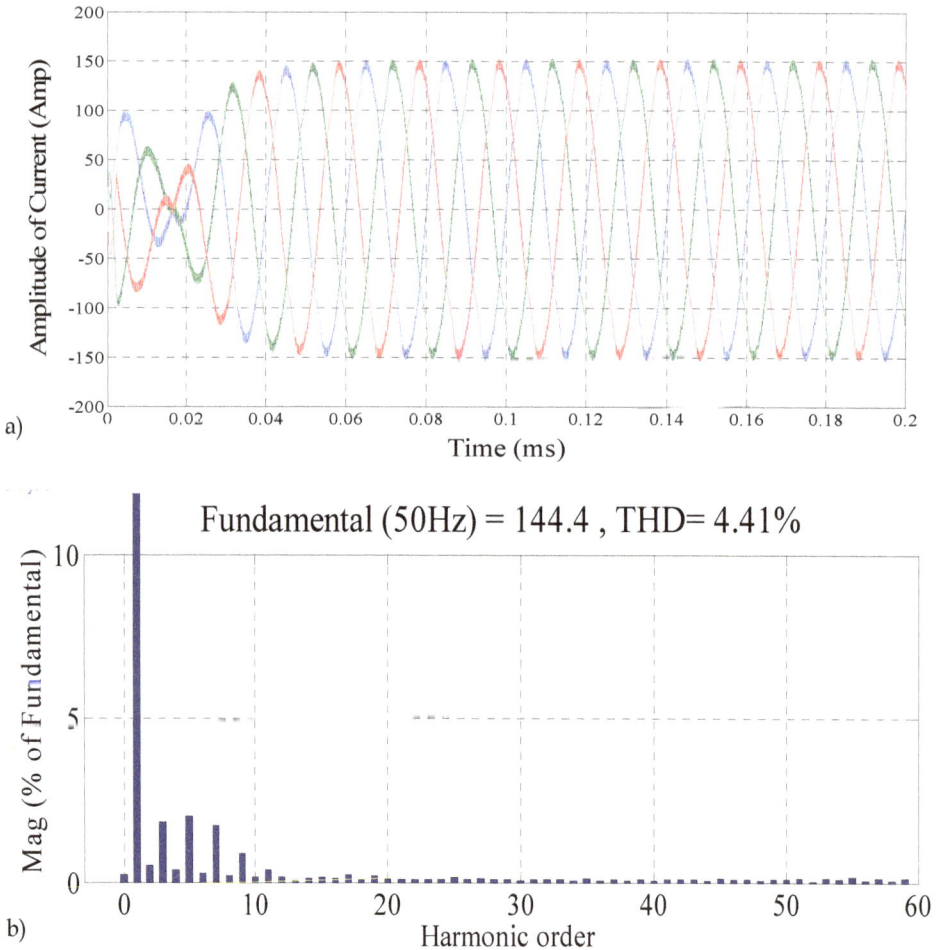

a)

b)

Fig. 12. (a) Three phase grid current waveform (b) Harmonic spectrum of grid current

Above mentioned figures shows the waveforms of output line voltage of VSI before filter in fig. 13(a) and grid line voltage waveform after filter inductor in fig. 13(b). Fig. 13(a) is output of inverter which is having some oscillations in the edges of inverter voltage around 2200V. This oscillation can be decreased by decrease ripple in DC link voltage. The inductor filter eliminates all harmonics present in output line voltage of VSI and produces pure sinusoidal voltage as that of grid voltage. And it ensures the proper synchronism between the VSI and Grid. We need to design the filter inductor and capacitor properly to synchronize inverter output with grid and drop across the filter elements should be very less. And these filter elements cannot make resonance with transmission line parameters like series line inductance and line to ground capacitance otherwise there is large power oscillation in the grid which leads to power quality problem in the system.

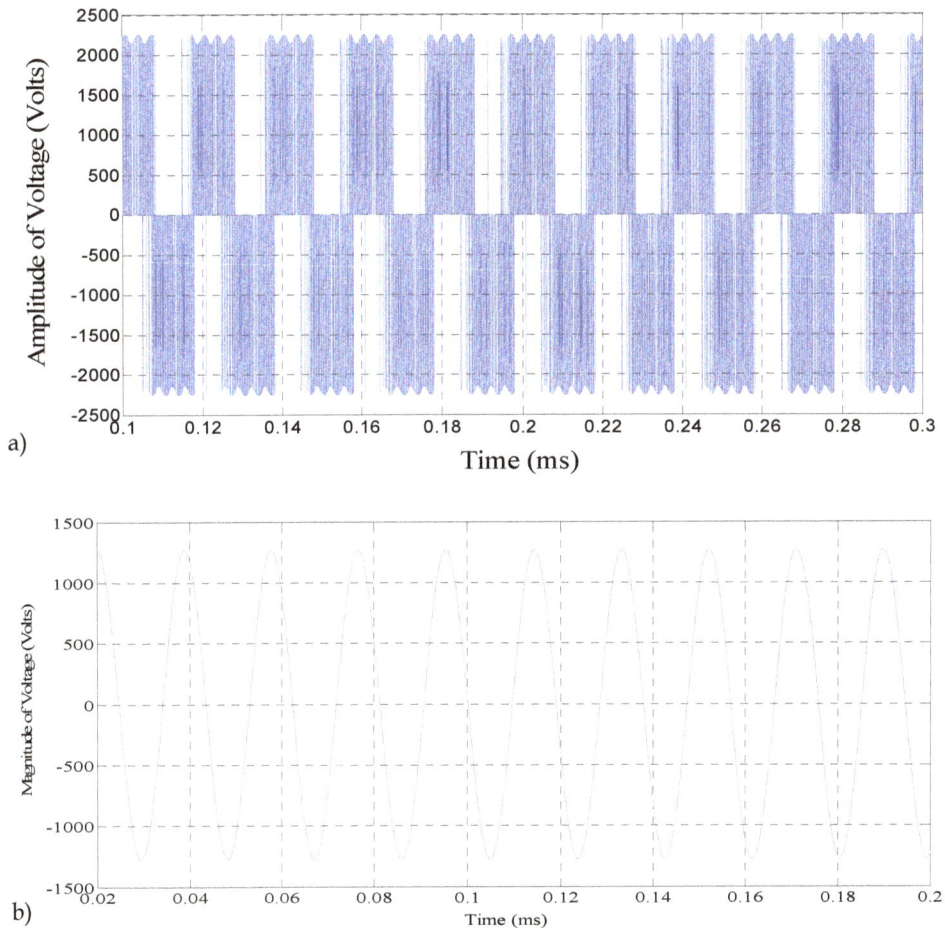

a)

b)

Fig. 13. (a) Output voltage of VSI before the filter (b) Grid voltage waveform

Figure 14 shows the waveforms of DC link capacitor current, inverter output voltage and displacement factor. Selection of capacitor is choice on basis of less ripple current in DC link capacitor. In fig. 14(a) capacitor is having fewer current ripples. And fig. 14(b) is displacement power factor, which is having value of 0.999999 almost unity power factor. This is because almost zero-phase angle difference between reference currents and Grid voltages. (i.e. reference currents fallows the same phase as grid voltages).

Power factor

Distortion factor (DF) is given by formula

$$DF = \frac{1}{\sqrt{1+THD^2}} \qquad (14)$$

Total power factor = DF*DPF   (15)

$$DF = \frac{1}{\sqrt{1+0.0441^2}} = 0.999029011$$

Total power factor = 0.999029011*0.999999 = 0.9989.

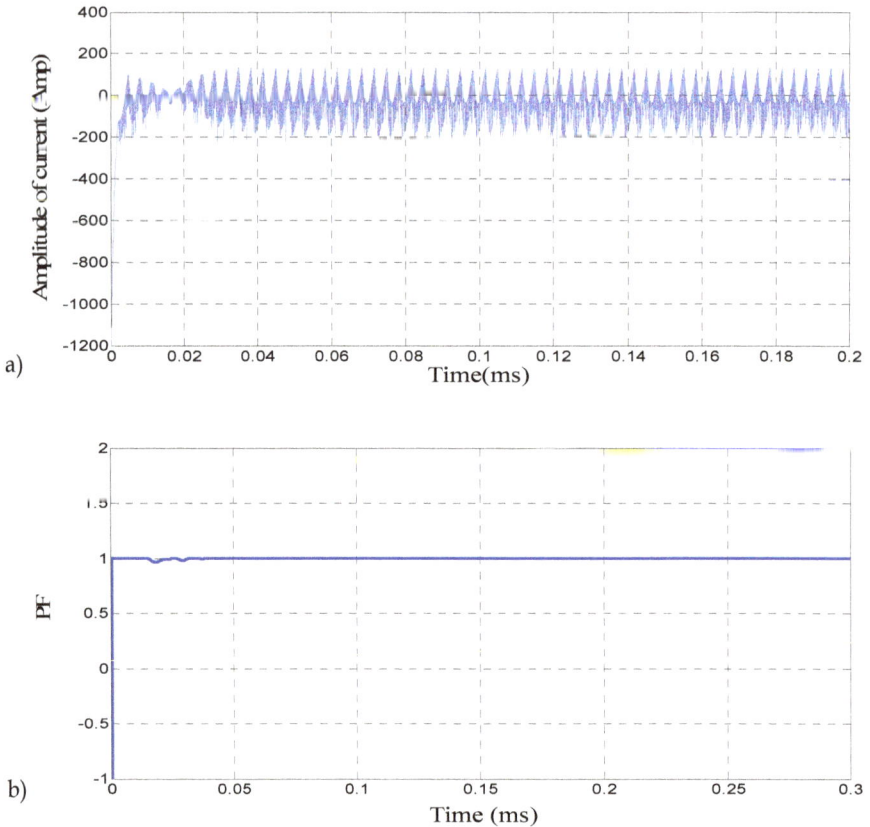

a)

b)

Fig. 14. (a) DC link Capacitor Current (b) Displacement power factor

Fig. 15(a) shows single phase instantaneous power and fig. 15(b) shows three phase average active power flowing into grid. The instantaneous power 'S' which is equal to active power flowing into grid this is due to zero phase angle difference between grid currents and grid voltages, from fig. 15(a) we can easily observe this fact (i.e. instantaneous power will not crossing zero that means reactive component of current flowing into grid is zero, only active component of current flowing). Fig. 15(b) shows the three phase active power flowing into grid which is around 160 KW. The oscillatory nature of the power is because of the harmonics present in currents which are flowing into the grid. The harmonic content (oscillatory nature of power) can be reduced by reducing the hysteresis band width. But by reducing band switching frequency in hysteresis current controller is going high.

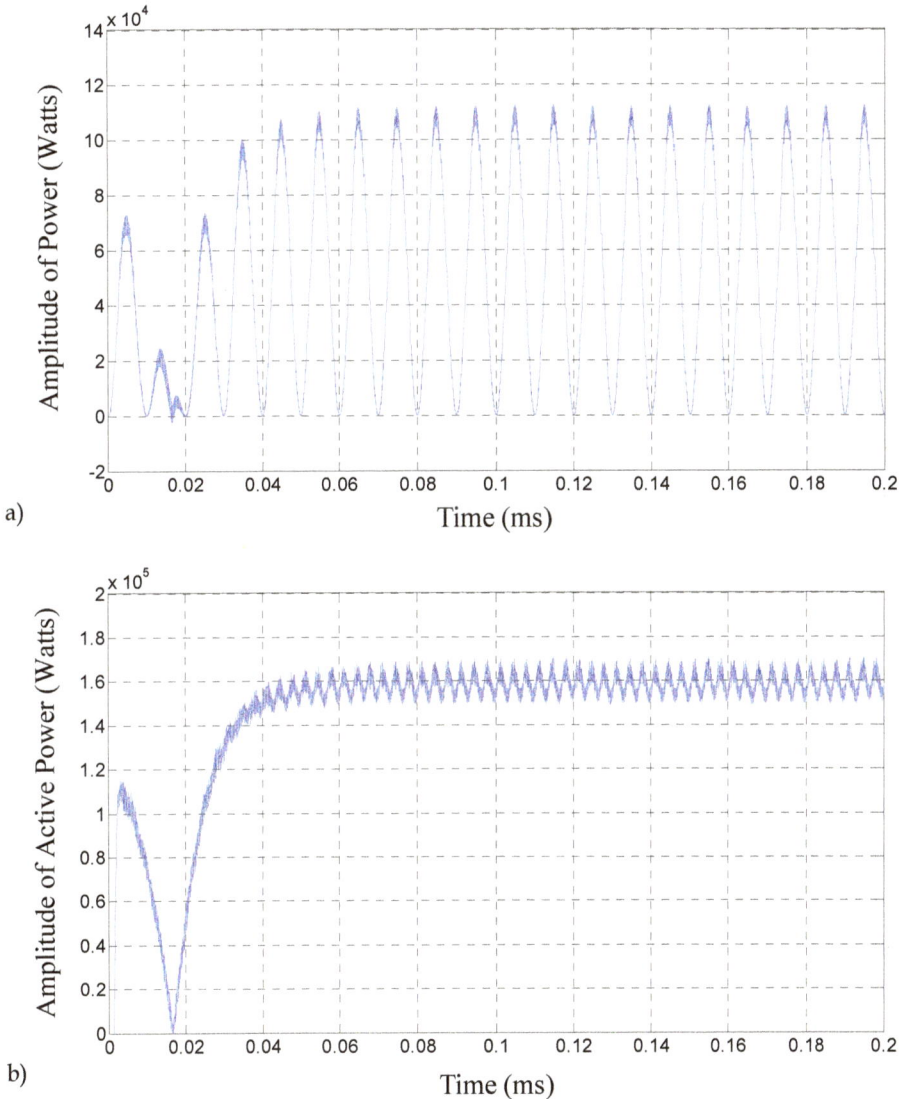

a)

b)

Fig. 15. (a) Single Phase Instantaneous Active Power (b) Three Phase Active Power

Figure 16 shows single phase instantaneous reactive and three phase reactive power flowing into grid. Fig. 16(a) shows instantaneous reactive power flowing into grid which is oscillating around zero. Fig. 16(b) shows three phase reactive power flowing into grid which having average zero value, this because of reference set value of reactive power is zero (i.e. Q* = 0 VAr). This ensures unity power factor operation of grid connected inverter. In such a case we are supplying only active power to the grid. The reactive power needed by the loads which are connected to the grid can be supplied from other generating stations or bulk capacitors connected to grid to maintain grid power factor almost unity.

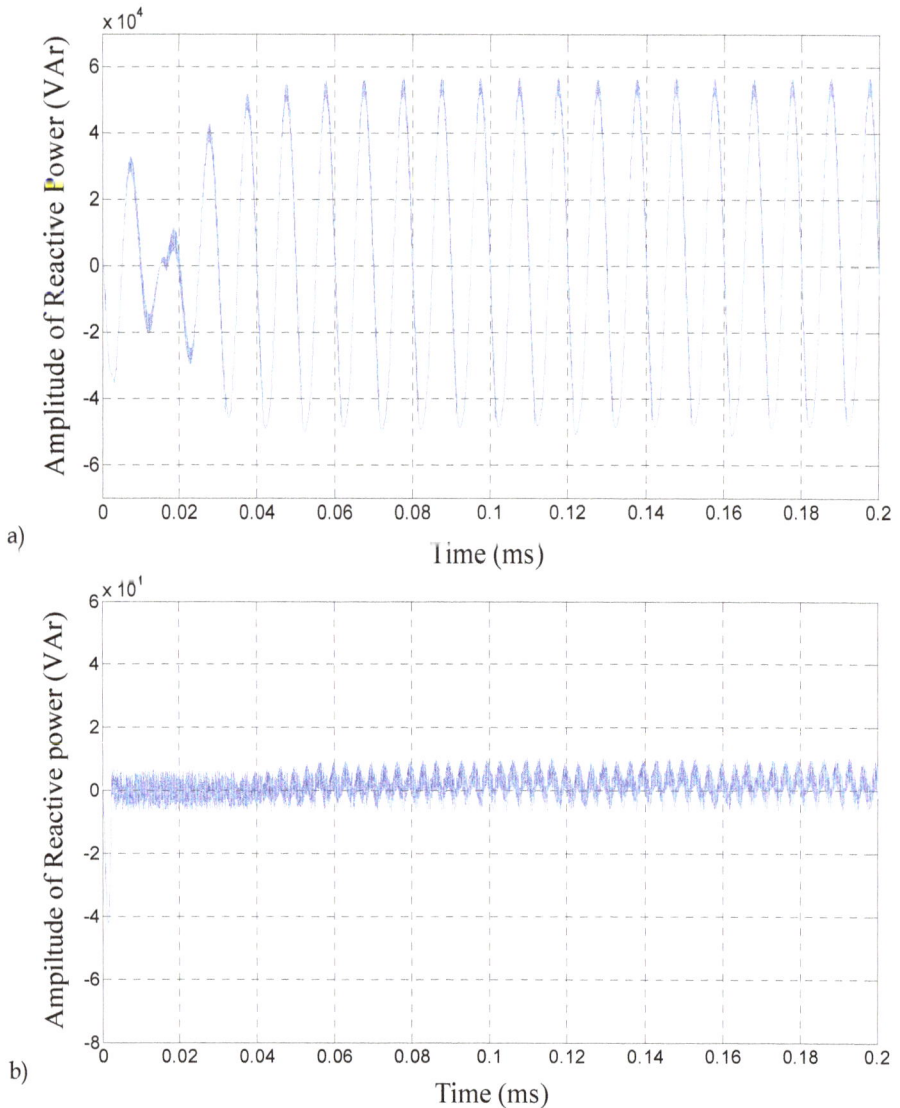

Fig. 16. (a) Single phase instantaneous reactive power (b) Three phase reactive power

5.2 Simulation results of current regulated delta modulator

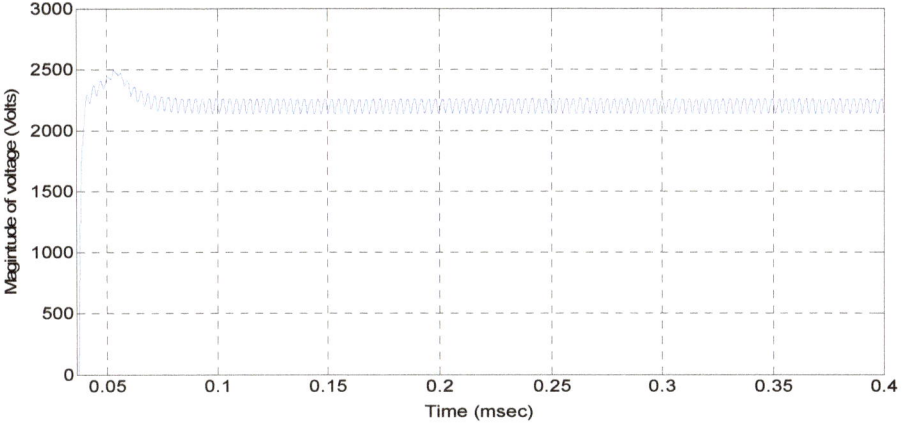

Fig. 17. DC link voltage

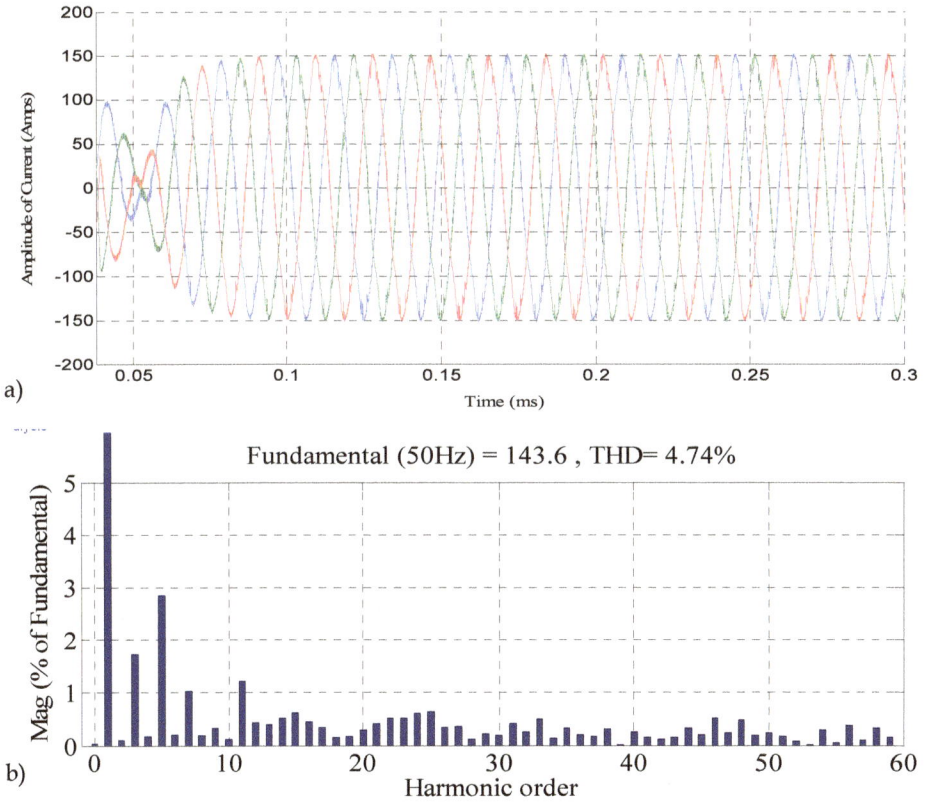

a)

Fundamental (50Hz) = 143.6 , THD= 4.74%

b)

Fig. 18. (a) Three phase grid current waveform (b) Harmonic spectrum of grid current

5.3 Simulation results for ramp type current controller

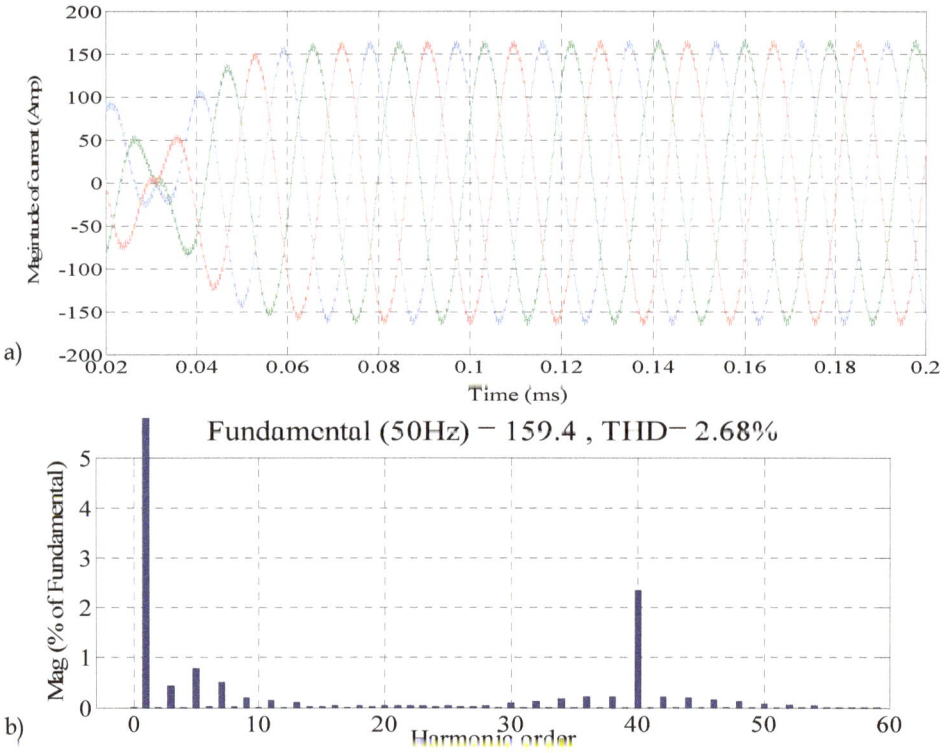

a)

Fundamental (50Hz) − 159.4 , THD− 2.68%

b)

Fig. 19. (a) Three phase grid current waveform (b) Harmonic spectrum of grid current

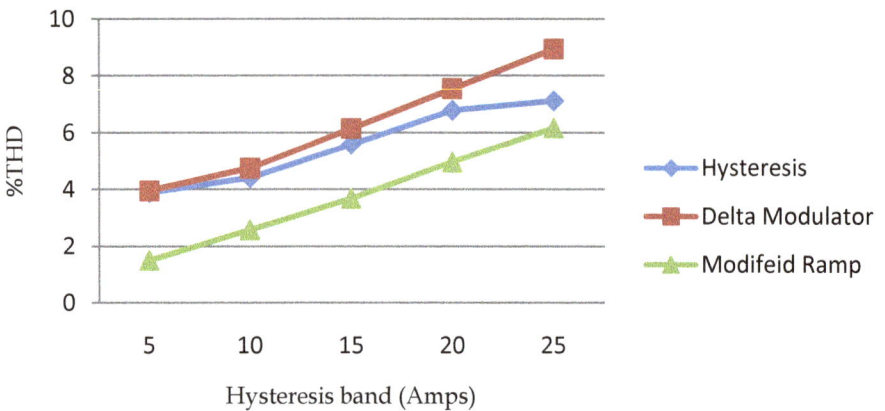

Fig. 20. Variation of % THD with hysteresis band

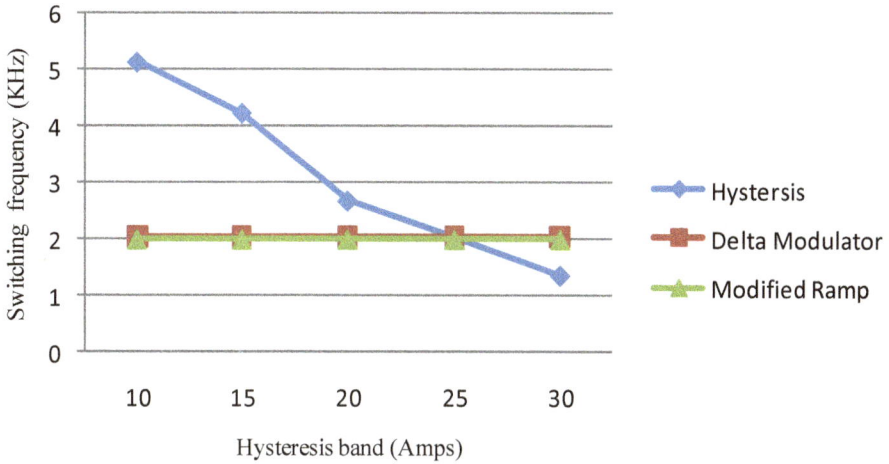

Fig. 21. Variation of switching frequency with hysteresis band

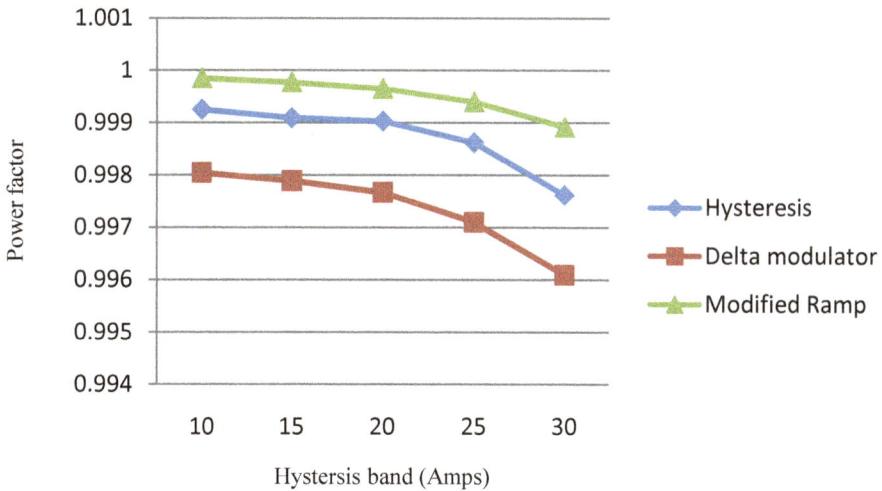

Fig. 22. Variation of power factor with hysteresis band

Above graphs shows the variations in %THD, Switching frequency, power factor, dynamic response, error current with hysteresis band. From the above graphs we could say that % THD variation is less in modified ramp type current controller. Switching frequency is constant in modified ramp type current controller, delta modulator as limited switching frequency. Switching frequency is varying more with hysteresis band in hysteresis current controller. Modified ramp type current controller is giving good system power factor compared to other controllers.

6. Discussion

Vector control of grid connected voltage source inverter is implemented in MATLAB/simulink. Simulation results are obtained for different current controllers. The discussion of the results as follows: vector control in virtual grid flux oriented reference frame is having an capability to decouple the active and reactive powers flowing into grid, which we can seen from the waveforms of active and reactive powers for three current controllers. Reactive power flowing into grid is almost zero in all the controllers to ensure unity power factor operation of grid. Total harmonic distortion of three current controllers is as follows 1. For hysteresis current controller percentage of THD is 4.41 2. For current regulated delta modulator percentage of THD is 4.74 and 3. For modified ramp type current controller percentage of THD is 2.68. Among all the current controllers modified ramp type current controller is having fewer harmonics in grid current, which ensures fewer ripples in three phase active power following into grid, and current regulated delta modulator having more harmonics in grid current which leads to more ripple is three phase active power following into grid. The switching frequency of modified ramp type current controller is 2 KHz it is a constant value at its ramp generator frequency. In Current regulated delta modulator switching frequency is limited to 2 KHz, but the variation of frequency with in fundamental period cannot be controlled. In hysteresis current controller the switching frequency is varies with load parameters and hysteresis band, switching frequency when hysteresis band 20 is equal to 2.72 KHz. Form the above discussion modified ramp type current controller has constant switching frequency. The power factor by using modified ramp type current controller is very good then compared to other current controllers. The following table 1 shows the switching frequency, THD, power factor for three current controllers.

Current control	THD	Switching frequency	Power factor
Hysteresis	4.41	2.72 KHz	0.9989
Delta modulator	4.74	2 KHz	0.9973
Modified ramp type	2.68	2 KHz	0.9996

Table 1. Comparison of three current control techniques of grid connected VSI

7. Conclusion

Analysis of different current control techniques for synchronous grid flux oriented control of grid connected voltage source inverter is presented in this chapter. For effectiveness of the study MATLAB/simulink is used here in GUI environment. Vector control in grid flux oriented reference frame is having capable of decoupling active and reactive powers following into grid, which we could see form figures of active and reactive powers for three current controllers. Reactive power following into grid is zero for all current control techniques to ensure the grid at unity power factor operation. There is a slight variation in power factors of three current controllers which is due to variation of percentage of THD in three current controllers. The DC link voltage is maintained at 2200V which is the set value of DC link voltage by using DC link voltage controller which controls the active current reference flows in the grid. The total harmonic distortion is less in modified ramp type current controller compared to other two current controllers. The switching frequency of modified ramp type current controller is maintained at 2 KHz which decreases the switching losses of power semiconductor devices compared to other current controllers where the switching frequency varies with load parameters. There is a less ripple in three phase active power for modified ramp type current controller compared to other two current controllers. Form the above discussion modified ramp type current controller is more advantages then other two current controller in grid connected voltage source inverter.

8. References

Azizur Rahman M., Osheiba Ali M. Analysis of current controllers for voltage-source inverter, *IEEE Transaction on industrial electronics*, vol.44, no.4, Augest 1997.

Hansen Lars Henrik and Bindner Henrik, Power Quality and Integration of Wind Farms in Weak Grids in India, Risø National Laboratory, Roskilde April 2000.

Hornik T. and Zhong Q.-C. , Control of grid-connected DC-AC Converters in Distributed Generation: Experimental comparison of different schemes, power electronic converters for power system, *compatibility and power electronics*, 2009. Page: 271-278.

Ibrahim Ahmed, Vector control of current regulated inverter connected to grid for wind energy applications. *International journal on renewable energy technology*, vol.1, no.1, 2009.

Kazmierkowski Marian P., and Malesani Luigi, Current Control Techniques for Three-Phase Voltage-Source PWM Converters: A Survey, *IEEE transactions on industrial electronics*, vol. 45, no. 5, October 1998.

Kohlmeier Helmut and SchrÄoder Dierk F., Control of a double voltage inverter system coupling a three phase mains with an ac-drive, in *Proc. IAS 1987*, Atlanta, USA, Oct. 18-23 1987, vol. 1, pp. 593-599.

Malesani L. and Tenti P. A novel hysteresis control method for current-controlled voltage-source PWM inverters with constant modulation frequency, *IEEE Trans. Ind. Electron.*, vol.26, pp.321-325.1998.

Milosevic Mirjana, Hysteresis current control in three phase voltage source inverter, ETH publications, power systems and high voltage laboratories.

Pongpit Wipasuramonton, Zi Qiang Zhu, Improved current regulated delta modulator for reduced switching frequency and low- frequency current error in permanent magnate brushless AC Drives. *IEEE Transactions on Power Electronics*,vol.20, no.2, march 2005.

4

Optimal Location and Control of Multi Hybrid Model Based Wind-Shunt FACTS to Enhance Power Quality

Belkacem Mahdad
Department of Electrical Engineering, Biskra University
Algeria

1. Introduction

Modern power system becomes more complex and difficult to control with the wide integration of renewable energy and flexible ac transmission systems (FACTS). In recent years many types of renewable source (Wind, solar,) and FACTS devices (SVC, STATCOM, TCSC, UPFC) integrated widely in the electricity market. Wind power industry has been developing rapidly, and high penetration of wind power into grid is taking place, (Bent, 2006), (Mahdad.b et al., 2011). According to the Global Wind Energy Council, GWEC, 15.197 MW wind turbine has been installed in 2006 (Chen et al. 2008), in terms of economic value, the wind energy sector has now become one of the important players in the energy markets, with the total value of new generating equipment installed in 2006 reaching US 23 billion.

FACTS philosophy was first introduced by (Hingorani, N.G., 1990), (Hingorani, N.G., 1999) from the Electric power research institute (EPRI) in the USA, although the power electronic controlled devices had been used in the transmission network for many years before that. The objective of FACTS devices is to bring a system under control and to transmit power as ordered by the control centers, it also allows increasing the usable transmission capacity to its thermal limits. With FACTS devices we can control the phase angle, the voltage magnitude at chosen buses and/or line impedances.

In practical installation and integration of renewable energy in power system with consideration of FACTS devices, there are five common requirements as follows (Mahdad. b et al., 2011):

1. What Kinds of renewable source and FACTS devices should be installed?
2. Where in the system should be placed?
3. How to estimate economically the number, optimal size of renewable source and FACTS to be installed in a practical network?
4. How to coordinate dynamically the interaction between multiple type renewable source, multi type of FACTS devices and the network to better exploit their performance to improve the reliability of the electrical power system?
5. How to review and adjust the system protection devices to assure service continuity and keep the indices power quality at the margin security limits?

Optimal placement and sizing of different type of renewable energy in coordination with FACTS devices is a well researched subject which in recent years interests many expert engineers. Optimal placement and sizing of renewable source in practical networks can result in minimizing operational costs, environmental protection, improved voltage regulation, power factor correction, and power loss reduction (Munteau et al., 2008). In recent years many researches developed to exploit efficiently the advantages of these two technologies in power system operation and control (Adamczyk et al., 2010), (Munteau et al., 2008).

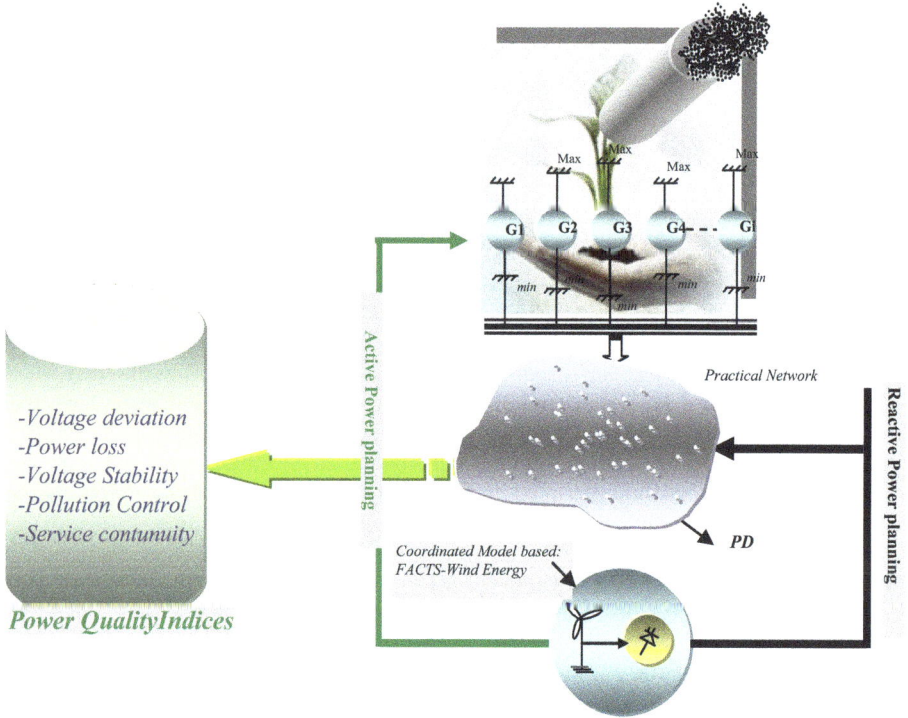

Fig. 1. Optimal power flow control strategy based hybrid model: wind and Shunt FACTS

In this study a combined flexible model based wind source and shunt FACTS devices proposed to adjust dynamically the active power delivered from wind source and the reactive power exchanged between the shunt FACTS and the network to enhance the power quality. Wind model has been considered as not having the capability to control voltages. Dynamic shunt compensators (STATCOM) modelled as a PV node used to control the voltage by a flexible adjustment of reactive power exchanged with the network.

2. Review of optimization methods

The optimal power flow (OPF) problem is one of the important problems in operation and control of large modern power systems. The main objective of a practical OPF strategy is to determine the optimal operating state of a power system by optimizing a particular objective while satisfying certain specified physical and security constraints. In its most general formulation, the optimal power flow (OPF) is a nonlinear, non-convex, large-scale, static optimization problem with both continuous and discrete control variables. It becomes even more complex when various types of practical generators constraints are taken in consideration, and with the growth integration of new technologies known as Renewable source and FACTS Controllers. Fig. 2 sumuarizes the basic optimization categories used by researchers to analysis and enhance the optimal power flow solution.

In first category many conventional optimization techniques have been applied to solve the OPF problem, this category includes, linear programming (LP) (Sttot et al., 1979), nonlinear programming (NLP) (Wood et al., 1984), quadratic programming (Huneault et al., 1991), and interior point methods (Momoh et al., 1999).

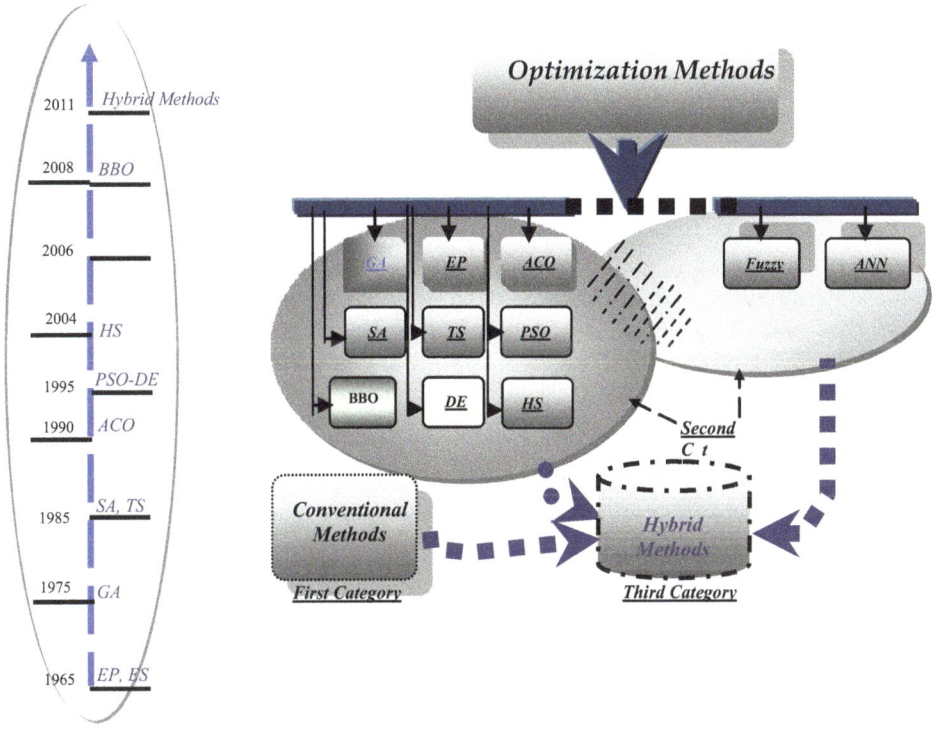

Fig. 2. Presentation of optimization methods: *Global, Conventional*, and *hybrids* methods

All these techniques rely on convexity to find the global optimum; the methods based on these assumptions do not guarantee to find the global optimum when taking in consideration the practical generators constraints (Prohibited zones, Valve point effect), (Huneault et al., 1991) present a review of the major contributions in this area. During the last two decades, the interest in applying global optimization methods in power system field has grown rapidly.

The second category includes many heuristique and stochastic optimization methods known as Global Optimization Techniques. (Bansal, 2005) represents the major contributions in this area. (Chiang, 2005) presents an improved genetic algorithm for power economic dispatch of units with valve-point effects and multiple fuels. (Chien, 2008) present a novel string structure for solving the economic dispatch through genetic algorithm (GA). To accelerate the search process (Pothiya et al., 2008) proposed a multiple tabu search algorithm (MTS) to solve the dynamic economic dispatch (ED) problem with generator constraints, simulation results prove that this approach is able to reduce the computational time compared to the conventional approaches. (Gaing, 2003) present an efficient particle swarm optimization to solving the economic dispatch with consideration of practical generator constraints, the proposed algorithm applied with success to many standard networks. Based on experience and simulation results, these classes of methods do not always guarantee global best solutions. Differential evolution (DE) is one of the most prominent new generation EAs, proposed by Storn and Price (Storn et al., 1995), to exhibit consistent and reliable performance in nonlinear and multimodal environment (price et al., 2005) and proven effective for constrained optimization problems. The main advantages of DE are: simple to program, few control parameters, high convergence characteristics. In power system field DE has received great attention in solving economic power dispatch (EPD) problems with consideration of discontinuous fuel cost functions.

The third category includes, a variety of combined methods based conventional (mathematical methods) and global optimization techniques like (GA-QP), artificial techniques with metaheuristic mehtods, like 'Fuzzy-GA', 'ANN-GA', 'Fuzzy-PSO'. Many modified DE have been proposed to enhance the optimal solution, (Coelho et al., 2009) present a hybrid method which combines the differential evolution (DE) and Evolutionary algorithms (EAs), with cultural algorithm (CA) to solve the economic dispatch problems associated with the valve-point effect. Very recently, a new optimization concept, based on Biogeography, has been proposed by Dan Simon (Simon, D., 2008), Biogeography describes how species migrate from one island to another, how new species arise, and how species become extinct.

To overcome the drawbacks of the conventional methods related to the form of the cost function, and to reduce the computational time related to the large space search required by many methaheuristic methods, like GA, (Mahdad, B. et al., 2010) proposed an efficient decomposed GA for the solution of large-scale OPF with consideration of shunt FACTS devices under severe loading conditions, (Mahdad, B. et al., 2009) present a parallel PSO based decomposed network to solve the ED with consideration of practical generators constraints.

This chapter presents a hybrid controller model based wind source and dynamic shunt FACTS devices (STATCOM Controller) to improve the power system operation and control. Choosing the type of FACTS devices and deciding the installation location and control of multi shunt FACTS coordinated with multi wind source is a vital research area. A simple algorithm based differential evolution (DE) proposed to find the optimal reactive power exchanged between shunt FACTS devices and the network in the presence of multi wind source. The minimum fuel cost, system loadability and loss minimization are considered as a measure of power system quality. The proposed methodology is verified on many practical electrical network at normal and at critical situations (sever loading conditions, contingency). Simulation results show that the optimal coordination operating points of shunt FACTS (STATCOM) devices and wind source enhance the power system security.

2.1 Standard Optimal Power Flow formulation

The OPF problem is considered as a general minimization problem with constraints, and can be written in the following form:

$$\text{Min } f(x,u) \tag{1}$$

$$\text{Subject to: } g(x,u) = 0 \tag{2}$$

$$h(x,u) \leq 0 \tag{3}$$

$$x_{min} \leq x \leq x_{max} \tag{4}$$

$$u_{min} \leq u \leq u_{max} \tag{5}$$

Where; $f(x,u)$ is the objective function, $g(x,u)$ and $h(x,u)$ are respectively the set of equality and inequality constraints. The vector of state and control variables are denoted by x and u respectively.

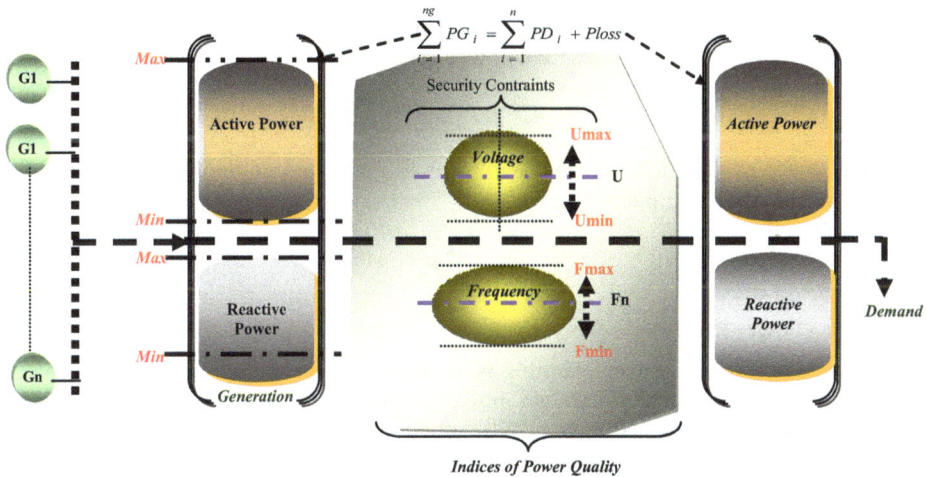

Fig. 3. Optimal power flow (OPF) strategy

In general, the state vector includes bus voltage angles δ, load bus voltage magnitudes V_L, slack bus real power generation $P_{g, slack}$ and generator reactive power Q_g. Fig. 3 shows the optimal power flow strategy. The problem of optimal power flow can be decomposed in two coordinated sub problemes:

a. Active Power Planning

The main role of economic dispatch is to minimize the total generation cost of the power system but still satisfying specified constraints (generators constraints and security constraints).

$$\sum_{i=1}^{Ng} Pg_i = P_D + P_{loss} \qquad (6)$$

For optimal active power dispatch, the simple objective function f is the total generation cost expressed as follows:

$$Min \ f = \sum_{i=1}^{N_g} \left(a_i + b_i P_{gi} + c_i P_{gi}^2 \right) \qquad (7)$$

where N_g is the number of thermal units, P_{gi} is the active power generation at unit i and a_i, b_i and c_i are the cost coefficients of the ith generator.

In the power balance criterion, the equality constraint related to the active power balance with consideration of wind power should be satisfied expressed as follow:

$$\sum_{i=1}^{Ng} P_{gi} + \sum_{i=1}^{NW} P_{wi} - P_D - P_{loss} = 0 \qquad (8)$$

Where; ng represents the total number of generators, nw the number of wind source integrated into the system, P_{wi} represents the active power of wind units, P_D is the total active power demand, P_{loss} represent the transmission losses.

The inequality constraints to be satisfied for this stage are given as follows:

- Upper and lower limits on the active power generations:

$$P_{gi}^{min} \le P_{gi} \le P_{gi}^{max} \qquad (9)$$

- Wind power availability: the total wind power generated, is limited by the available amount from the wind park P_w^{av},

$$P_{loss} + P_D - \sum_{i=1}^{Ng} P_{gi} \le P_w^{av} \qquad (10)$$

b. Reactive Power Planning

The main role of reactive power planning is to adjust dynamically the control variables to minimize the total power loss, transit power, voltages profiles, and voltage stability, individually or simultaneously, but still satisfying specified constraints (generators constraints and security constraints). Fig. 4 shows the structure of the control variable to be optimized using DE.

- Upper and lower limits on the reactive power generations:

$$Q_{gi}^{min} \le Q_{gi} \le Q_{gi}^{max}, i = 1,2,...,NPV \qquad (11)$$

- Upper and lower limits on the generator bus voltage magnitude:

$$V_{gi}^{min} \le V_{gi} \le V_{gi}^{max}, i = 1,2,...,NPV \qquad (12)$$

- Upper and lower limits on the transformer tap ratio (t).

$$t_i^{min} \le t_i \le t_i^{max}, i = 1,2,...,NT \qquad (13)$$

- Upper transmission line loadings.

$$S_{li} \leq S_{li}^{\max}, \; i = 1, 2, \ldots, NPQ \tag{14}$$

- Upper and lower limits on voltage magnitude at loading buses (PQ buses)

$$V_{Li}^{\min} \leq V_{Li} \leq V_{Li}^{\max}, \; i = 1, 2, \ldots, NPQ \tag{15}$$

- Parameters of shunt FACTS Controllers must be restricted within their upper and lower limits.

$$X^{\min} \leq X_{FACTS} \leq X^{\max} \tag{16}$$

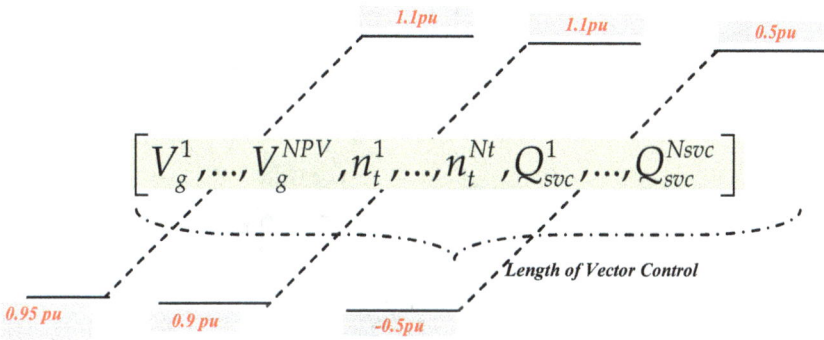

Fig. 4. Vector control structure based DE for reactive power planning

Fig. 5 shows the strategy of FACTS controllers integrated in power system to improve the power quality. In general these FACTS devices are classified in three large categories as follows (Mahdad et al, 2010):

1. Shunts FACTS Controllers (SVC, STATCOM): Principally designed and integrated to adjust dynamically the voltage at specified buses.

$$V^{des} - V_i = 0 \tag{17}$$

2. Series FACTS Controllers (TCSC, SSSC): Principally designed and integrated to adjust dynamically the transit power at specified lines.

$$\tag{18}$$

3. Hybrid FACTS Controllers (UPFC): Principally designed and integrated to adjust dynamically and simultaneously the voltage, the active power, and the reactive power at specified buses and lines.

$$\begin{cases} V^{des} - V_i = 0 \\ P_{ij}^{des} - P_{ij} = 0 \\ Q_{ij}^{des} - Q_{ij} = 0 \end{cases} \tag{19}$$

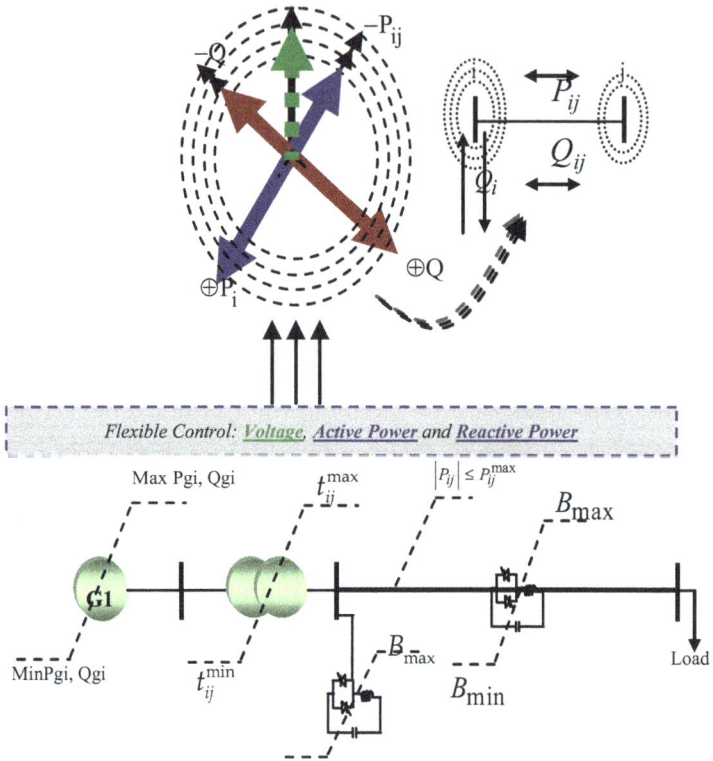

Fig. 5. Basic strategy of FACTS technology integrated in power system

In this study we are interested in the integration of hybrid model based shunt FACTS controller (STATCOM) and wind energy to enhance the indices of power quality at normal and at critical situations.

3. Hybrid model based wind energy and shunt FACTSController

The proposed approach requires the user to define the number of wind units to be installed, in this study voltage stability used as an index to choose the candidate buses. The differential evolution (DE) algorithm generates and optimizes combination of wind sources sizes. Minimum cost, and power losses, used as fitness functions. Wind units modelling depend on the constructive technology and their combined active and reactive power control scheme.

In this study wind has been considered as not having the capability to control voltages. Dynamic shunt compensators (STATCOM) modelled as a PV node used in coordination with wind to control the voltage by a flexible adjustment of reactive power exchanged with the network (Mahdad.b et al., 2011). Fig. 6 shows the proposed combined model based wind source and STATCOM Controller.

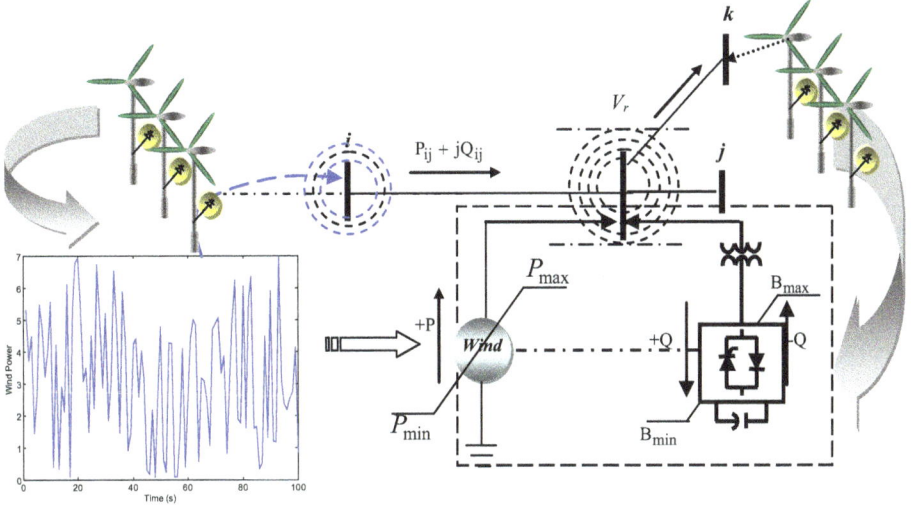

Fig. 6. The proposed combined model based wind/STATCOM Compensators integrated in power flow algorithm

3.1 Wind energy

The principle of wind energy transformation based aerodynamic power can be formulated using the following equations:

$$
P_w = \begin{cases} 0 & V < V_D \\ \dfrac{1}{2}.\rho.S.V^3.C_p(\lambda,\beta) & V_D \leq V < V_N \\ P_N & V_N \leq V < V_A \\ 0 & V \geq V_A \end{cases}
\tag{20}
$$

Where;

ρ : is the air density,

S : the surface swept by the turbine

V : Wind speed

V_D : Critical wind speed

V_A : Wind stopping speed

$$
\lambda = \frac{Q_t.R_t}{v} \; ;
$$

λ : tip speed ratio

R_t : is the blade length

Q_t : is the angular velocity of the turbine

Detailed descriptions about various types of aero-generators are well presented in many references (Bent, S., 2004), (Chen et al. 2008).

3.2 Steady state model of Static Compensator (STATCOM)

The first SVC with voltage source converter called STATCOM commissioned and installed in 1999 (Hingorani, N.G., 1999). STATCOM is build with Thyristors with turn-off capability like GTO or today IGCT or with more and more IGBTs. The steady state circuit for power flow is shown in Fig. 7, the V-I characteristic presented in Fig. 8.

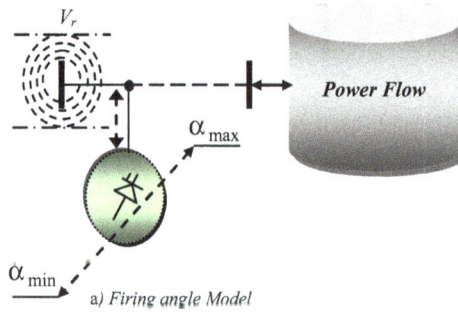

a) *Firing angle Model*

Fig. 7. STATCOM steady state circuit representation

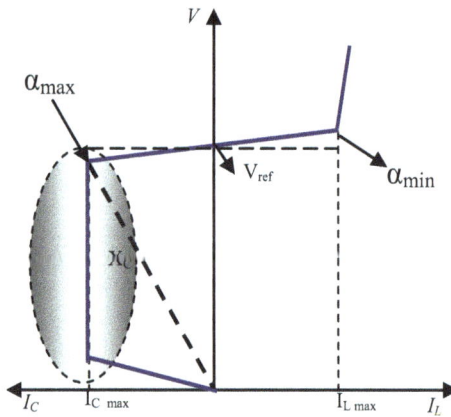

Fig. 8. V-I Characteristic of the STATCOM

3.2.1 Advantages of STATCOM

The advantages of a STATCOM compared to SVC Compensators summarized as follows (Hingorani, N.G., 1999):

- The reactive power to be exchanged is independent from the actual voltage on the connection point. This can be seen in the diagram (Fig. 9) for the maximum currents being independent of the voltage in comparison to the SVC. This means, that even during most severe loading conditions, the STATCOM keeps its full capability.
- Reduced size is another advantage of the STATCOM as compared to the SVC Controller, sometimes even to less than 50%, and also the potential cost reduction achieved from the elimination of many passive components required by SVC, like capacitor and reactor banks.

3.2.2 STATCOM modelling based power flow

In the literature many STATCOM models have been developed and integrated within the load flow program based modified Newton-Raphson, the model proposed by (Acha, et al, 2004), is one of the based and efficient models largely used by researchers. Fig. 9 shows the equivalent circuit of STATCOM, the STATCOM has the ability to exchange dynamically reactive power (absorbed or generated) with the network.

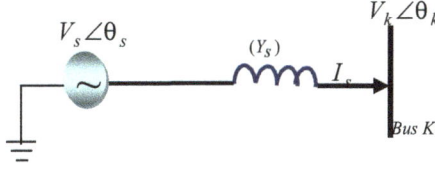

Fig. 9. STATCOM equivalent circuit

Based on the simplified equivalent circuit presented in Fig. 9, the following equation can be formulated as follows:

$$I_s = Y_s.(V_s - V_k);$$

(21)

Where,
$Y_s = G_s + jB_s$; is the equivalent admittance of the STATCOM;
The active and reactive power exchanged with the network at a specified bus expressed as follows:

$$S_s = V_s.I_s^* = V_s.Y_s^*.(V_s^* - V_k^*);$$

(22)

After performing complex transformations; the following equations are deduced:

$$P_s = |V_s|^2 G_s - |V_s||V_k|\{G_s \cos(\theta_s - \theta_k) + B_s \sin(\theta_s - \theta_k)\}$$

(23)

$$Q_s = -|V_s|^2 B_s - |V_s||V_k|\{G_s \sin(\theta_s - \theta_k) - B_s \sin(\theta_s - \theta_k)\}$$

(24)

The modified power flow equations with consideration of STATCOM at bus k are expressed as follows:

$$P_k = P_s + \sum_{i=1}^{N} |Y_{ki}|.|V_k|.|V_i|\cos(\theta_k - \theta_i - \theta_{ki})$$

(25)

$$Q_k = Q_s + \sum_{i=1}^{N} |Y_{ki}|.|V_k|.|V_i|\sin(\theta_k - \theta_i - \theta_{ki})$$

(26)

4. Overview of Differential Evolution technique

Differential Evolution (DE) is a new branch of EA proposed by (Storn and Price, 1995). DE has proven to be promising candidate to solve real and practical optimization problem. The strategy of DE is based on stochastic searches, in which function parameters are encoded as floating point variables. The key idea behind differential evolution approach is a new mechanism introduced for generating trial parameter vectors. In each step DE mutates vectors by adding weighted, random vector differentials to them. If the fitness function of the trial vector is better than that of the target, the target vector is replaced by trial vector in the next generation.

4.1 Differential evolution mechanism search

The differential evolution mechanism search is presented based on the following steps (Gonzalez et al., 2008):

Step 1. Initialize the initial population of individuals: Initialize the generation's counter, $G=1$, and also initialize a population of individuals, $x(G)$ with random values generated according to a uniform probability distribution in the n-dimensional space.

$$X_i^{(G)} = x_{ij}^{(L)} + rand[0,1] * \left(x_{ij}^{(U)} - x_{ij}^{(L)} \right) \qquad (27)$$

Where:

G :   is the generation or iteration

$rand[0,1]$: denotes a uniformly distributed random value within $[0, 1]$.

$x_{ij}^{(L)}$ and $x_{ij}^{(U)}$ are lower and upper boundaries of the parameters x_{ij} respectively for $j = 1,2,...,n$.

Step 2. the main role of mutation operation (or differential operation) is to introduce new parameters into the population according to the following equation:

$$v_i^{(G+1)} = x_{r3}^{(G)} + f_m * \left(x_{r2}^{(G)} - x_{r1}^{(G)} \right) \qquad (28)$$

Two vectors $x_{r2}^{(G)}$ and $x_{r1}^{(G)}$ are randomly selected from the population and the vector difference between them is established. $f_m \succ 0$ is a real parameter, called mutation factor, which the amplification of the difference between two individuals so as to avoid search stagnation and it is usually taken from the range $[0,1]$.

Step 3. Evaluate the fitness function value: for each individual, evaluate its fitness (objective function) value.

Step 4. following the mutation operation, the crossover operator creates the trial vectors, which are used in the selection process. A trial vector is a combination of a mutant vector and a parent vector which is formed based on probability distributions.

For each mutate vector, $v_i^{(G+1)}$, an index $rnbr(i) \in \{1,2,...,n\}$ is randomly chosen using a uniform distribution, and a trail vector, $u_i^{(G+1)} = \left[u_{i1}^{(G+1)}, u_{i2}^{(G+1)},...,u_{i2}^{(G+1)} \right]$ is generated according to equation:

$$u_{ij}^{(G+1)} = \begin{cases} v_{ij}^{(G+1)} \; if \left(rand[0,1] \le CR \right) or \left(j = rnbr(i) \right) \\ x_{ij}^{(G)} \; otherwise \end{cases} \qquad (29)$$

Step 5. the selection operator chooses the vectors that are going to compose the population in the next generation. These vectors are selected from the current population and the trial population. Each individual of the trial population is compared with its counterpart in the current population.

Step 6. Verification of the stopping criterion: Loop to step 3 until a stopping criterion is satisfied, usually a maximum number of iterations, G_{max}.

4.2 Active power dispatch for conventional source

The main objective of this first stage is to optimize the active power generation for conventional units (>=80% of the total power demand) to minimize the total cost, Fig 9 shows the three phase strategy based deferential evolution (DE). Fig. 10 shows the structure of the control variables related to active power dispatch for conventional source. In this stage the fuel cost objective J_1 is considered as:

$$J_1 = \sum_{i=1}^{NG} f_i \tag{30}$$

$$Pd1 = \sum_{i=1}^{NG} Pg_i \tag{31}$$

Where;

f_i : is the fuel cost of the *ith* generating unit.

$Pd1$: the new active power associated to the conventional units;

$Pd2$: the new active power associated to the wind source;

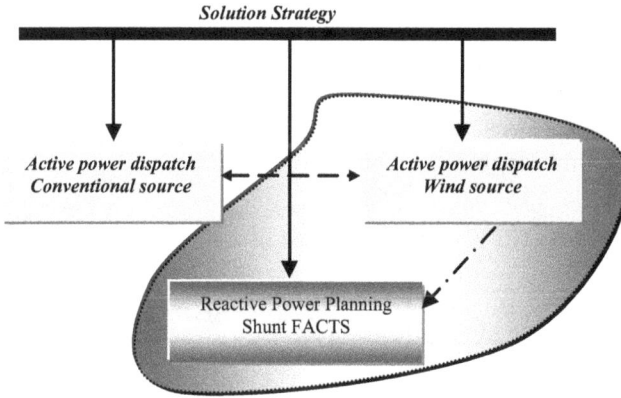

Fig. 10. Three phase strategy based differential evolution (DE)

4.3 Combined active and reactive power planning based hybrid model

The main objective of this second stage is to optimize the active power generation for wind source (<=20% of the total power demand) in coordination with the STATCOM installed at the same specified buses, the objective function here is to minimize the active power loss (P_{loss}) in the transmission system. It is given as:

$$J_2 = Min \ P_{loss} \tag{32}$$

$$P_{loss} = \sum_{k=1}^{N_l} g_k \left[(t_k V_i)^2 + V_j^2 - 2t_k V_i V_j \cos \delta_{ij} \right] \tag{33}$$

The equality constraints to be satisfied are given as follows:

$$Pd_2 = \sum_{i=1}^{NW} Pw_i \tag{34}$$

$$Pd1 + Pd2 - \sum_{i=1}^{NG} Pg_i - \sum_{i=1}^{NW} Pw_i = P_{loss} \tag{35}$$

$$Pd1 + Pd2 = PD \tag{36}$$

Where, N_i is the number of transmission lines; g_k is the conductance of branch k between buses i and j; t_k the tap ration of transformer k; V_i is the voltage magnitude at bus i; δ_{ij} the voltage angle difference between buses i and j.

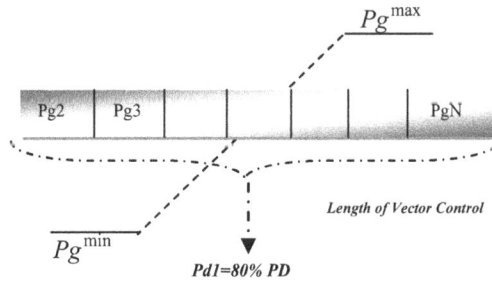

Fig. 11. Vector control structure: conventional source

The inequality constraints to be satisfied are all the security constraints related to the state varaibles and the control variables mentioned in section 2.1.

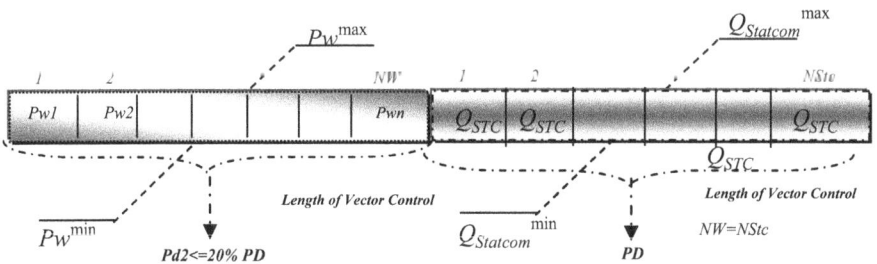

Fig. 12. Coordinated vector control

Fig. 12 shows the two coordinated vectors control structure related to this stage, the individual of the combined vector control denoted by $X_{pq} = [P_{w1}, ..., P_{wN}, Q_{STC1}, ..., Q_{STCN}]$
Where $[P_{w1}, ..., P_{wN}]$ indicate active power outputs of all units based wind source, $[Q_{STC1}, ..., Q_{STCN}]$ represent reactive power magnitude settings of all STATCOM controllers exchanged with the network.

5. Simulation results

The proposed algorithm is developed in the Matlab programming language using 6.5 version. The proposed approach has been tested on many practical electrical test systems (small size:

IEEE 14-Bus, IEEE 30-Bus, and to large power system size). After a number of careful experimentation, following optimum values of DE parameters have been settled for this test case: population size = 30, mutation factor =0.8, crossover rate = 0.7, maximum generation =100. Due to the limited chapter length, details results related to values of control variables (active power generation, voltage magnitudes, reactive power compensation ($Q_{STATCOM}$) for other practical network test will be given in the next contribution.

5.1 Test system 1

The first test system has 6 generating units; 41 branch system, the system data taken from (). It has a total of 24 control variables as follows: five units active power outputs, six generator-bus voltage magnitudes, four transformer-tap settings, nine bus shunt FACTS controllers (STATCOM). The modified IEEE 30-Bus electrical network is shown in Fig 13.

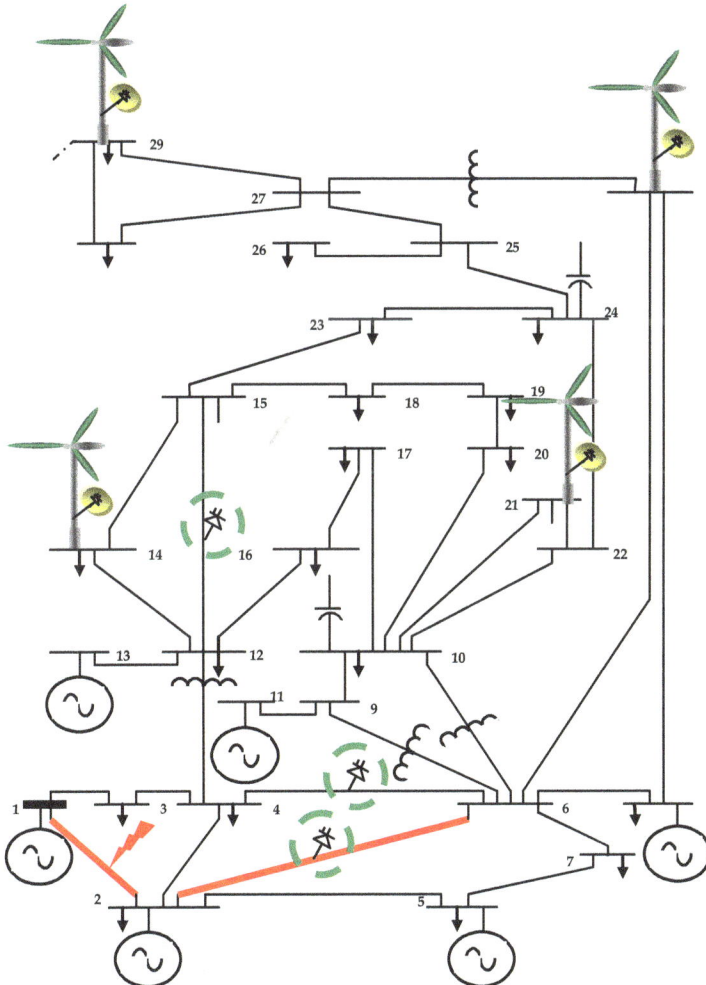

Fig. 13. Single line diagram for the modified IEEE 30-Bus test system (with FACTS devices)

Case1: Normal Condition

	Buses								
STATCOM	10	12	15	17	20	21	23	24	29
Q (MVAR)	35.92	-16.27	-9.82	-19.34	-1.96	-19.94	1.25	4.82	-4.25
Pw (MW)	3.9194	3.9202	4.0070	4.0615	4.1781	4.1956	3.9707	3.8700	3.8776
V (p u)	1.02	1.0	1.0	1.0	1.0	1.0	1.0	1.0	1.0
$\sum_{i=1}^{NW} P_w$ (MW)	36 (12.7%), PD =283.4 MW								
Ploss (MW)	7.554								
Pg1 (MW)	149.92								
Pg2 (MW)	46.53								
Pg5 (MW)	20.64								
Pg8 (MW)	15.81								
Pg11(MW)	10.05								
Pg13(MW)	12	Case1 Normal Condition							
Qg1	5.39								
Qg2	21.67								
Qg5	23.04								
Qg8	45.77								
Qg11	15.43								
Qg13	39.09								
$\sum_{i=1}^{NG} P_G$ (MW)	254.95 (89.96%)								
Cost ($/h)	676.4485								

Table 1. Power Quality Results based Hybrid Model: Wind Source: STATCOM: IEEE-30Bus: Normal Condition

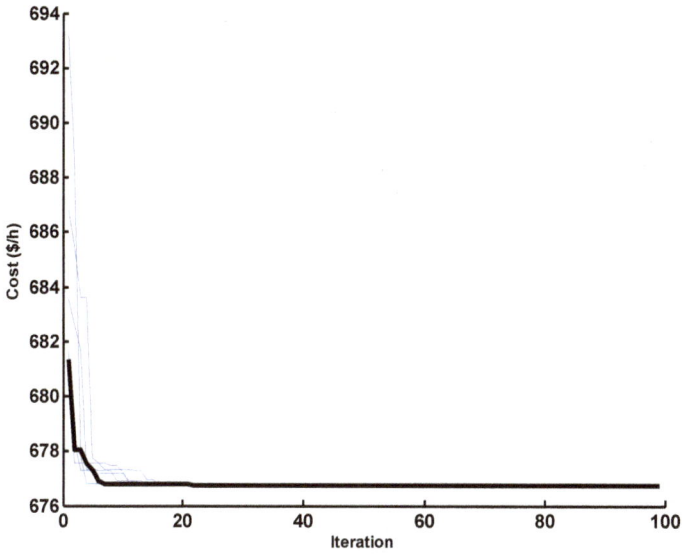

Fig. 14. Convergence characteristic of the 6 generating units with consideration of wind source and STATCOM

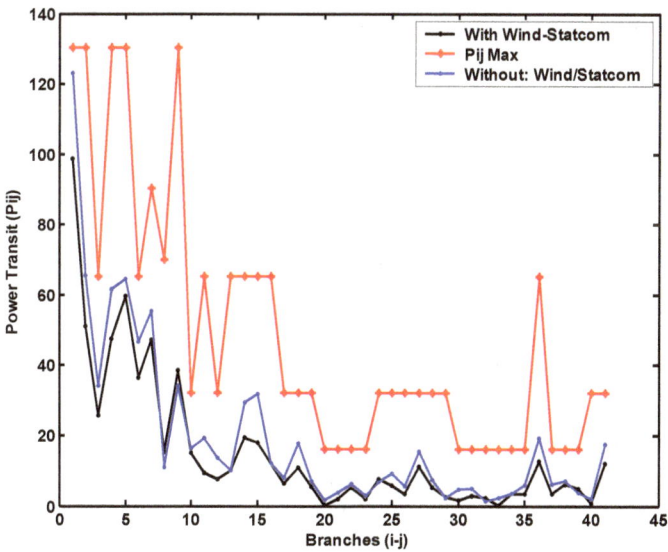

Fig. 15. Active power transit (Pij) with and without wind and STATCOM, Case1: Normal Condition: IEEE 30-Bus

Table 1 shows the results based on the flexible integration of the hybrid model, the goal is to have a stable voltage at the candidate buses by exchanging the reactive power with the network, the active power losses reduced to 7.554 MW compared to the base case: 10.05 MW, without integration of the hybrid controllers, the total cost also reduced to 676.4485 $/h compared to the base case (802.2964 $/h), Fig. 14 shows the convergence characteristic of fuel cost for the IEEE 30-Bus with consideration of the hybrid models, Fig. 15 shows the distribution of power transit in the different branches at normal condition, Fig. 17 shows the distribution of power transit in the different branches at contingency situation (without line 1-2).

The active power transit reduced clearly compared to the case without integration of wind source which enhance the system security. Fig. 16 shows the improvement of voltage profiles based hybrid model. Results at abnormal conditions (contingency) are also encouragement.

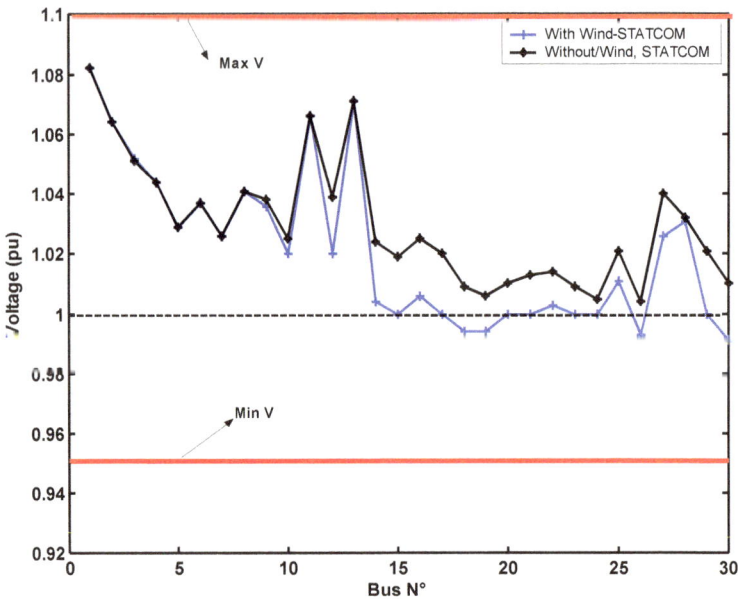

Fig. 16. Voltage profiles with and without hybrid model (wind and STATCOM): IEEE 30-Bus

Case2: Under Contingency Situation

The effeciency of the integrated hybrid model installed at different critical location is tested under contingency situation caused by fault in power system, so it is important to maintain the voltage magnitudes and power flow in branches within admissible values. In this case a contingency condition is simulated as outage at different candidate lines. Table 2 shows sample results related to the optimal power flow solution under contingency conditions (*Fault at line 1-2*).

	Buses								
STATCOM	10	12	15	17	20	21	23	24	29
Q (MVAR)	42.76	-15.65	-11.00	-20.10	-2.90	-20.83	0.28	4.09	-4.52
Pw (MW)	5.8791	5.8803	6.0105	6.092	6.2671	6.2934	5.9560	5.8050	5.8164
V (p u)	1.02	1.0	1.0	1.0	1.0	1.0	1.0	1.0	1.0
$\sum_{i=1}^{NW} P_w$ (MW)	54 MW (19.05%), PD =283.4 MW								
Ploss (MW)	5.449								
Pg1 (MW)	64.12								
Pg2 (MW)	67.98								
Pg5 (MW)	26.86								
Pg8 (MW)	34.65								
Pg11(MW)	21.00								
Pg13(MW)	20.24	Abnormal Condition Without line 1-2							
Qg1	1.76								
Qg2	41.3								
Qg5	20.98								
Qg8	35.55								
Qg11	8.08								
Qg13	39.26								
$\sum_{i=1}^{NG} P_G$ (MW)	235.610MW (83.14%)								
Cost ($/h)	686.1220								

Table 2. Power Quality Results based Hybrid Model: IEEE-30Bus: Abnormal Condition

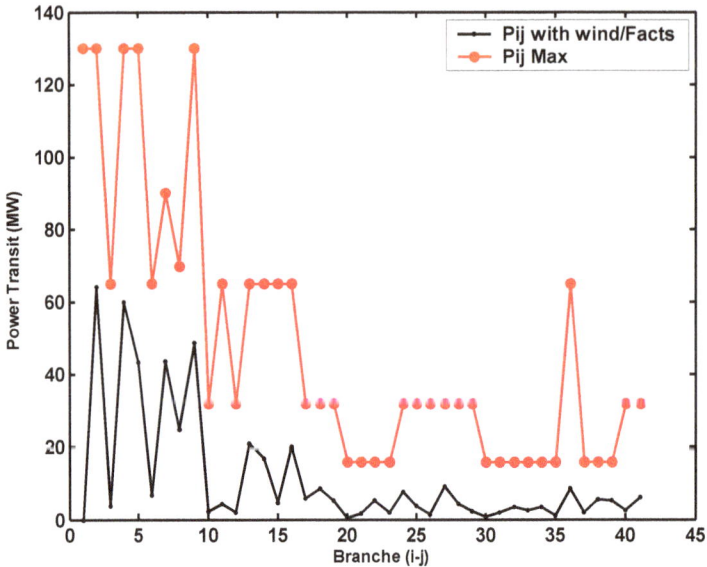

Fig. 17. Active power transit (Pij) with hybrid model: Case 2: Abnormal Condition: without line 1-2: IEEE 30-Bus

6. Conclusion

A three phase strategy based differential evolution (DE) method is proposed to enhance the power quality with consideration of multi hybrid model based shunt FACTS devices (STATCOM), and wind source. The performance of the proposed approach has been tested with the modified IEEE 30-Bus with smooth cost function, at normal condition and at critical loading conditions with consideration of contingency. The results of the proposed hybrid model integrated within the power flow algorithm compared with the base case with only conventional units (thermal generators units). It is observed that the proposed dynamic hybrid model is capable to improving the indices of power quality in term of reduction voltage deviation, and power losses.

Due to these efficient properties, in the future work, author will still to apply this algorithm to solve the practical optimal power flow of large power system with consideration of multi hybrid model under severe loading conditions and with consideration of practical constraints.

7. References

Acha E, Fuerte-Esquivel C, Ambiz-Perez (2004) FACTS Modelling and Simulation in Power Networks. John Wiley & Sons.

Adamczyk, A.; Teodorescu, R.; Mukerjee, R.N.; Rodriguez, P., Overview of FACTS devices for wind power plants directly connected to the transmission network, *IEEE International Symposium on Industrial Electronics (ISIE)*, Page(s): 3742– 3748, 2010.

Bansal, R. C., Otimization methods for electric power systems: an overview, *International Journal of Emerging Electric Power Systems*, vol. 2, no. 1, pp. 1-23, 2005.

Bent, S., Renewable energy: its physics, use, environmental impacts, economy and planning aspects, 3rd ed. UK/USA: Academic Press/Elsevier; 2004.

C. Chien Kuo, A novel string structure for economic dispatch problems with practical constraints, *International Journal of Energy Conversion and management*, ,vol. 49, pp. 3571-3577, 2008.

Chen, A., Blaadjerg, F, Wind farm-A power source in future power systems, *Renewable and Sustainable Energy Reviews*. pp. 1-13, 2008.

Chiang C.-L., Improved genetic algorithm for power economic dispatch of units with valve-point effects and multiple fuels, *IEEE Trans. Power Syst.*, vol. 20, no. 4, pp. 1690-1699, Nov. 2005.

Coelho, L. S., R. C. Thom Souza, and V. Cocco Mariani, (2009) Improved differential evoluation approach based on clutural algorithm and diversity measure applied to solve economic load dispatch problems, *Journal of Mathemtics and Computers in Simulation*, Elsevier, 2009.

Gaing, Z. L., Particle swarm optimization to solving the economic dispatch considering the generator constraints, *IEEE Trans. Power Systems*, vol. 18, no. 3, pp. 1187-1195, 2003.

Gonzalez, F. D., M. M. Rojas, A. Sumper, O. Gomis-Bellmunt, L. Trilla, Strategies for reactive power control in wind farms with STATCOM,

Gupta, A., Economic emission load dispatch using interval differential evolution algorithm, *4th International Workshop on reliable Engineering Computing* (REC 2010).

Hingorani NG, Gyugyi L (1999) Understanding FACTS: Concepts and Technology of Flexible A Transmission Systems. IEEE Computer Society Press.

Hingorani, N.G., FACTS: flexible ac transmission systems, *EPRI Conference on Flexible AC Transmission System*, Cincinnati, OH, November 1990.

Huneault, M., and F. D. Galiana, A survey of the optimal power flow literature, *IEEE Trans. Power Systems*, vol. 6, no. 2, pp. 762-770, May 1991.

Mahdad, B., K. Srairi, T. Bouktir, and and M. EL. Benbouzid, Fuzzy Controlled Parallel PSO to Solving Large Practical Economic Dispatch, Accepted and will be Published at *IEEE IECON Proceeding* , 2010.

Mahdad, B., T. Bouktir, K. Srairi, and M. EL. Benbouzid, Dynamic Strategy Based Fast Decomposed GA Coordinated with FACTS devices to enhance the Optimal Power Flow, *Intenational Journal of Energy Conversion and Management*(IJECM), vol. 51, no. 7, pp. 1370-1380, July 2010.

Mahdad, B., T. Bouktir, K. Srairi, OPF with Environmental Constraints with SVC Controller using Decomposed Parallel GA: Application to the Algerian Network. *Journal of Electrical Engineering & Technology*, Korea, Vol. 4, No.1, pp. 55~65, March 2009.

Mahdad, B., T. Bouktir, K. Srairi, Optimal Location and Control of Multi Hybrid Model Based Wind-Shunt FACTS to Enhance Power Quality. Accepted at *World Renewable Energy Congress -Sweden*, 8-11 May 2011, Linköping, Sweden, Mai 2011.

Mahdad, B., T. Bouktir, K. Srairi, Optimal Power Flow for Large-Scale Power System with Shunt FACTS using Efficient Parallel GA, *Intenational Journal of Electrical Power & Energy Systems* (IJEPES), vol. 32, no. 4, pp. 507- 517, Juin 2010.

Momoh, J. A., and J. Z. Zhu, Improved interior point method for OPF problems, *IEEE Trans. Power Syst.* , vol. 14, pp. 1114-1120, Aug. 1999.

Munteau, I., AI. Bratcu, N-A. Cutululis, E. Ceaga , Optimal control of wind energy, towards a global approach, London: Springer-Verlag: 2008.

Nikman,T., (2010) A new fuzzy adaptive hybrid particle swarm optimization algorithm for non-linear, non-smooth and non-convex economic dispatch, *Journal of Applied Energy*, vol. 87, pp. 327-339.

Pothiya, S., I. Nagamroo, and W. Kongprawechnon, Application of multiple tabu search algorithm to solve dynamic economic dispatch considering generator constraints, *International Journal of Energy Conversion and Management*, vol. 49, pp. 506-516, 2008.

Price, K., R. Storn, and J. Lampinen, Differential Evolution: A Practical Approach to Global Optimization. Berlin, Germany: Springer- Verlag, 2005.

Simon, D., Biogeography-based optimization, *IEEE Trans. Evol.Comput.*, vol. 12, no. 6, pp. 702–713, Dec. 2008.

Storn, R. and K. Price, Differential Evolution-A Simple and Efficient Adaptive Scheme for Global Optimization Over Continuous Spaces, International Computer Science Institute, Berkeley, CA, 1995, Tech. Rep. TR-95-012.

Sttot, B., and J. L. Marinho, Linear programming for power system network security applications, *IEEE Trans. Power Apparat. Syst.*, vol. PAS-98, pp. 837-848, May/June 1979.

Wood, J. , and B. F. Wollenberg, Power Generation, Operation, and Control, 2nd ed. New York: Wiley, 1984.

Wood, J., and B. F. Wollenberg, Power Generation, Operation, and Control, 2nd ed. New York: Wiley, 1984.

Yankui, Z., Z. Yan, B. Wu, J. Zhou, Power injection model of STATCOM with control and operating limit for power flow and voltage stability analysis, *Electic Power Systems Researchs*, 2006.

Zhang, X.P., Energy loss minimization of electricity networks with large wind generation using FACTS, *IEEE Power and Energy Society General Meeting-Conversion and Delivery of Electrical Energy in the 21st Century*, 2008.

5

Reliability Centered Maintenance Optimization of Electric Distribution Systems

Dorin Sarchiz[1], Mircea Dulau[1], Daniel Bucur[1] and Ovidiu Georgescu[2]
[1]Petru Maior University,
[2]Electrica Distribution and Supply Company
Romania

1. Introduction

A basic component of the power quality generally and electricity supply in particular is the management of maintenance actions of electric transmission and distribution networks (EDS). Starting from this fact, the chapter develops a mathematical model of external interventions upon a system henceforth, called Renewal Processes. These are performed in order to establish system performance i.e. its availability, in technical and economic imposed constraints. At present, the application of preventive maintenance (PM) strategy for the electric distribution systems, in general and to overhead electric lines (OEL) in special, at fixed or variable time intervals, cannot be accepted without scientifically planning the analysis from a technical and economic the point of view. Thus, it is considered that these strategies PM should benefit from mathematical models which, at their turn, should be based, on the probabilistic interpretation of the actual state of transmission and distribution installations for electricity. The solutions to these mathematical models must lead to the establishment objective necessities, priorities, magnitude of preventive maintenance actions and to reduce the life cycle cost of the electrical installations (Anders et al., 2007).

1.1 General considerations and study assumptions

Based on the definition of IEC No: 60300-3-11 for RCM: „method to identify and select failure management policies to efficiently and effectively achieve the required safty, availability and economy of operation", it actually represents a conception of translating feedback information from the past time of the operation installations to the future time of their maintenance, grounding this action on:

- statistical calculations and reliability calculations to the system operation;
- the basic components of preventive maintenance (PM), repair/renewal actions.

So, Reliability Based Maintenance (RCM) implies planning the future maintenance actions (T^+) based on the technical state of the system, the state being assessed on the basis of the estimated reliability indices of the system at the planning moment (T^0). At their turn, these reliability indices are mathematically estimated based on the *record of events*, that is, based on previously available information, related to the behaviour over period (T^-), i.e. to the (T^0) moment, concordant with Figure 1.

$$T^- \qquad\qquad\qquad\qquad T^+$$

$$\text{System Reliability} \quad T^0 \quad \text{System Maintenance}$$

-t «_____|_____» +t

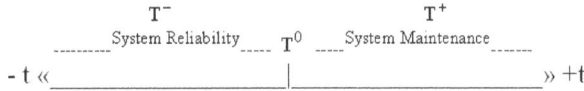

Fig. 1. Assessment and planning times of the RCM

Even if in case of the OEL, the two actions are apparently independent, as they take place at different times, they influence each other through the model adopted for each of them. Thus:
- the analytical expression of the reliability function adopted in while (T^-), depends directly by the basis of specific physical phenomena (such as wear, failure, renewal, etc.) of equipment during maintenance actions in time interval (T^+);
- in turn, the actions of the preventive maintenance in time interval (T^+), depend directly on the reliability parameters at the time (T^0) and by the time evolution of the *reliability function* of the system over this period of time. The interdependence of the two actions can be expressed mathematically given by the analytical expression by the *availability of system*, which is the most complex manifestation of the quality of the system's operation, because it includes both: the reliability of the system and its maintainability.

Accepting that the expression of the availability at a time *t*, the relationship given by the form (Baron et al., 1988):

$$A(t) = R(t) + \left[1 - R(t)\right] \cdot M(t') \qquad\qquad (1)$$

where:

A(t) – Availability;

R(t) – Reliability;

M(t') – Maintainability

It can be inferred that the availability of an element or of a system is determined by two probabilities:
a. the probability that the product to be in operation without failure over a period of time (*t*); R(t) – is called the *reliability function*;
b. the probability that the element or the system, which fails over time interval (*t*), to be reinstated in operation in time interval (*t'*) ; M(t') - is called the *Maintainability function*.

Thus, for components of EDS which are submissions by RCM actions, a distinctive study is required to model their availability under the aspect of modelling those two components:
1. The *reliability function* R(t) of EDS, respectively of the studied component, in this case a OEL;
2. The *Maintainability function* M(t'), under the aspect of their specific constructive and operation conditions, maintenance.

As the mathematical models of the two components R(t) & M(t') interdependent, we will continue by exposing some considerations regarding the establishment of M(t'), and section 3 will be devoted to developing the component R(t) of the EDS.

1.2 EDS-Degradable systems subjected to wear processes

EDS in general and the OEL in particular, contain parts with mechanical and electrical character and their operation is directly influenced by different factors, being constantly subjected to the requirement of mechanical, electrical, thermal processes, etc. The literature shows the percentage of causes which determin interruptions of OEL, (Anders et al., 2007).

- Weather 55%; Demage from animals 5%; Human demage 3%; Trees 11%; Ageing 14%. Thus, one can say with certainty that these failures of systems are due to wear and can be defined as NBU type degradable systems *(New Better than Used)*, whose operation status can improve through the preventive maintenance actions, called in the literature as *renewal actions*. As OEL maintainability is directly conditioned by the evolution in time of maintenance actions and it is imposed a defines mathematics, from the point of view of reliability and of the notions of: *wear, degradable system and renewal actions*.

1.3 Modelling of systems wear

In theory of reliability, the concept of the *wear* has a wider meaning than in ordinary language. In this context, the wear includes any alteration in time of characteristics of reliability, for the purposes of worsening or improving them (Catuneanu & Mihalache 1983). Considering the function of the reliability of equipment for a mission with a duration x, initialized at the t moment, of the type:

$$R(t,t+x) = \frac{R(t+x)}{R(t)} \qquad (2)$$

- the equipment is characterized by the *positive wear* if the reliability function $R(t, t+x)$ is decreases in the range $t \in (0,\infty)$ for any value of $x \geq 0$, i.e. for any mission of the equipment, its reliability decreases with its age.
- the equipment is characterized by the *negative wear* if the reliability function $R(t, t+x)$ is increases in the range $t \in (0,\infty)$ for any value of $x \geq 0$, i.e. for any mission of the equipment, its reliability increases with its age.

From a the mathematical point of view, we can express the wear through *failure rate (FR)* of the equipment. From the relationship (2) and from the defining relationship of failure rate we obtain:

$$z(t) = \lim_{x \to 0} \frac{R(t) - R(t+x)}{x \cdot R(t)} = \lim_{x \to 0} \frac{R(t,t+x)}{x} \qquad (3)$$

If the failure rate of the equipment with positive wear increases in time, then these systems will be called IFR type systems (Increasing Failure Rate) and if the failure rate of the equipment with negative wear decreases in time, these systems will be called DFR type systems (Decreasing Failure Rate).

An equipment is of *NBU degradable type* if the reliability function associated to a mission of the duration x, initialized at the t age of equipment, is less than the reliability function in interval *(0, t)*, regardless of the age and duration of x mission.

$$R(t,t+x) < R(t); \quad t,x \geq 0 \qquad (4)$$

From the previous definition, results that a degradable equipment which was used is inferior that to a new equipment.

The notion of degradable equipment is less restrictive than that of equipment with positive wear, which supposed the decreasingd character of the function $R(t, t+x)$ with t age of equipment.

A particular problem consist of identifying the type of wear that characterizes an equipment/system, which can be obtained from reliability tests or from the analysis of the moments of failure in operation, which mathematically modelled, give the wear function, denoted by T_s (F) .

With:

- $F(t)$, *probability of failure function*, representing the probability that the equipment should enter the failure mode over time interval *(0, t)* and is complementary to reliability function *R(t):*

$$F(t) = 1 - R(t) \tag{5}$$

- $R(t)$ represent the *reliability function* of the equipment over time interval *(0, t).*

According to system reliability $T_s = f(F(t))$, the graph function of the wear inscribed in a square with l side, can indicate the type of equipment wear by its form:

    concave for equipment with positive wear, of IFR type;
- convex for equipment with negative wear, of DFR type;
- first bisectrix for equipments characterized by lack of wear.

Thus, the literature allow for the study maintenance of OEL, the following assumptions:

- the OEL are degradable systems, by the of NBU type;
- the OEL are systems with positive wear, of IFR type;
- the OEL are repairable systems;
- the OEL allowed external interventions (the renewal actions), to restore the performances.

We mention that all models specific to these assumptions/behaviours of the OEL are dependent on the reliability function of the system in study and their effects model this function.

2 RCM – Evolutive process of renewal

In the analysis presented in section 1, aimed to observe the natural process of degradation of the system performance, without taking into account the capacity of external interventions, to oppose this degradation through the partial or total reconditioning system, in a word's through its renewal. Starting from these considerations, in this section we will try to present the specific models of external interventions, defined as a renewal strategy, performed to restore systems performance and to modelling the reliability function of systems in study, (Catuneanu & Mihalache, 1983).

2.1 Renewal process

We define the action of *renewal*, as: *the external intervention performed on a system that restores the system operating status and/or changes the level of its wear, respectively of the system reliability.* From the definition we can distinguish two types of renewal actions that can be performed on systems:

- *Failure Renewal (FR),* is performed only at the appearance of some failures and its purpose is to restore the system operation. They are random events generated by the failure of the system;
- *Preventive renewals (PR),* have the purpose of renewing the system before its failure. They can be random or deterministic events, according to the way in which they design

their strategies. Thus, the random or deterministic strategy of preventive renewals is added to a random process of failure renewals.

The classification of renewal actions can be performed on several criteria, of which we remind a few: the purpose, timing and costs of their occurrence, distribution and frequency and last but not least, the effects on the system safety, (Anders et al., 2007).

In this work, we analyze and study only preventive renewal processes, through their modelling influence on the system in the following cases:

- the system is known due to the reliability function and reliability indices, based on data received from operation;
- PR actions have the purpose of reducing the influence of the system wear and thus improve its reliability, which are considered preventive maintenance actions (PM) from this point of view;
- PR does not entirely change the characteristics of the system, and after the renewal action, the evolution of the system follows the same law of reliability until a new renewal;
- determining the PR frequency on the system will be based on technical criteria and/or economic ones;
- PR of system elements, will be performed considering the same assumptions of study as the system as a whole.

The evolution of an equipment will thus be represented by the succession of renewal moments $t_1, t_2, \ldots, t_n$, and the intervals between them: $x_1, x_2, \ldots, x_n \ldots$ presented in the Figure 2.

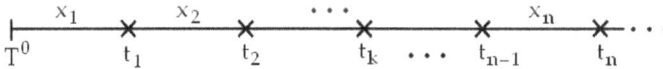

Fig. 2. The evolution of an equipment with renewal

If it is considered a certain time interval (0, t), the number of PR denoted by N_t, performed during this time, is a discrete deterministic process, called preventive renewal process, as the basic component of preventive maintenance, which in turn determine planning, development and the effects of RCM on system. This, requires the development of PR strategies, enabling knowledge of behaviour of renewal equipment, used in the development of PM programs.

2.2 Renewal strategies

Studies classify and model the renewal strategies according to technical and economic parameters, which determine these strategies, in two distinct categories: non-periodic and periodic, (Catuneanu & Mihalache, 1983), (Andres et al., 2007).

- *Non-periodic renewal strategies*, are external interventions on one equipment of the system, which take place only when accidental failure of the equipment, in order to restore its performance and are performed in negligible time; are characteristic of the failure renewals.
- *Periodic renewal strategies*, are characteristic to the prophylactic renewals or preventive PR, providing the renewal of equipment before their failure and providing increased efficiency of equipments in operation. These types of strategies are characterized by a constant/variable and known duration between two consecutive preventive renewals,

based on a preventive maintenance program. In conclusion, that these strategies have a deterministic character.

According to the mathematical model adopted in the calculation of specific parameters of renewals non-periodic or periodic renewal, we can distinguish the following types of strategies:

- *BRP (Block Replacement Policy)*,

    It represents the simplest strategy of periodic renewal and consists of the renewal of the equipment either at the moment of its failure or at equal time intervals $\{ k\Delta T, k = 1, 2, ...\}$.

    A first criterion used to implement this strategy, i.e. the calculation of period ΔT or the number of k renewals, over the imposed period T, consists of imposing a certain minimum level of the reliability function over the interval between two successive renewals, or at the end of a period T. The strategies by BRP type have the disadvantage of inflexible schedules of renewals, by the fact that preventive renewals are performed at predetermined moments of time, without taking into account the failure renewals imposed by the failures occurring in the system.

- *FRP (Failure Replacement Policy)*, define the situation for which T is not running the preventive renewals.

- *DRP (Delayed Replacement Policy)*, eliminates the disadvantages of BRP- type strategies by inserting deterministic preventive renewals between the Failure Renewal random type strategies.

- *ARP (Age Replacement Policy)*, represents the simplest strategy, in which preventive renewal is determined by the moment when the equipment reaches a certain age. In order to establish the age of the system at which the preventive renewal can be performed various technical criteria or economic criteria can be used.

These preventive strategies renewal operating at some predetermined time points and lead to total or partial elimination of accumulated wear, each time bringing equipment in operating condition characterized by lack of wear or with negligible wear. Establishing the renewal moment is made using the reliability model of equipment, based on the information referring to the behavior of the equipment in conditions of operating data, principles which are the foundation of the RCM design.

The option between the different RCM preventive renewal strategies must rely on uniform criteria leading to the best assessment. These criteria must be the result of technical and safety factors imposed to the system, as well as the result of economic considerations imposed to the preventive maintenance actions.

We mention only some of these criteria which are valid for EDS, these criteria developed in the following sections:

- imposing a certain minimum level of reliability function in the interval between two successive renewals or after a fixed period of time;
- imposing a maximum number of unplanned interruptions during a certain period;
- optimizing the operating costs and maintenance costs.

Different studies present additional evolutionary renewal strategies used in the renewal equipment, which are part of the *CRP type (Continuous Replacement Policy)* strategies.

Implementing the CRP type strategy by requires a continuous monitoring of equipment through measured quantities. Determining the moment of the next preventive renewal is made according to the evolution equipment parameters, established through diagnosis techniques.

This type of evolutionary strategy of renewal is underlying the maintenance actions planning by *CBM (Condition Based Maintenance) type*.

3. Electric Distribution Systems reliability assessment

In this section, we try to exemplify an algorithm for the RCM planning, through the optimizing the PM strategies for OEL 20kV, through his components, as a subsystem of the EDS.

3.1 Reliability assessment stages

Knowing the *availability* of EDS implies statistical processing of the database containing historical events, with the final purpose of determining the parameters and the analytical shape of the *reliability function R(t)*. This process is applied to each component of OEL, in accordance with corrective or preventive maintenance actions performed over time.

The analytical expression of the reliability function is a central issue of the reliability theory, in general and especially in RCM. The procedure for assessing the unknown parameters of a distribution function, is called an estimate and she performed in the selective treatment of data resulting from the analysis of reaction time in operation of the system with its components.

Estimating the reliability of OEL, based on statistical data processed during the operation, in this case from the $(0 \div T^0)$, presented in Figure 1 is based on modeling the subsystems' behavior with random processes.

Estimation modeling involves the following steps, for a OEL:

a. the selection and processing of data resulting from the network operation until (T^0), i.e. of random variables given by the behaviour of the system;
b. establish the type of the theoretical distribution function which model the best random variables;
c. calculation of distribution function parameters;
d. verifying the correspondence between the adopted theoretical law and the database, with regard to the behavior of the system in operation. The regression method is given by the statistical information;
e. reliability indices of OEL at the time T^0;
f. availability estimation function of OEL.

To illustrate these steps, we study the case of OEL by 20 kV, has the following construction parameters:

-   Transmission length L = 90.589 km;
-   1097 concrete pillars and 9 metal pillars;
-   1107 porcelain and glass insulators;
-   Aluminium conductor steel reinforced (ACSR) by 35/50/95 mm².

3.2 Random variables

The OEL operation time was selected as random variable form the database of the beneficiary of the 20 kV power line, which includes the sheets of incidents and interventions over a period of about five years. The OEL operation time is expressed according to:

-   moments of time t_i (expressed in days), when the OEL stopped functioning. Only those failures/interruptions in electricity supply were considered, which were followed by corrective actions in installation to bring the facility into operation of OEL;

- the period of corrective actions, tr_i (expressed in minutes), to restore in operation OEL, the values are presented in Table 1.

Nr	1	2	3	4	5	6	7	8	9	10	11	12	13	14
t_i	62	68	80	200	210	212	254	290	407	418	483	497	510	539
t_{ri}	213	118	352	209	339	640	165	305	229	215	154	1970	204	303

...	45	46	47	48	49	50	51	52	53	54
...	1272	1370	1399	1483	1578	1596	1597	1608	1638	1712
...	482	1350	270	150	343	270	200	403	178	841

Table 1. Incidents database

3.3 Distribution function

Establishing a law distribution used in reliability implies good knowledge of the physical bases of the phenomenon of wear, of the specific ways in which these phenomena manifest themselves and of the type of wear to which each OEL component and the entire system has been subjected. Considered as an EDS component, the OEL contains, at its turn, components of a mechanical character, whose operations are directly influenced by mechanical actions, electrical ones (e.g. overvoltages, over currents), temperature, environmental pollution, etc. We can say with certainty that the OEL failures are due to wear and slow aging and, from the standpoint of their reliability, they are treated as IFR and NBU type, with an increasing failure rate.

Given these findings, *the law of Weibull distribution* is adopted as *theoretical law* for modelling such survival processes. This law is specific for positive wear systems, being also characteristic for overhead electrical lines.

In the case of an OEL in operation whose components are characterized by the absence of hidden defects, but show a striking phenomenon of aging in time while the intensity of failures increases monotonically, *the law of Weibull distribution* is adopted as theoretical law, which is specific for positive wear systems, being also characteristic for overhead electrical lines.

Of all the known forms of the Weibull distribution law (two and three parameters, normalized) let us accept the form with two parameters for modeling the reliability of the electrical line in study. This form has the mathematical expression (Baron et al., 1988), (IEC 61649, 2008):

$$R(t,\alpha,\beta) = e^{-\alpha \cdot t^{\beta}} \tag{6}$$

Where:

$\alpha > 0$ - is a scale parameter;

$\beta > 0$ - is a shape parameter, $\beta > 1$ for components of IFR type;

t $(0, +\infty)$ - time variable.

The relationship (6) expresses the probability that the event will occur in time interval $(0, t)$ or as they say in the theory of reliability is the probability of the OEL functioning without fault until t moment.

3.4 The parameters of Weibull distribution function

There are several studies which present a lot of techniques and methods to evaluate the Weibull function parameters, depending on: the number of function parameters chosen, the scope of the application, the available statistical data, etc.

In (Dickey, 1991), (IEC 61649, 2008), are mentions statistical methods for assessing the parameters for Weibull distribution of failures type and constant repair time. The Statistics Toolbox MATLAB programming environment allows the evaluation of parameters of the Weibull function, with the instructions, (Blaga, 2002):

$$[parmhat,parmci]=wblfit\ (t)$$

$$parmhat\ (1);\ \ parmhat\ (2);$$

A synthesis of the calculation methods and accuracy of the Weibull parameters is presented in (IEC 61649, 2008).

In the paper (Baron et al.,1988), the authors presents a practical mathematical method of setting the a and β parameters with two parameters of Weibull law by the form of relationship (6), based on statistical data obtained from the analysis of operating arrangements of the OEL, as well as in the following assumptions:

a.   there are significant records concerning the concerning to the number of defects and the operating times between them;

b.   the average recovery times are negligible compared with the operating times.

The value of the parameters 'α' and 'β' is derived from the relationship (6), which by logarithm application, results as:

$$\lg R_N(t_{fi}) = -\alpha \cdot t_{fi}^{\beta} \cdot \lg e \ \ or: \ \ \lg\left[1 / R_N(t_{fi})\right] = \alpha \cdot t_{fi}^{\beta} \cdot \lg e \tag{7}$$

repeating the operation of logarithms, result:

$$\lg\left\{\lg\left[1 / R_N(t_{fi})\right]\right\} = \lg(\lg e) + \lg\alpha + \beta\lg t_{fi} \tag{8}$$

with notations: $R_N(t_{fi}) = 1 - t_{fi} / t_{fi}^{max}$; $a = \lg(\lg e) + \lg\alpha$; $b_i = \lg\left\{\lg\left[1 / R_N(t_{fi})\right]\right\}$ (9)

is obtained from (8) equation of a straight line: $b_i = a + \beta\lg t_{fi}$ (10)

for which using the method of least squares system is obtained the following equations system is obtained:Cititi fonetic

$$\sum_{i=1}^{n} b_i = n \cdot a + \beta \cdot \sum_{i=1}^{n} \lg t_{fi}$$

$$\sum_{i=1}^{n} b_i \lg t_{fi} = a \sum_{i=1}^{n} \lg t_{fi} + \beta \sum_{i=1}^{n} (\lg t_{fi})^2 \tag{11}$$

or :
$$A = n \cdot a + \beta \cdot B$$
$$C = B \cdot a + \beta \cdot D \tag{12}$$

which solved in relation with a and β unknown, result:

$$a = \frac{C \cdot B - A \cdot D}{B^2 - n \cdot D} \; ; \; \beta = \frac{n \cdot C - A \cdot B}{n \cdot D - B^2} \quad \text{and} \quad \alpha = 10^{[a - \lg(\lg e)]} \tag{13}$$

For the OEL 20 KV in study, by replacing the parameters from Table 3.1 in the previous relations, the following values of parameters of the Weibull function distribution result:

$$\alpha = 1.2669 \cdot 10^{(-4)} \quad \text{and} \quad \beta = 1.2939 \tag{14}$$

which allows modeling the reliability function $R(t)$ of OEL the through the relationship:

$$R(t) = e^{-1.2669 \cdot 10^{(-4)} \cdot t^{1.2939}} \tag{15}$$

and the nonreliability function $F(t)$, through the relationship: $F(t) = 1 - R(t)$ $\qquad$ (16)

and the graph presented in Figure 3.

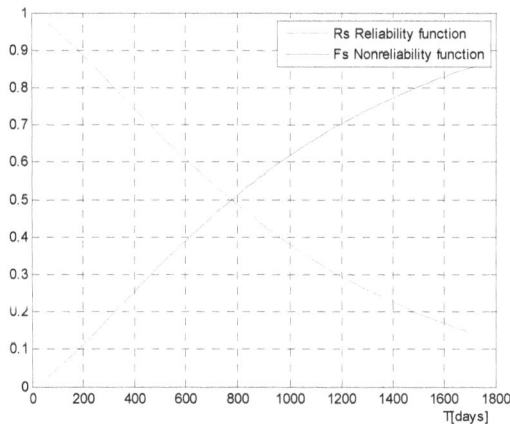

Fig. 3. Variation of reliability functions and failure probability

3.5 Concordance test

The fundamental criterion in adopting the distribution law is the concordance between the theoretical law, which in our case is Weibull distribution and experimental data, which in this case is the recorded database.

For this study, the validation of the chosen distribution law it is imposed by concordance study between: Weibull distribution by the form (15), with two parameters α, β, calculated with formulae (13), with experimental data presented in Table 1.

This can be achieved using the chi square concordance test which, in MATLAB, Statistics Toolbox is achieved with the procedure, (Blaga, 2002) :

$$[h,p] = chi2gof(t,@cdfweib\_OEL)$$

where: - t is line matrix of experimental data, and
- @cdfweib_OEL is the cumulative density function F(t)=1-R(t).

The following values resulted are h=0, p=0.935. This means the acceptance of the null hypothesis of concordance between the observed data and the theoretical Weibull distribution of the parameters α, β, with the level of significance of 6.5%.

3.6 Reliability indices

Determining the reliability indices of an OEL facilitates the knowledge of the safety level in the operation of the OEL analyzed and the whole essembly which composes the OEL. The OEL 20 KV is studied as a reparable simple element, which regains its operating ability after failure, through repair, and then it can continue operation until the next failure. The evolution in time of such an element is a sequence of tf_i operating times with tr_i and repair times, for which, the following indices of reliability can be defined and calculated (Dub, 2008):

- MTBF, Mean Time Between Failures, given by the relationship:

$$MTBF = \frac{\sum_{i=1}^{n} t_{fi}}{n} \ [day] \quad with \ t_{fi} = t_i - t_{i-1}; \ MTBF = 31.7037 \ [day] \tag{17}$$

- λ, failure rate, $\quad \lambda = \frac{1}{MTBF} \ \left[day^{-1}\right]; \ \lambda = 0.031542 \ \left[day^{-1}\right]$ (18)

$$MTTR = \frac{\sum_{i=1}^{n} tr_i}{n-1} \ [day]; \ MTTR = 0.2506 \ [day] \tag{19}$$

- MTTR, Mean Time Repair, (as above)

- μ, repair rate, $\quad \mu = \frac{1}{MTTR} \ \left[day^{-1}\right]; \ \mu = 3.9897 \ \left[day^{-1}\right]$ (20)

For the case of steady state, indices P and Q are defined. In our case, which chose the Weibull model for shaping the reliability function, and considering that ($\beta = 1.2939$), $\beta \cong 1$, we used the relationships of availability associated with the exponential distribution with operating times and repair times.

- P, success probability, $\quad P = \frac{\mu}{\lambda + \mu}; \ P = 0.9922$ (21)

- Q, failure probability, $\quad Q = \frac{\lambda}{\lambda + \mu} \quad or \quad Q = 1 - P$ (22)

- $M(\alpha(t))$, Total average duration of operation of the OEL for a time interval (0,T)

$$M(\alpha(t)) = P \cdot T \tag{23}$$

- $M(\beta(t))$, Total average duration of failure of the OEL for a time interval (0,T)

$$M(\beta(t)) = Q \cdot T = (1 - P) \cdot T \tag{24}$$

-   $M(\gamma(t))$, the average probably number of interruptions in operation of the OEL for a time interval $(0,T)$

$$M(\gamma(t)) = \lambda \cdot P \cdot T = \mu \cdot Q \cdot T \tag{25}$$

3.7 The availability function

In the case of an exponential variation of *reliability function R(t)*, the following indices are specified, (Dub, 2008):

-   A(t), operating probability at a time *t*, called *availability*, which corresponds to the relation (1)

$$A(t) = \frac{\mu}{\lambda + \mu} + \frac{\lambda}{\lambda + \mu} \exp\left[-(\lambda + \mu)t\right] \tag{26}$$

-   $\bar{A}(t)$, probability to be in course of repairing at a time *t*, called *unavailability*

$$\bar{A}(t) = \frac{\lambda}{\lambda + \mu} \cdot \left[1 - \exp\left[-(\lambda + \mu) \cdot t\right]\right] \tag{27}$$

For the OEL in study, based on the known indices λ and μ and on relations (26) and (27), Figure 4 presents a variation of the availability/unavailability function of the OEL.

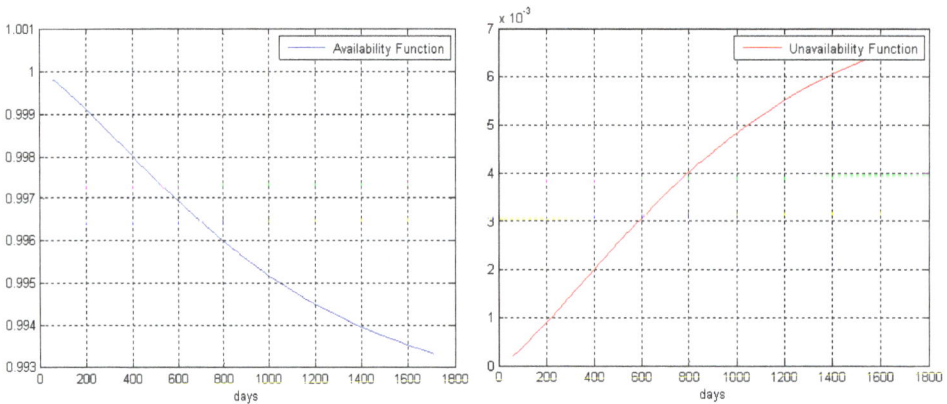

Fig. 4. Availability and unavailability functions for OEL 20 kV

The proposed algorithm allows to establishing the parameters and reliability indices for each component of OEL: pillar, conductor, insulator, assuming that we have a specific database for each component.

Considering these parameters and indices as the intrinsic dimensions of the OEL at a time, we can move on to preventive maintenance planning, namely the establishment of preventive maintenance strategies for a future period.

4. Modelling strategies by preventive maintenance with renewals

In accordance with the information presented in section 2 and section 3, it follows that the prophylactic strategy related to a system with RCM represents the combination of two

defining elements: *the type of wear* and intervention operations, *the renewal type and moment.*
In the cases in which at the time t = T^0 presented in Figure 1:

a.   The system is in operation, and has the probability function P_0 (T^0), known by the relationship (21) ;
b.   the law of variation of system reliability is know, being expressd by he relationship (15);
c.   The system and its components will follow the same law of variation of reliability including during the periods of preventive maintenance (T^+), presented in Figure 1; we can shape the preventive renewal process for the following situations:

4.1 Modeling on a time interval

In the conditions of preventive maintenance actions for a period of [0, ΔT], with a number of renewals r performed during that period, we can express the system reliability function, on the relationship given by (Catuneanu & Popentiu, 1998), (Georgescu et al., 2010). and the nonreliability function *F(t)*, through the relationship:

$$R(r, \Delta T) = \exp\left(-\alpha(r+1)^{(1-\beta)} \cdot \Delta T^{\beta}\right) \tag{28}$$

where: α ; β - Weibull parameters, the relationship (14);
r – the number of preventive renewals during the study ΔT.
Figure 5 illustrates the graph in the case of the OEL in study, with the parameters given in the relations (15) and (21) with he purpose of exemplifying the variation of the reliability function R(r,ΔT), under the influence of PMA, carried out by varying the number of renewals r for [0,10,50,100], assuming a reliability of the OEL, at the beginning of the study period by P_0 (t) = 0.9922.

$$R(r, \Delta T) = P_0(t) \cdot \exp\left(-\alpha(r+1)^{(1-\beta)} \cdot \Delta T^{\beta}\right) \tag{29}$$

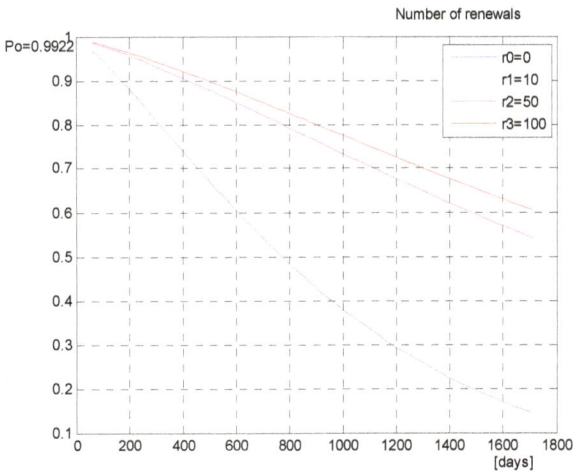

Fig. 5. The influence of the renewals r on the reliability of the OEL 20 kV

From the graphic of Figure 5, we note an increase reliability of the OEL at the end of the period ΔT, under the influence of renewals r0, r1, r2, r3.

4.2 Modeling over n time intervals

Let us study the variation probability for a system operating with PM, over a period of time T_n, composed of n time intervals ΔT_i. A number r_i of renewals is performed over each interval ΔT_i for i=1,n, while the reliability of system at the beginning of the interval ΔT_1 is known $Po \mid _{t=0}$.

$$R(T_n) = P_o(t) \prod_{i=1}^{n} R_i(r_i, \Delta T_i) \tag{30}$$

where:
$$R_i(r_i, \Delta T_i) = \exp\left[-\alpha(r_i + 1)^{(1-\beta)} \cdot \Delta T_i^{\beta}\right] \quad \mid i = 1, n \tag{31}$$

or:
$$R(T_n) = P_0(t) \prod_{i=1}^{n} \exp\left[-\alpha(r_i + 1)^{(1-\beta)} \cdot \Delta T_i^{\beta}\right] \tag{32}$$

In the case of OEL in study, the variation of reliability was studied for:

- $T = \sum \Delta T_i$, with equal periods of time of $\Delta T_i = 600$ days $\mid$ i=1,2,3;
- for each period i having a number $r_{ij} \mid j = 1, 2, 3$ renewals, or:
- $\Delta T_1 \mid r_1 = 0,5,20,50$; $\Delta T_2 \mid r_2 = 0,3,15,30$; $\Delta T_3 \mid r_3 = 0,1,10,20$;

With the parameters given in relations (15) and (21), we obtain the graph of variation presented in Figure 6:

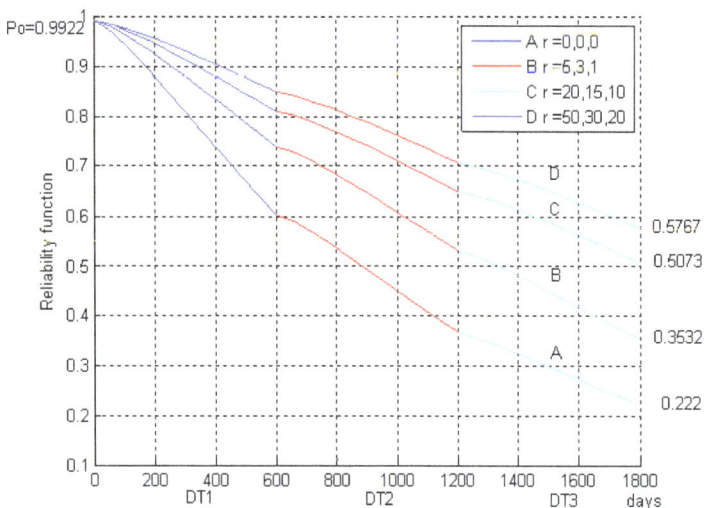

Fig. 6. Variation of reliability over time periods $\Delta T \mid i = 1, 2, 3$

From the graph of Figure 6, we note an increased reliability of OEL at the end of the three periods $\Delta T_i \mid i = 1, 2, 3$, under the influence of renewals.

4.3 Modeling on the components of the OEL

For the purpose of studying the PM of the system components, we consider the OEL with basic components: pillars, conductors, insulators, with the following assumptions:

- the exploitation data is known, due to corrective/preventive maintenance actions, for each component, for a certain period of operation;
- each component follows the same law of distribution reliability as the entire OEL, of form (6);
- using the model presented in section 3, it is possible to calculate the parameters and the reliability indicators for each component;

Through the OEL components, the structural model of the OEL reliability is the *series model*, such as the one presented in Figure 7. The reliability parameters and indicators of each component are known.

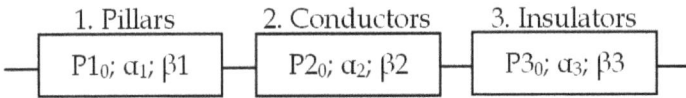

1. Pillars	2. Conductors	3. Insulators
$P1_0$; α_1; $\beta1$	$P2_0$; α_2; $\beta2$	$P3_0$; α_3; $\beta3$

Fig. 7. Equivalent scheme of reliability on components of the OEL

In the line of these assumptions we can model the probability of operating with renewals, on each component over an interval ΔT, knowing the operating probability for each component in the early of period Pi_0, under the form:

$$R_i(r, \Delta T) = Pi_0 \cdot \exp\left(-\alpha_i (r_i + 1)^{(1-\beta_i)} \cdot \Delta T^{\beta_i}\right) \quad \big| i = 1,2,3 \qquad (33)$$

where: r_i - the number of renewals on the component $i = 1, 2, 3$.
In this case, the model of reliability of the OEL with renewals on components will take the form:

$$R(r, \Delta T) = \prod_{i=1}^{3} R_i(\Delta T) = \prod_{i=1}^{3} Pi_0 \cdot \exp\left[-\alpha_i (r_i + 1)^{(1-\beta_i)} \cdot \Delta T^{\beta_i}\right] \qquad (34)$$

Let us study the variation of the OEL reliability parameters and of its components for a period of time $\Delta T = 350$ days, assuming that the reliability parameters of the components and the number of preventive maintenance actions on each component are known, according to Table 2.
If we have the number of renewals $r_i | i=1,2,3$, each in four variants, the results are presented in Table 3. and their variation is presented in Figure 8.

i	Element	R_{0i}	$\alpha_i * 10^{-4}$	β_i	r_i[buc]
1	Pillars	0.99888	2.66883	1.29	0/5/10/20
2	Conductor	0.99750	9.56379	1.21	0/10/20/25
3	Insulators	0.99610	15.45002	1.33	0/10/15/30

Table 2. Reliability parameters of OEL components
** The data are estimates, because there is no information on the component maintenance

	Ri_0	$r_i = 0/0/0$ $i=1,2,3$	$r_i = 5/10/10$ $i=1,2,3$	$r_i = 10/20/15$ $i=1,2,3$	$r_i = 20/25/30$ $i=1,2,3$
R1	0.9988	0.8762	0.9230	0.9344	0.9447
R2	0.9975	0.7300	0.8264	0.8465	0.8527
R3	0.9961	0.4055	0.6632	0.6954	0.7464
R_{OEL}	0.9922	0.2593	0.5058	0.5501	0.6012

Table 3. Renewal influence on OEL components reliability

From the study and comparison of the values presented in Table 3, we note an increase of reliability for each component and for the whole OEL, when the number of renewals is increased. Figure 8., graphically presents the influence of renewals on the components and on the OEL.

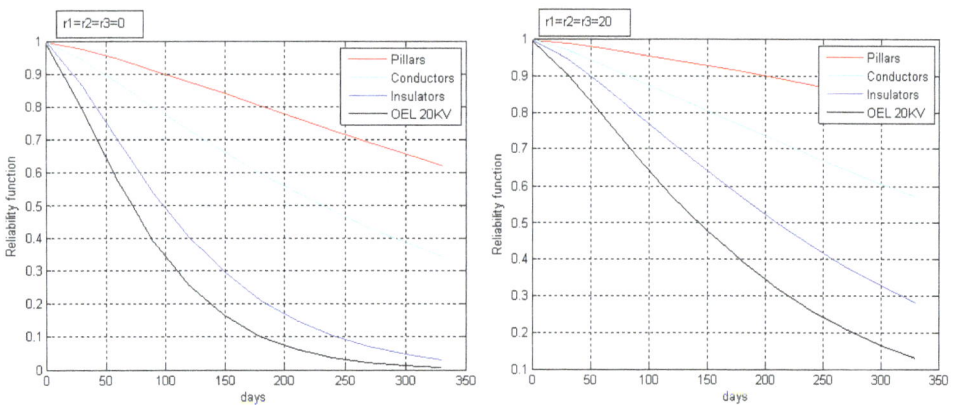

Fig. 8. The influence of renewals on the components and on the OEL

5. The economic modeling of preventive maintenance with renewals

The second component of the actions of the PM, after *strategy modeling*, is the *cost management* involved in the prophylactic strategy associated to the OEL systems with maintenance. In turn, the action of management and control of the maintenance plan of the electricity provider and in particular the implications of these actions on the financial relationship between the *electricity supplier* and *electricity consumers*, impose realizing an *economic model of PM*. Such a model should summarize all the costs and effects generated by the actions of the preventive and corrective maintenance, in all their aspects: planning, execution, management, etc., including the effect of inflation through the update method (Sarchiz et al., 2009), (IEC 60300-3-11, 2009).

To optimize the effect of RCM through the action and through the effect of the renewal processes for a period ΔT, belonging to the interval (T^+), presented in Figure 1, we propose an economic model, based on the state parameters of the OEL and the number of r renewals.

5.1 Total costs

The economic model of the PMA, from the perspective of the total cost (TC), has three basic components, (Anders et al., 2007):

- C_F - the costs due to corrective maintenance actions;
- C_{PM} - the costs due to preventive maintenance actions;
- C_{INT} - the costs due to unplaned intrerruptions.

Or:
$$TC(r) = C_F(r) + C_{PM}(r) + C_{INT}(r) \quad [\text{cost}/\text{year}] \tag{35}$$

We express these costs depending on the parameters of reliability and on the number of planned interventions r during the period ΔT of by PMA for an OEL belonging to the EDS, through the following components.

5.2 Costs with unplanned interuptions

The costs due to unplanned interruptions, due to system failure, can be expressed as follows:

$$C_F(r) = N_F \cdot c_F \tag{36}$$

where: N_F, average number of failures for the reference interval ΔT from the relation (25),

or:
$$N_F = R_S \cdot \lambda_S \cdot \Delta T \tag{37}$$

where:
$$R_S = P_0 \cdot \exp\left(-\alpha(r+1)^{(1-\beta)} \cdot \Delta T^\beta\right) \quad \text{from the relationship (28)} \tag{38}$$

a ; β - Weibull parameters, the relationship (14);
λ_S - the failure rate of a EDS, considered constant during the study ΔT, OEL being half way through its life cycle, the relationship (18);
c_F - the average cost to fix a failure.

or:
$$C_F(r) = P_0 \cdot \exp\left(-\alpha(r+1)^{(1-\beta)} \cdot \Delta T^\beta\right) \cdot \lambda_S \cdot \Delta T \cdot c_F \tag{39}$$

5.3 Renewal costs

The cost of preventive maintenance activities performed on a system or system element is determined on the basis of existing statistical information available at each EDS operating unit. This cost can be appreciated as a function directly proportional to the number of interventions, i.e. renewals r on the system or on the system element, of the form:

$$C_{PM}(r) = r \cdot c_{PM} \tag{40}$$

with: c_{PM} - average cost of a preventive renewal.

5.4 Penalty costs

The penalties suported by the *electricity supplier*, due unplanned interruptions, for duration ΔT, can be by two types:
- PENs - proportional with the time of interruption T_F, for undelivered electricity at the average power on OEL by P_{OEL}, and
- PEN_U - for unrealized production during interruption by the electricity consumer connected to the OEL.

$$C_{INT}(r) = PEN_S + PEN_U \tag{41}$$

with:
$$PEN_S = P_{OEL} \cdot c_w \cdot T_F \tag{42}$$

$$PEN_U = \sum_{k=1}^{K} P_k \cdot c_k \cdot T_F \tag{43}$$

where:
P_{OEL} - average power on OEL; P_k - average power at the k consumer; c_W - electricity cost; c_k - production cost unrealized for a kilowatt hour of electricity undelivered, at the k consumer; T_F - total average duration of OEL nonoperation during the reference period ΔT, relation (24); K - number of consumers connected to the OEL in the study.

with:
$$T_F = (1 - R_S) \cdot \Delta T \tag{44}$$

By replacing the relations established in (41), it follows:

$$C_{INT}(r) = \left(P_{OEL} \cdot c_w + \sum_{k=1,K} P_k \cdot c_k \right) \cdot \left(1 - P_0 \cdot \exp\left(-\alpha(r+1)^{(1-\beta)} \cdot \Delta T^\beta\right) \right) \cdot \Delta T \tag{45}$$

By replacing the relations (39), (40) and (45) in relation (35), we obtain the expression of the total costs over a period of time ΔT, depending on the safety operating parameters of OEL and depending on the renewal actions of the PM.

$$TC_{(r)} = P_0 \cdot \exp\left(-\alpha(r+1)^{(1-\beta)} \cdot \Delta T^\beta\right) \cdot \lambda_S \cdot \Delta T \cdot c_F + r \cdot c_{PM} +$$
$$+ \left(P_{OEL} \cdot c_w + \sum_{k=1,K} P_k \cdot c_k \right) \cdot \left(1 - P_0 \cdot \exp\left(-\alpha(r+1)^{(1-\beta)} \cdot \Delta T^\beta\right) \right) \cdot \Delta T \tag{46}$$

With the following specifications on the relashionsip (46)
1. The term (43) is included into the optimization calculations for situations where an OEL provides electricity for consumer, which can calculate the costs of production according to the electric energy supplied, which allows the calculation of the coefficient c_k;
2. To certain categories of electricity consumers, the higher production losses occur depending on the number of interruptions N_F during ΔT and less influence on time of interruption T_F;
3. In a system with multiple components, each having a specific number of renewals r_i, the total costs are the sum of costs on each system element.

6. The optimization of RCM strategies

The literature in this field approaches a wide range of classifications, according to different criteria and parameters, used in the design and optimization of RCM strategies (Hilbert, 2008), (Anders et al., 2007). Further on, we will give examples of RCM strategies for an OEL belonging to EDS, for the following cases and mathematical models.

6.1 Study assumptions

The strategies of optimizing PM of OEL, depending on the optimal number of preventive renewals r over a given period (0, T), can be approached from the perspective of the consequences it has on the relationship between electricity supplier and electricity consumer, based on two different criteria, which from the standpoint of the electricity supplier are (Sarchiz, 1993, 2005), (Georgescu, 2009):

a. *The economic criterion:* through the total cost involved in providing a safe supply of electricity to consumers. This optimization model is:

$$\min\{TC(T,r)\} \tag{47}$$

in presence of technical constraints imposed on safety criteria:

$$R_S(t,r) \geq R_S^{min} \quad \text{or} \quad N_F(t,r) \leq N_F^{max} \tag{48}$$

where: - R_S^{min}, the minimum reliability imposed on the study interval, and

- N_F^{max}, the maximum number of failures permitted on the study interval.

b. *The tehnical criterion,* aims to maximize safety in operation, respectively of system reliability on interval (0, T).

$$\max\{R_S(T,r)\} \quad \text{or} \quad \min\{1 - R_S(T,r)\} \tag{49}$$

in presence of economic constraints due to maintenance actions:

$$TC(T,r) \leq TC^{max} \tag{50}$$

where, TC^{max}, the maximum cost allocated to the exploitation of the OEL on the studied interval.

The study of the strategies used to optimize the RCM based on models (47) and (49), can be performed depending on the degree of safety imposed to ensure electricity and/or depending on the degree of assurance of financial resources during the PM during (0,T).

We will exemplify the application of the two models to the OEL 20KV in study, for the duration, T, a year, on these assumptions:

- we consider as action of renewal *r*, one or more specific PMA, or BRP type defined in section 2, for an OEL component and it can take one of the following forms (Mahdavi & Mahdavi 2009), (Teresa Lam & Yeh 1993):

Simple/minimal preventive maintenance and/or;

Maximal preventive replacement;

Sequential and/or continuous inspection strategies.

- we admit a periodic distribution of preventive renewals on the interval (0, T), with a constant time between two renewals for:

OEL $\Delta t = T / r$ or

OEL component $\Delta t_i = T / r_i \quad | i = 1, 2, 3$

-   the technical parameters of OEL are given in the relations: (15); (18); (21)
-   we admit an average power on OEL per year: P_{OEL} = 5.16 MW;
-   the values of the economic parameters are only calculation values given in monetary units [m.u.];

$$c_F = 1000 \text{ m.u.} ; \quad c_{PM} = 17500 \text{ m.u.}; \quad c_W = 530 \text{ m.u.}/\text{MWh.}$$

Note: The values of costs used in the program are not real, that is why the obtained results are only *demonstrative theoretical results*.

-   we ignore the PEN_U component from the relation (43), because we do not have data regarding the technical and economic parameters of the consumers connected to the electric line.

The solution of the mathematical optimization models (47) or (49) in relation with *r variable optimization*, impose the use of nonlinear optimization techniques in relation with *criterion functions* and the *restrictions of the models*. In order to solve the RCM strategy models presented below, we used the software package MATLab 7.0\Optimization Toobox\procedure *fmincon*.

The design of preventive strategies for renewal by type (47) or (49) can be done based on different criteria, which will be listed below within each RCM model optimization strategies.

6.2 The model: The minimum costs and imposed reliability

We will determine the optimum number of renewals, in the situation of minimum total costs, in such away that OEL relaibility does not drop below the imposed value R_S^{min} .

The *fmincon* procedure imposes the following structure of mathematical models to be optimized:

1. Vector of decision variable: $x \equiv r$ the number of renewals
2. Objective function: $\min\{TC(x)\}$

$$\text{where}: \quad TC(x) = P_0 \cdot \exp\left(-\alpha(x+1)^{(1-\beta)} T^\beta\right) \cdot \lambda_S \cdot T \cdot c_F + x \cdot c_{PM} + \\ + P_{OEL} \cdot c_w \cdot \left(1 - P_0 \cdot \exp\left(-\alpha \cdot (x+1)^{(1-\beta)} T^\beta\right)\right) \cdot T \qquad (51)$$

3. Constraints of the model:

$$R_S(x) \geq R_S^{min}; \quad 0 \leq x \leq x^{max} \qquad (52)$$

$$\text{or}: \quad P_0 \cdot \exp\left(-\alpha(x+1)^{(1-\beta)} T^\beta\right) \geq R_S^{min}; \quad -x \leq 0; \quad x \leq x^{max} \qquad (53)$$

with x^{max} – the maximum imposed number of renewals.

By running the application program for different values imposed to the minimum reliability R_S^{min} , we obtain the optimum number of renewals r^{opt}, for PM of OEL during a year, in conditions of the minimum total cost TC (T, r), presented in Figure 9, i.e. the graph $r^{opt} = f(R_S^{min})$. From the graph of variation we remark that in order to ensure reliability of 0.95, it is required to perform a number of 16 preventive renewals per year and for a reliability of 0.96, it is impose a number of 41 renewals. Also, we can extract the variation costs with reliability R_S^{min} imposed to the OEL or with the optimum number of renewals, i.e.: $TC = f(R_S^{min})$ or $TC = f(r^{opt})$.

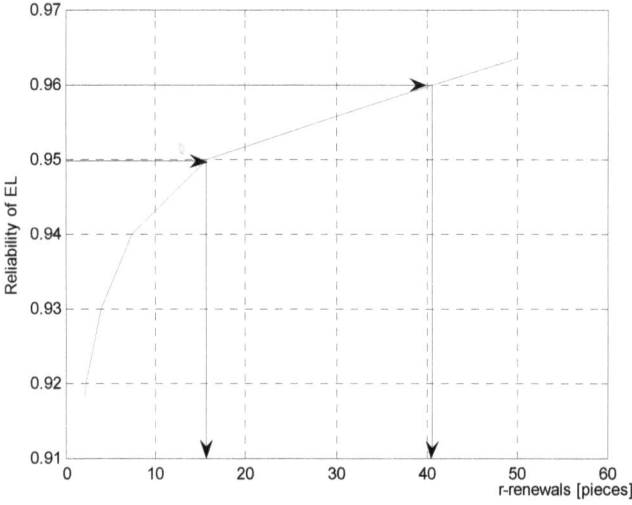

Fig. 9. Variation of the optimum number of renewals with reliability imposed to the OEL

6.3 The model: Minimum costs and number of interruptions imposed

We determine the optimum number of renewals with minimum total cost, in conditions in which the number of failures (unplanned interruptions) N_F does not exceed a maximum number imposed per year, N_F^{max}. In this hypothesis, the structure of the mathematical model to be optimized is:

1. Vector of decision variable: $x \equiv r$ the number of renewals
2. Objective function: $\min\{TC(x)\}$

$$where: \quad TC(x) = P_0 \cdot exp\left(-\alpha(x+1)^{(1-\beta)} T^\beta\right) \cdot \lambda_S \cdot T \cdot c_F + x \cdot c_{PM} +$$
$$+ P_{OEL} \cdot c_w \cdot \left(1 - P_0 \cdot exp\left(-\alpha(x+1)^{(1-\beta)} T^\beta\right)\right) \cdot T \tag{54}$$

3. Constraints of the model:

$$N_F(x) \le N_f^{max}; \quad 0 \le x \le x^{max} \tag{55}$$

or: $$P_0 \cdot exp\left(-\alpha(x+1)^{(1-\beta)} T^\beta\right) \cdot \lambda_S \cdot T \le N_F^{max}; \quad -x \le 0; \quad x \le x^{max} \tag{56}$$

with x^{max} – the maximum imposed number of renewals.

By running the application program, for different maximum values imposed to the numbers of unplanned interruptions on the OEL, we obtain the optimum number of renewals that are required to apply to the OEL during a year, in conditions of minimum total cost TC(T,r), presented in Figure 10, i.e. $r^{opt} = f(N_F)$.

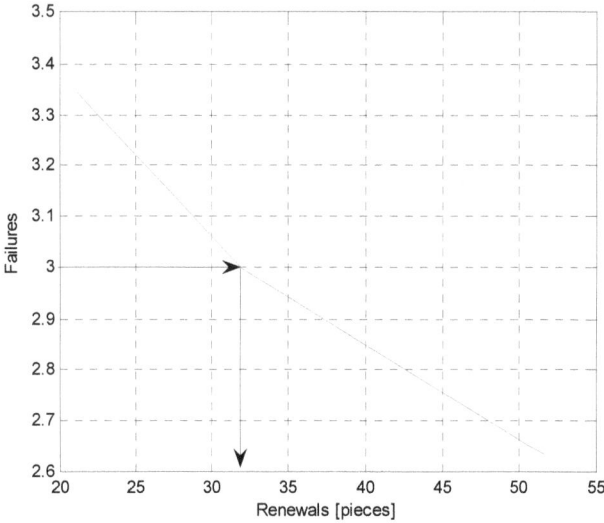

Fig. 10. Variation of the optimum number of renewals with the number of unplanned interruption

6.4 The model: Maximum safety and imposed costs

We determine the optimum number of renewals, to maximize the reliability of the OEL, in conditions of total costs TC(T,r) does not exceed a maximum value imposed TC^{max}.

In this hypothesis, the structure of mathematical model to be optimized is:

1. Vector of decision variable: $x \equiv r$ the number of renewals

2. Objective function: $\min\{F(x) = 1 - R_S(x)\}$

where:

$$F(x) = 1 - P_0 \cdot \exp\left(-\alpha(x+1)^{(1-\beta)} T^\beta\right) \tag{57}$$

3. Constraints of the model:

$$TC(x) \le TC^{max}; \quad 0 \le x \le x^{max} \tag{58}$$

or : $P_0 \cdot \exp\left(-\alpha(x+1)^{(1-\beta)} T^\beta\right) \cdot \lambda_s \cdot T \cdot c_F + x \cdot c_{PM} +$

$$+ P_{OEL} \cdot c_w \cdot \left(1 - P_0 \cdot \exp\left(-\alpha(x+1)^{(1-\beta)} T^\beta\right)\right) \cdot T \le TC^{max}; \quad -x \le 0; \ x \le x^{max}; \tag{59}$$

with, x^{max} – the maximum imposed number of renewals.

By running the application program, for different values imposed for TC^{max}, we obtain the maximum reliability of the OEL, presented in Figure 11 and the optimum number of renewals required to achieve that reliability, presented in Figure 12.

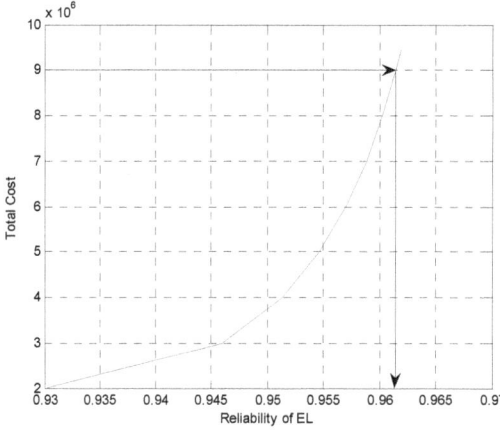

Fig. 11. Variation of the reliability of the OEL with TC^{max} imposed

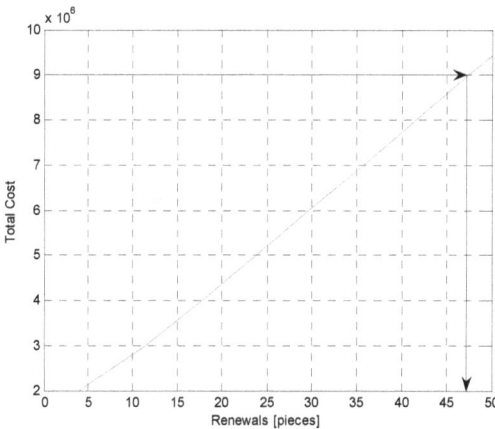

Fig. 12. Variation of the number of renewals with TC^{max} imposed

6.5 The model on components: The minimun costs and reliability imposed

We determine the optimum number of renewals on the three components of the OEL (pillars, conductors, and insulators), in the assumption of minimum total costs and on condition that the reliability of the OEL does not fall below an imposed value R_S^{min}.

In this case, the structure of the mathematical model to be optimized will include the total costs $TC(T, r_i)$ corresponding to those three components.

1. In this case, the vector of variables to be optimized will have three components, corresponding to the number of renewals on the three components of the electric line: 1-pillars; 2-conductors; 3-insulators.

$$X = \left[x_1 x_2 x_3\right]^T \quad \text{with} \quad x_i \equiv r_i \left| i = 1, 2, 3 \right.$$

2. Objective function: $\min\{TC(X)\}$
    where:

$$TC(X) = \sum_{i=1}^{3} TC_i(X) \quad \text{where:}$$

$$TC_i(X) = P_{0i} \exp\left(-\alpha_i(x_i+1)^{(1-\beta i)} T^{\beta i}\right) \cdot \lambda_i \cdot T \cdot c_{Fi} + x_i \cdot c_{PMi} + \qquad (60)$$

$$+ P_{OEL} \cdot c_w \cdot \left(1 - P_{0i} \exp\left(-\alpha_i(x_i+1)^{(1-\beta i)} T^{\beta i}\right)\right) \cdot T$$

3. Constraints of the model:

$$R_S(x) \geq R_S^{min}; \quad 0 \leq x_i \leq x_i^{max} \qquad (61)$$

$$\text{or:} \quad \prod_{i=1,2,3} P_{0i} \exp\left[-\alpha_i(x_i+1)^{(1-\beta i)} T^{\beta i}\right] \geq R_S^{min}; \quad -x_i \leq 0; \ x_i \leq x_i^{max}; \qquad (62)$$

with: x_i^{max} – the maximum imposed number of renewals on the i component for $|$ i=1, 2, 3;

$\alpha_i; \beta_i; \lambda_i$ - the reliability parametres of components;

$c_{Fi}; c_{PMi}$ - the cost with the corrective and preventive maintenance on components.
where:
The reliability parameters of the components have the estimated values presented in the Table 2.
The maintenance costs of the components can be assessed as a percentage of the costs related to the maintenance on the electric line, in the absence of a database with the costs on the components.
By running the application program, for different values imposed to the reliability of the OEL, we obtain the optimum number of renewals $r_i^{optim} |$ i=1,2,3; on each component of the OEL, to ensure the minimum reliability imposed R_S^{min}, in the conditions of the minimum total cost TC, presented in Figure 13.

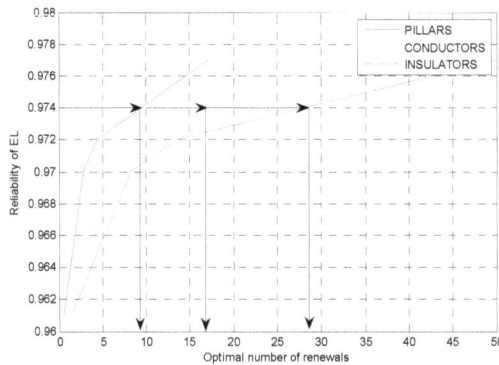

Fig. 13. Variation of the optimum number of renewals with the reliability imposed to the OEL

In conclusion, through the models of RCM optimization strategies that have been developed, we can obtain technical and economic information on the analysis, policy and planning maintenance actions for a period of time. This information pertains to:
- the optimum number of PMA on the components;
- the time interval between two actions;
- the optimum degree of safety in the electricity supply to consumers;
- the costs with preventive maintenance and/or corrective maintenance;
- the penalty costs due to improper maintenance.

7. RCM - maintenance management integrated software

Based on the algorithm for calculating and optimizing PM, presented in the previous sections, we propose an integrated model of software for the implementation and exploitation of the RCM. This model is specifically designed for OEL belonging to EDS.

To achieve the proposed information system we used the Matlab environment, because it offers the possibility to write and add programmes to the original files, allowing the development of the application characteristic to the EDS domain. The choice of Matlab was decisively influenced by the facilities offered by this environment in terms of achieving interactive user interfaces in the form of windows and menus, with the toolbox GUI (Graphical User Interface) and statistical processing with the Statistics Toolbox (Dulau et al., 2007), (Dulau et al., 2010).

For interactive control of various representations, we do the following steps:
- the application startup;
- choice and study of the OEL in operation according to the type of interruption;
- initialization and/or completing the database with the record of incidents;
- the reliability evaluation;
- the maintenance estimate by the preventive maintenance actions.

7.1 The application startup

The user interface that opens as a result of startup operations identifies two types of overhead electric lines, of 20 kV and 110 kV, and allows choosing the overhead electric line that will be examined by pressing the corresponding button, as in Figure 14 (Dulau et al., 2010).

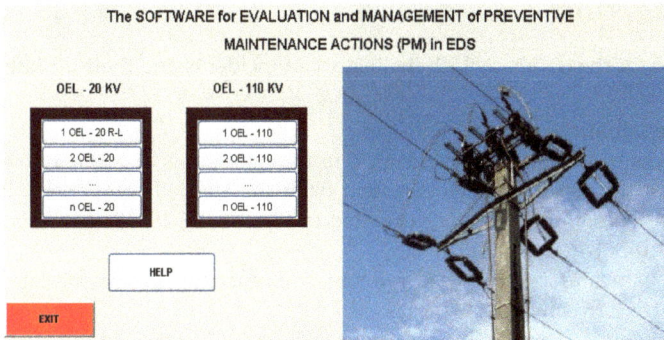

Fig. 14. User interface

7.2 Database of the events

Example: by selecting an OEL 20 KV, from the user interface, we obtain the technical and economical parameters of the OEL in a window. Further on, one is required to select the type of interruption to complete the database, corresponding to corrective maintenance or preventive maintenance, presented in the Figure 15.

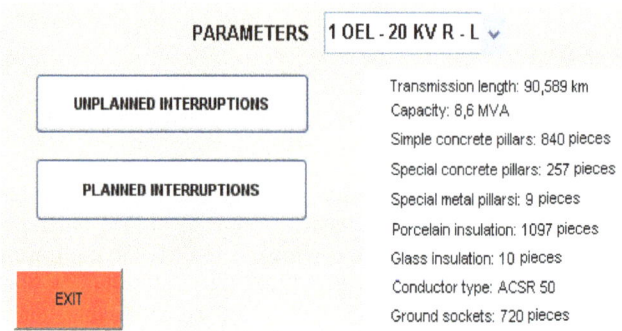

Fig. 15. User interface for line 1 OEL 20 kV

For each of the two possible directions, one must access the database for OEL and/or OEL component and fill in with technical and economic data corresponding to the type of interruption, presented in the Figure 16.

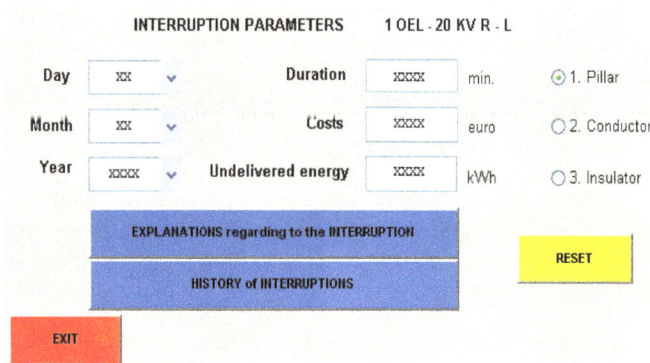

Fig. 16. Interface for the database with the history of incidents and their update

7.3 Reliability evaluation

The database thus created allows moving on to the operations of estimation and graphical representation of the reliability of the system or of the studied component, based on data from historical events until that date presented in Figures 17, 18, 19, 20 (Dulau et al., 2007), (Dulau et al., 2010).

The numerical values associated data are saved in specific files of Matlab environment, respectively matrix or vectors.

All vectors are initialized and any subsequent call will add new data to existing ones and will save the new content.

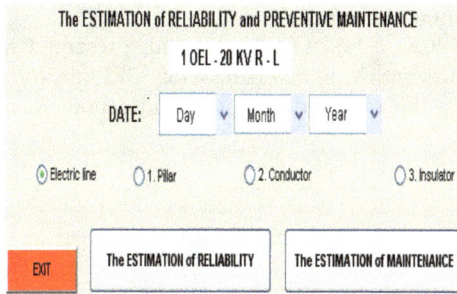

Fig. 17. RCM interface

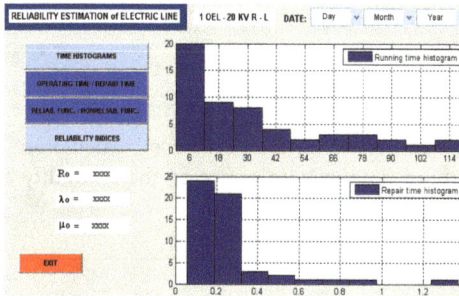

Fig. 18. Interface of the time histograms

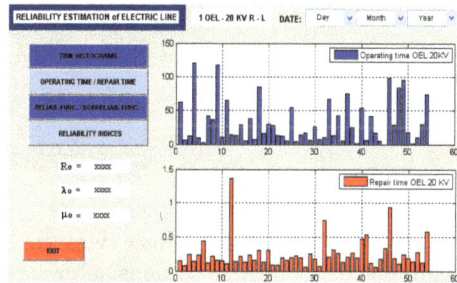

Fig. 19. Interface distribution of the operation times and repair times

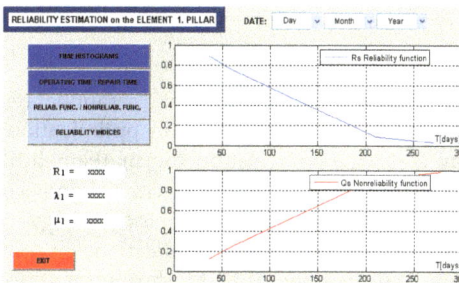

Fig. 20. Interface for the reliability and nonreliability functions

7.4 Preventive maintenance planning

The button MAINTENANCE ESTIMATION comand, presented in Figure 17, opens the interface for planning preventive maintenance for OEL or for its components, for a determined period and in the required technical and economic conditions, presented in Figure 21.

Fig. 21. Interface for planning PM on the OEL or its components

The influence of the number of renewals r [pieces], calculated on the OEL or on its components, allows a study of renewals influence on the reliability at the end of the period of study, presented in Figure 22.

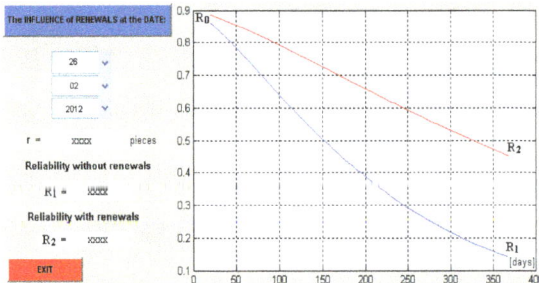

Fig. 22. Influence of the number of renewals

Do to its complexity, the developed program, which was conducted on the basis of theoretical considerations presented in previous sections, allows a much easier assessment of the reliability of the studied system and allows an optimization in planning PM, in different technical and/or economic situations.

8. Conclusions

The paper is the result of basic research in the field of operational research and maintenance management, with contributions and applications in optimization strategies of RCM for EDS. The contributions refer to the formulation the mathematical models of preventive maintenance strategies belonging to RCM, solving technical and economic objectives of the exploitation distribution systems and systems use electrical energy.

The optimal solutions to these models, with applications on a OEL 20KV as an EDS subsystem, allow a fundamental planning of RCM, through: setting the optimal number of future actions for the preventive maintenance PM, on overhead electric line or its components; the optimal interval between actions; the optimum degree of safety in electricity supply; the optimal management of financial resources for RCM.

The results are summarized by an integrated software for the maintenance management, to manage the database regarding the history of events, as well as the RCM design and analysis. We consider the models presented can be developed in the following research directions: failure rate λ was considered constant throughout the PMA, although in reality it changes value after every action; the time between two successive renewals was considered constant, although it may be placed in the model, as a new variable to be optimized; in evaluation of costs did not take into account the influence of inflation, which could influence the results; last but not least, the lack of real databases, on technical and maintenance events, and their costs in relation to the components of the OEL.

9. Acknowledgment

The excellent collaboration that my colleagues and I have with the publisher & editors was an experience and an honor for the professional manner in which they coordinated the accomplishment of this book.
Dorin Sarchiz
"Petru Maior" University of Targu Mures, Romania

10. References

Anders, G., Bertling, L., & Li, W. (2007). *Tutorial book on Asset Management – Maintenance and Replacement Strategies at the IEEE PES GM 2007*, KTH Electrical Engineering, Stockholm, Sweden, Available from
http://eeweb01.ee.kth.se/upload/publications/reports/2007/IRE-EE-ETK_2007_004.pdf

Baron, T., Isaia-Maniu, A., & Tővissi, L. (1988). *Quality and Reliability*, Vol. 1, Technical Publisher, Bucharest, Romania

Blaga, P. (2002). *Statistics with Matlab*. "Babes-Bolyai" University, Cluj University Press, ISBN 973-610-096-0, Cluj-Napoca, Romania

Catuneanu, V.M., & Popentiu, F. (1998). *Optimization of Systems Reliability*, Romanian Academy Publisher House, ISBN 973-27-0057-2, Bucharest, Romania

Catuneanu, V.M., & Mihalache, A. (1983). *Theoretical Fundamentals of Reliability*, Romanian Academy Publisher House, Bucharest, Romania

Dickey, J.M. (1991). The renewal function for an alternating renewal process, wich has a Weibull failure distribution and a constant repair time, In: *Reliability Engineering & System Safety*. Vol. 31, Issue 3, 1991, pp.321-343

Dub, V. (2008). *System Reliability*, Didactic and Pedagogic Publisher, ISBN 978-973-30-2371-5, Bucharest, Romania

Dulau, M., Dub, V., Sarchiz, D., & Georgescu, O. (2007). Informatic system for preventive maintenance actions in electric systems, In: *Proceedings of the 3rd International symposium on modeling, simulation and system's identification SIMSIS 13*, September 21-22, 2007, "Dunarea de Jos" University of Galati, ISBN 978-973-88413-0-7, pp.203-208, Galati University Press, ISSN 1843-5130

Dulau, M., Sarchiz, D., & Bucur, D. (2010). Expert system for Maintenance Actions in Transmission and Distribution Networks, In: *Proceedings of the 3rd International*

Conference on Power Systems MPS 2010., Acta Electrotehnica Journal, Academy of Techinal Sciences of Romania, Technical University of Cluj-Napoca, Vol. 51, No. 5, 2010, Mediamira Science Publisher, pp.134-137, ISSN 1841-3323

Georgescu, O. (2009). *Contributions to maintenance of electric power distribution.* PhD Thesis, "Transilvania" University, Brasov, Romania, Available from http://www.unitbv.ro/biblio

Georgescu, O., Sarchiz, D., & Bucur, D. (2010). Optimization of reliability centered maintenance (RCM) for power transmission and distribution networks, In: *CIRED Workshop – Lyon, 7-8 June 2010, Sustainable Distribution Asset Management & Financing.*, ISSN 2032-9628, Available from http://www.cired2010-workshop.org/pages/012/title/Home.en.php

Hilber, P. (2008). *Maintenance Optimization for Power Distribution Systems.* PhD Thesis, defended 18th of April 2008, KTH, Stockholm, Sweden, ISBN 978-91-628-7464-3

IEC 61649, International Standard, Edition 2.0, 2008-08, *Weibull analysis*, ISBN 2-8318-9954-0

IEC 60300-3-11, International Standard, Edition 2.0, 2009-06, *Application Guide-Reliability centred maintenance*, ISBN 2-8318-1045-3

Mahdavi, M., Mahdavi, M. (2009). Optimization of age replacement policy using reliability based heuristic model, In: *Journal of Scientific & Industrial Research*, Vol.68, August 2009, pp. 668-673

Sarchiz, D. (1993). *Optimization of the Electric Power Systems,* Multimedia System Publisher, ISBN 973-96197-9-7, Targu-Mures, Romania

Sarchiz, D. (2005). *Optimization of the Electric System Reliability,* Matrixrom Publisher, ISBN 973-685-990-8, Bucharest, Romania

Sarchiz, D., Bica, D., & Georgescu, O. (2009). Mathematical model of reliability centred maintenance (RCM)-Power transmission and distribution networks application, In: *2009 IEEE Bucharest Power Tech Conference.*, ISSN 978-1-4244-2235-7, Available from http://ieeexplore.org/search/freesearchresult.jsp?newsearch=true&queryText=sarchiz&x=39&y=4

Teresa Lam, C., Yeh, R.H. (1993). Comparison of Sequential and Continous Inspection Strategies for Deteriorating Systems, In: *Department of Industrial and Operations Engineering,* The University of Michigan. Technical Report No: 92-22, Available from http://www.jstor.org/pss/1427444

Part 2

Disturbances and Voltage Sag

6

Electrical Disturbances from High Speed Railway Environment to Existing Services

Juan de Dios Sanz-Bobi, Jorge Garzón-Núñez,
Roberto Loiero and Jesús Félez
Research Centre on Railway Technologies – CITEF,
Universidad Politécnica de Madrid
Spain

1. Introduction

New high speed lines are becoming more and more frequent. These new lines due to their power needs are constructed with a new electric system in order to be able to meet the power demands of the rolling stock and to lessen the electrical losses of the system. There are special cases like the Spanish one where new lines are built in parallel with the existing ones due to lack of space when entering populated zones. This line deployment is more usual in the access to the railway stations in big cities. In this case both lines interfere with each other for several kilometres. The Spanish case is especially interesting because both railway systems work together and there are also points of rolling stock transference between both lines due to the gauge changing facilities and train units that allow trains to operate in both systems.

The main cause of electrical disturbances for the already existing services is the electrification system of the new high speed lines. These lines are electrified with 1x25 kV or 2x25 kV 50 Hz systems. These systems, since they use AC, provoke electromagnetic disturbances in the nearby environment.

The most prone-to-be-affected systems among the existing lines are the track circuits that use 50 Hz for train detection and the wired signalling and telecommunications systems that run in parallel to the railway line.

Track circuits are the most used elements in order to detect train presence in a track section of the line. There are other systems like axle counters that are also used to detect train presence in a section. Most of the deployed track circuits use a 50Hz signal in order to detect the train. Modern track circuits use audiofrequency signals for this purpose and this kind of elements are used in the new built lines, but since is a very reliable safety item there's no need to replace them unless new disturbances can affect this safety related equipment.

The effects of these electrical disturbances can be also affect people. Induced voltages or currents can exceed the limits of the standards or laws. For example **EN 50122: Railway applications. Fixed installations. Protective provisions relating to electrical safety and earthing** defines the limits for AC and DC transients and continuous voltage in the railway environment.

The presence of these high speed lines have two main ways of electrical interfering with the so called 'conventional lines'. These interferences can be split into two main blocks:

- Disturbances in the existing lines due to 'direct contact' between both systems. This occurs when a train transits between both systems. This usually takes place in the gauge changing environment.
- Disturbances in the existing lines due to the presence nearby of a high speed line. This electrical disturbance is related to magnetic induction and electrostatic electric fields.

The Centro de Investigación en Tecnologías Ferroviarias, CITEF (Research Centre on Railway Technologies) that belongs to Universidad Politécnica de Madrid has a long dated expertise dealing with this disturbance matters that affect existing railway lines in the Spanish case. CITEF expertise is required in order to analyse and calculate the determined probable disturbed sections so preventive actions can be carried on well ahead before disturbance can affect those facilities. Once the new railway lines are totally functional and before open them to commercial services, CITEF engineer are required to perform electrical in-field measurements in the disturbance-susceptible sections in order to validate that the disturbance levels can't act against the safety of the railway line. There are singularly important measurement points that are those that confirm the border between the protected railway line zone and the unprotected against these electrical disturbances zone.

This chapter is structured following a usual approach. First of all there is a description of the disturbance problem, why it is important to detect and apply measures in order to mitigate or eliminate those effects. After that there is a section where there is a description of the physical principles that intervene originating the disturbances and those physical principles that can help to solve them. This section is the most academic part of the chapter with the use of formulations and equations. The next section gives a guideline referring what to do in order to lessen the effects of the disturbances.

The following section shows the methodology used in order to measure the effect of the disturbances in the 'conventional' facilities and equipments. Some results of these measurements are showed in the next section of the chapter, precisely called 'results'. To finalise the chapter there is one section to cope with the conclusions that summarises the chapter most important aspects and the final section that shows a list of references in this field.

2. Description of the problem

Disturbances in electric and electronic circuits are problems that appear while designing and implementing a system. Bearing in mind the presence of electrical disturbances (radiated or driven) systems are design in order to avoid or lessen the effects of these phenomena.

In the case of existing railway services there are not effective preventive measures since the electrical 50 Hz disturbance problem was not present while designing those equipments and facilities for the railway line. So preventive measures have to be developed and implemented long after the line is in service. The main measures that are carried out in order to lessen or avoid these problems are:

- Exchange the old equipment that can be affected by new equipment immune against disturbances.
- Provide ways of protecting the existing services changing only parts of it (usually wires) immunizing the system against disturbances.

There are different effects of the disturbances that affect different victims. Victim is the usual term while talking about electrical disturbances for the affected element. Disturbances can be grouped into two main groups:

- Radiated disturbances: these disturbances appear when there's a source of disturbance emits a electromagnetic wave that can be coupled by the victim with the air acting as means of transportation. This is usually called radiated noise in EMC terminology.
- Conducted disturbances: these disturbances occur when there's a direct electric contact between the source of disturbance and the victim. This is usually called conducted noise in EMC terminology.

The term noise is used when the effects of the disturbance can cause malfunction of the equipments. When the disturbance can affect human health or can provoke a fail against safety, the usual term is safety hazard. The victims of these disturbances are also different ones:

- Communications and signalling equipment: these are mainly affected by sections of parallelism and voltage inductions in the wires.
- Detection equipments: these equipments are basically track circuits and these can be affected both by radiated and conducted disturbances.
- Rail workers and passengers: people in general can be affected if induced or conducted voltages in the metallic elements that can be accessible are above the limits established by norms.

Usually radiated disturbances are related to quite high frequencies in usual EMC treatises but in this case and due to the high power consumption – up to 8MVA for each high speed train – and the antenna like structure of the railway line that make it possible to emit 50 Hz electromagnetic disturbances and to be affected by them in the case of the conventional lines fed in DC.

Crossings between high speed lines and conventional lines are another special disturbance point. The angle of the crossing have a special relevance since the small this angle is, the more effect will have in disturbing the already existing systems, while if the cross is right angled the effects of the disturbances will be almost negligible.

Since a way to lessen current return by earth in DC configurations is to isolate the rail from earth, high voltages can appear between the rail and earth. This problem can also appear if railway lines are fenced. If the fence is not properly earthed, high induced voltages can appear between fence and earth. In DC lines shielded cables are used, but they are often earthed in just one edge in order to avoid DC return through them finding a less impedance way back to the feeder station. This way of earthing wires is effective against electrostatic disturbances (capacitive coupling). Electrostatic field does not disturb the wires inside the shielding, but this way of shielding does not protect against induced electromagnetic fields (inductive coupling). This problem has to be carefully identified because a compromise between inductive coupling and DC current return by the shielding earthed in both extremes of the wire.

Another problem in order to determine the effects of the disturbances is to identify the currents that can flow through the high speed line overhead wire system. This is not an easy task since signalling and electrification branches in major railway infrastructure management companies or administrations are not related and obtaining information from the electrification branch from the signalling part can be difficult. So trying to obtain shortcut currents or fault duration is not an easy task for signalling engineers. Determine these currents will affect – as can be easily seen – to the determination of the boundaries between affected and not affected equipments, facilities or systems. This status quo is changing in the last years, at least in Spain, so current limits are defined taking into account electrification systems information.

Conducted disturbances have a major impact in two aspects: the rise of the voltage between the rail and earth and the disturbances that may cause to the 50 Hz track circuits that might be present in the line. The effects on the track circuit are highly affected by the impedance balance of the two rails. If the impedances of both rails are unbalanced a 50 Hz voltage will appear between the rails and depending on the relative phase between the 50Hz signal of the track circuit and the conducted one the operation of the line (if a false track circuit occupation occurs) or even a failure against safety if a shunted track circuit (occupied) due to the disturbance signal will give a false unoccupied information. CITEF, ADIF and the major Spanish 50Hz track circuits suppliers have carried out test to determine these effects and both of the above related cases can occur depending on the rail impedance unbalance and the relative phase of the signals. The location of feeder stations has also an impact in the distribution of currents along the rail and also affects the propagation of the effects.

These conducted disturbances might appear when a train unit is contacting both lines while crossing through a gauge change facility or a transition zone from AC to DC fed lines or electrified to non electrified lines. The train unit might be fed by AC through the pantograph and some of the returning current might, depending on the train configuration, end returning through the DC rails. In order to avoid this undesired current return isolation rail joints and impedances are used.

There are other signalling systems – like for example electronic block – that might be disturbed by harmonics originated by traction units. Electronic block depending on the used technology can be established between stations using a modulated signal of few kilohertz. Rolling stock modern composition for high speed lines do usually use asynchronous electric engines controlled with frequency shifters in order to enhance the performance and lessen the losses. Asynchronous engines are more efficient for the same weight than other electric engines and very easy to maintain so they are thoroughly used in train units using electronic controllers. These units can emit a series of harmonics that can interfere with such systems as electronic block.

3. A few physical notations

As mentioned above, one of the drawbacks of AC railway system is the electromagnetic interference to other surrounding electric circuits, particularly if parallel. These interferences can disturbs communication lines, as well as overhead conductors that run parallel.

An example of these phenomena is showed in Figure 1. As can be seen, between the AC railway system and the victim system exist coupled inductors and coupled capacitors that are the cause of the appearance of low frequency disturbances.

So, near an electrified railway AC, two electrical phenomena exist that can disrupt the lower current circuits (González Fernández & Fuentes Losa, 2010. Perticaroli, 2001):

- electrostatic induction, whose importance is due to the high value of the voltage and to the capacitances between the system, the earth and the disturbed line;
- Electromagnetic induction due to the nature of alternating current and to the coupling loop between the earth and the catenary, and the one represented by the induced line and the earth.

The resulting induced voltages in the victim system, in absence of protection, can cause hazards to personnel, material deterioration, problems in the proper functioning as unexpected operation of railway signaling installations.

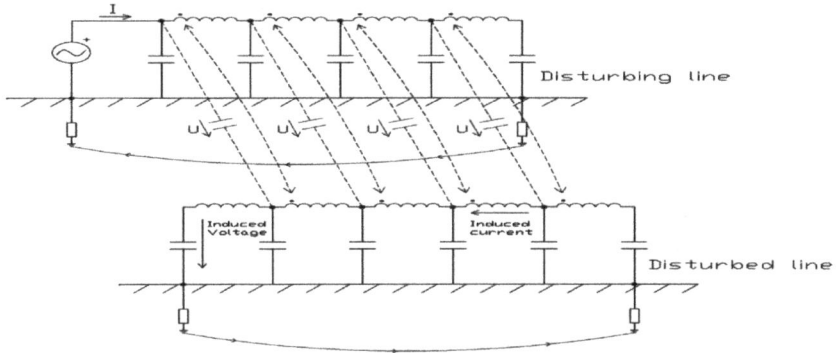

Fig. 1. Coupled inductors and coupled capacitors between the disturbing line and the disturbed line

To explain the phenomena the induced voltages in a victim wire will be studied. The development will be done considering the generic catenary pole of Figure 2, composed by:

- 2 catenary wires;
- 2 contact wires
- 2 negative feeders;
- 2 return current wires;
- 4 tracks (considered as active wires).

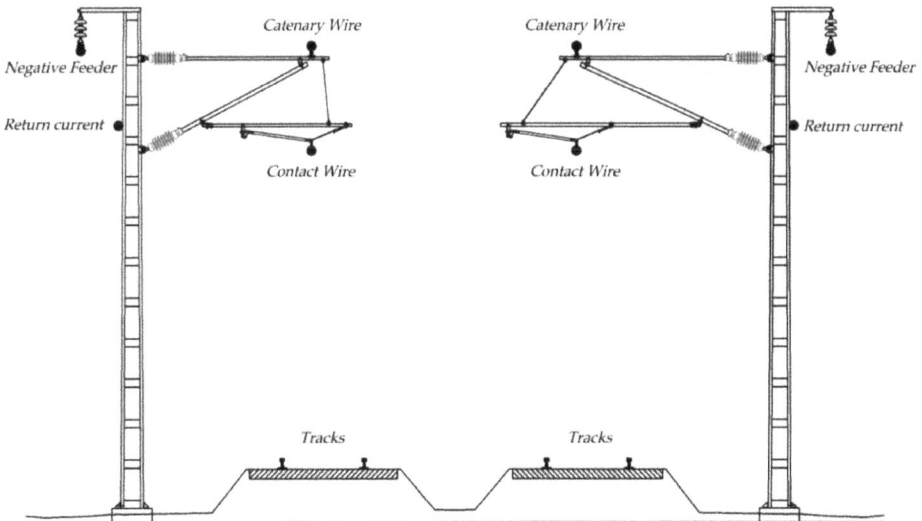

Fig. 2. Structure of a catenary pole

3.1 Electrostatic induction

The electric field consists of open field lines starting from the charge generating the field to other charges where field lines end. Considering a wire, the linear charge is calculated using the Gauss Law, stating that the electrical flux coming out of a closed surface is equal to the electrical charge contained in the surface (Arturi, 2008):

$$\psi = Q_{int} \tag{1}$$

Where the electrical flux in equal to the integral of the electrical flux density in a surface:

$$\Psi = \int_S d\psi = \oint D \cdot dS = D_\rho \int dS \tag{2}$$

Considering a cylindrical conductor of radius ρ and length l the electrical flux is:

$$\psi = D_\rho \cdot 2\pi\rho l \tag{3}$$

The electrical charge is the integral of the linear charge in a line:

$$Q_{int} = \int_l \lambda_l dl = \lambda_l \cdot l \tag{4}$$

Where λ_l is the linear charge density of the wire.
Comparing the equation (3) and (4) the electrical flux is:

$$D_\rho = \frac{Q}{2\pi\rho \cdot l} = \frac{\lambda_l \cdot l}{2\pi\rho \cdot l} = \frac{\lambda_l}{2\pi\rho} \tag{5}$$

And the electrical flux vector is:

$$\bar{D} = D_\rho \bar{a}_\rho = \frac{\lambda_l}{2\pi\rho} \bar{a}_\rho \tag{6}$$

The electrical field is proportional to the electrical flux vector through the permittivity of the material:

$$\bar{D} = \varepsilon_0 \bar{E} \tag{7}$$

Where $\varepsilon_0 = 8.85 \cdot 10^{-12} \frac{F}{m}$ is the permittivity of vacuum.
Consequently the electrical field is:

$$\bar{E} = \frac{\bar{D}}{\varepsilon_0} \tag{8}$$

$$\bar{E} = \frac{\bar{D}}{\varepsilon_0} = \frac{\lambda_l \cdot l}{2\pi\rho\varepsilon_0 \cdot l} \bar{a}_\rho = \frac{Q}{2\pi\rho\varepsilon_0 \cdot l} \bar{a}_\rho \tag{9}$$

Through the image theorem and considering the position of the catenary pole conductors, the electric field intensity inducted in a generic point (x,y) is due to the following relation expressing the influence of all the wires and its images to the point:

$$\bar{E}(x,y) = \frac{1}{2\pi\varepsilon_0 l} \sum_i \frac{Q}{\sqrt{(x-x_i)^2 + (y-y_i)^2}} \cdot \bar{u}_i \tag{10}$$

Where x_i and y_i are the coordinates of each wire.
Using the electrical field is possible to calculate the potential induced on a wire through the definition of electrical potential:

$$V = -\int_{r'}^{r} \bar{E} \cdot dl \tag{11}$$

Where r is the generic point where the potential has to be calculated and r' is the reference of the potential.
From (9) and (11) we obtain:

$$V = -\int_{r'}^{r} \frac{Q}{2\pi\varepsilon_0 l} \cdot \frac{1}{\rho} \cdot d\rho = -\frac{Q}{2\pi\varepsilon_0 l} \cdot \ln\frac{r}{r'} \tag{12}$$

To calculate the influence of all the wires of the catenary pole the image theorem must be applied. In Figure 3 are showed the principal conductor and the respective image conductors in a standard High Speed Line catenary Pole. Wires 1-14 are the main wires and wires 1'-14' are the images of the main wires. Wire 15 and 15' are the victim wire and its image. $a_{2,13}$ distance between wires 2-13. $b_{2,8}$ distance between wires 2 to image of wire 8. h_2 is the height of wire 2. The wire 15 is the hypothetical conductor exposed to the electrical field produced by all the catenary pole wires and by theirs images. The tracks are considered active wires.

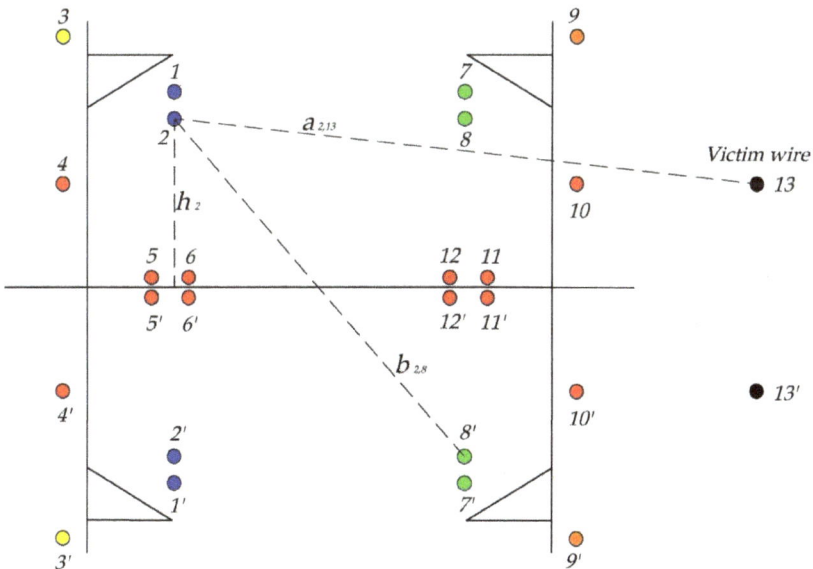

Fig. 3. Application of the image method to High Speed Lines

To apply the image method it is assumed that the earth is a perfect conductor. Consequently the earth should have an electrical charge to verify the following relation:

$$q_{earth} = -\sum_{i=1}^{n\,wires} q_i \tag{13}$$

Being the whole system neutral:

$$\sum_{i=1}^{n\,wires+earth} q_i = 0 \tag{14}$$

The following notation will be applied:
- a_{ij} : distance between main wires i and j;
- b_{ij} : distance between main wire i and the image of the main wire j;
- h_i : height of wire conductor i ;
- d_i : diameter of wire conductor i ;
- the main wires will have a charge of $+q_i$ and theirs images a charge of $-q_i$;
- the victim wire will have a charge equal to zero.

Applying the image theorem the following system is obtained:

$$V_i = \left(\frac{1}{2\pi\varepsilon_0 l} \ln \frac{2h_i}{d_i} \right) q_i + \sum_{\substack{j=1 \\ j\neq i}}^{n\,wires} \left(\frac{1}{2\pi\varepsilon_0 l} \ln \frac{b_{ij}}{a_{ij}} \right) q_j \tag{15}$$

Where $n\,wires$ includes also the victim conductor.
The system can be also rewritten in the following way:

$$V_i = s_{ii} \cdot q_i + \sum_{\substack{j=1 \\ j\neq i}}^{n\,wires} s_{ij} \cdot q_j \tag{16}$$

Where s_{ii} and s_{ij} are the Maxwell's potential coefficients.
From (16) it could be deduced the coupling capacitances of the systems:

$$[c] = [s]^{-1} \tag{17}$$

Consequently the system can be rewritten as a function of the capacitances:

$$Q_i = c_{ii} \cdot V_i + \sum_{\substack{j=1 \\ j\neq i}}^{n\,wires} c_{ij} \cdot V_j \tag{18}$$

Using the system (16) is easy to evaluate the inducted voltage in the victim conductor p, just applying the boundary condition that the victim wire has a null charge (floating conductor) and using the voltages of the catenary pole wires:

$$Q_p = c_{pp} \cdot V_p + \sum_{\substack{j=1 \\ j\neq i}}^{n\,wires} c_{ij} \cdot V_j = 0 \tag{19}$$

So the voltage induced in the victim wire can be calculated through the following relation:

$$V_p = -\frac{\left(\displaystyle\sum_{\substack{j=1 \\ j \neq i}}^{n\,wires} c_{ij} \cdot V_j \right)}{c_{pp}} \qquad (20)$$

In the equation (20) can be seen that the electrostatic induction relies only on the voltages of the wires and not on frequency power.

3.2 Electromagnetic induction

The magnetic field, unlike the electric field, consists of closed field lines that can pass through any material.

As for electrostatic induction, the various conductors which form the railway system have a magnetic coupling between them. For the development of the following method the Figure 2 will be used.

The study of inductive coupling refers to the characteristic impedance of a conductor with earth return and the mutual impedance between two insulated conductors that have both earth return. Instead of using the inductance value, the impedance value is used. The relation between impedance Z, self-inductance L_{ii} and mutual inductance M_{ij}, for alternating current of frequency f is given by:

- Self-impedance: $\qquad\qquad Z_{ii} = j\omega L_{ii}$ $\qquad\qquad$ (21)

- Mutual impedance: $\qquad\qquad Z_{ij} = j\omega M_{ij}$ $\qquad\qquad$ (22)

Expressions defining the impedance have real and imaginary components whose magnitudes vary greatly with the distance between the wires. Imaginary component of Z predominates near the inducing line. The electric field induced in this area along a parallel conductor and the current density at earth level has an offset of almost 90 ° respect the inducing current.

At great distance from the inducing line real component dominates Z. The electric field induced along a parallel conductor and the current flow at earth level, are almost in phase opposition with the inducing current.

The variance based on the phase difference between the magnitudes of the two currents is continuous and fairly uniform. For the model the following assumptions have been considered:

- inducing line is a straight horizontal conductor of infinite length in both directions;
- the mutual impedance is calculated between parallel lines;
- the earth is homogeneous, with finite resistivity and unitary relative magnetic permeability.

3.3 Calculation of self and mutual impedances

In Figure 4 is showed the circuit for the calculation of the self-impedance. Electrical parameters associated with self and mutual impedances of conductors with earth return are:

- Self-impedance of the wire.
    - I_i : Current in conductor i (inductor) with earth return.

- U_{0i} : Voltage in the outer surface of conductor i caused by the current i.
- U_{ei} : External voltage in the circuit formed by the conductor i and earth, caused by the current i.

The self-impedance can be defined as the sum of the following impedances:

$$Z_{sii} = Z_{0i} + Z_{ei} \tag{23}$$

Where the meaning of the terms is:

$$Z_{sii} = \frac{U_s}{I_i \cdot l} \tag{24}$$

$$Z_{0i} = \frac{U_{0i}}{I_i \cdot l} \tag{25}$$

$$Z_{ei} = \frac{U_{ei}}{I_i \cdot l} \tag{26}$$

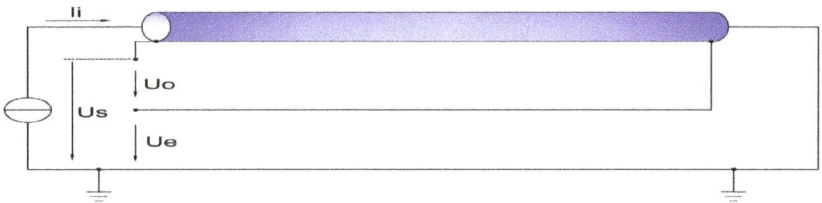

Fig. 4. Calculation of the self-impedance

- Mutual impedance between two wires.

In Figure 5 is showed the circuit for the calculation of the mutual impedance. I_i is the current for the conductor I and U_{ij} in the voltage in conductor j due to the current for the current I_i .

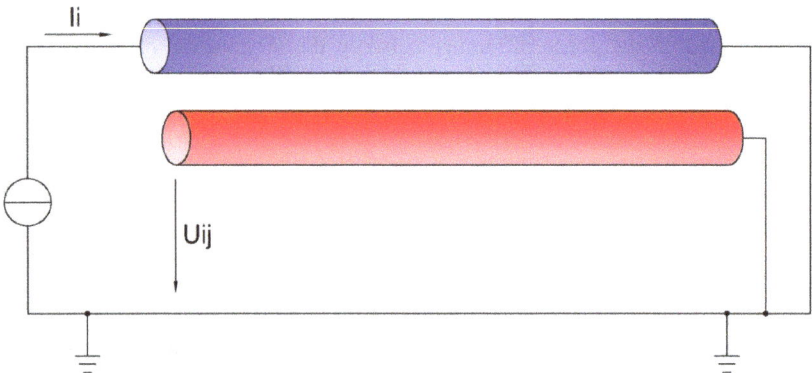

Fig. 5. Mutual impedance calculation

According to the law of reciprocity, the values of impedance for the wires i^{th} and j^{th} can be exchanged, i.e. $Z_{ij} = Z_{ji}$.

$$Z_{ij} = \frac{U_{ij}}{I_i \cdot l} \tag{27}$$

3.4 Calculation of the impedances of the basic Carson's formulas

The impedance of the conductors can the calculated easily using the expressions developed by Carson (Carson, 1926; Dommel et al., 1992). These expressions are based on series that depend on the position of these wires.

The starting point is the same image method used in 3.1 applied in Figure 3.

3.4.1 Internal impedance of the self-impedance

In a conductor flowing current flows is distributed over the entire surface of its cross section. Consequently magnetic fields exist inside and outside. To properly evaluate the internal field is necessary to know the geometric properties of the cross section of conductor and internal distribution of the current. For the study, the following hypotheses are assumed:

- current is distributed throughout the cross section of the conductor;
- the relative permeability of the material μ_r is constant.

The internal impedance of a conductor per unit length is defined by the formula:

$$Z_{oi} = \left(R_i + j \cdot 2 \cdot 10^{-4} \cdot \omega \cdot \frac{\mu_r}{4} \right) \frac{\Omega}{km} \tag{28}$$

Considering the following relations,

r Radius of the wire
r' GMR (geometric mean radius)

$$r_i' = r_i \cdot e^{-\mu/4} \tag{29}$$

relation (28) can be rewritten as:

$$Z_{oi} = \left(R_i + j \cdot 2 \cdot 10^{-4} \cdot \omega \cdot \ln\frac{r_i}{r_i'} \right) \frac{\Omega}{km} \tag{30}$$

3.4.2 External impedance of the self-impedance and mutual impedance

The external impedance of the self-impedance is due to the following relation:

$$Z_{ei} = \left(j \cdot \omega \cdot \frac{\mu_0}{2\pi} \cdot \ln\frac{2h_i}{r_i} + (\Delta R_{ii} + \Delta X_{ii}) \right) \frac{\Omega}{km} \tag{31}$$

The mutual impedance can be calculated using the equation:

$$Z_{ij} = \left(j \cdot \omega \cdot \frac{\mu_0}{2\pi} \cdot \ln\frac{D_{ij}}{d_{ij}} + 2(\Delta R_{ii} + \Delta X_{ii}) \right) \frac{\Omega}{km} \tag{32}$$

The correction terms ΔR and ΔX take into account the effect of earth return and they are a function of angle θ (between conductors) and the parameter p, which is:

- For self-impedance: $p = \alpha 2h$ (33)

- For mutual impedance: $p = \alpha b_{ij}$ (34)

Where $\alpha = 4\pi\sqrt{5} \cdot 10^{-4} \cdot \sqrt{\dfrac{f}{\rho}}$ (35)

with f : frequency of the current.

ρ : resistivity of the earth.

Expressions of ΔR and ΔX series given by Carson are:

$$\Delta R = \omega \cdot 2 \cdot 10^{-4} \left\{ \frac{\pi}{8} - b_1 p \cos\theta + b_2 p^2 \left[(c_2 - \ln p)\cos 2\theta + \theta \sin 2\theta \right] + \right.$$
$$\left. + b_3 p^3 \cos 3\theta - d_4 p^4 \cos 4\theta - b_5 p^5 \cos 5\theta + ... \right\}$$
(36)

$$\Delta X = \omega \cdot 2 \cdot 10^{-4} \left\{ \frac{1}{2}(0.6159315 - \ln p) + b_1 p \cos\theta - d_2 p^2 \cos 2\theta + b_3 p^3 \cos 3\theta - \right.$$
$$\left. - b_4 p^4 \left[(c_4 - \ln p)\cos 4\theta + \theta \sin 4\theta \right] - b_5 p^5 \cos 5\theta + ... \right\}$$
(37)

Where the coefficients a, b, c and d are:

$$b_1 = \frac{\sqrt{2}}{6}$$ (38)

$$b_i = b_{i-2}\frac{sign}{i(i+2)}$$ (39)

With the starting value:

$$b_2 = \frac{1}{16}$$ (40)

$$c_i = c_{i-2} + \frac{1}{i} + \frac{1}{(i+2)}$$ (41)

With the starting value:

$$c_2 = 1.3659315$$ (42)

$$d_i = \frac{\pi}{4}b_i$$ (43)

3.5 Voltage calculation

Once the matrix $[Z]$ containing the self-impedances and the mutual impedances has been calculated, the inducted voltage in the victim wire can be evaluated using the following relation:

$$[\Delta V] = [Z] \cdot [I] \tag{44}$$

$$\begin{bmatrix} \Delta V_1 \\ \vdots \\ \Delta V_{13} \end{bmatrix} = \begin{bmatrix} Z_{1,1} & \cdots & Z_{1,13} \\ \vdots & \ddots & \vdots \\ Z_{13,1} & \cdots & Z_{14,13} \end{bmatrix} \begin{bmatrix} I_1 \\ \vdots \\ I_{13} \end{bmatrix} \tag{45}$$

Voltage ΔV_{13} can be evaluated considering the border condition that $i_{13} = 0$:

$$\Delta V_{13} = \sum_{i=1}^{12} Z_{13,i} \cdot I_i \tag{46}$$

4. Field reduction

It is possible to prevent the problems caused by the electrostatic and electromagnetic coupling using some criteria that will be explained in next sections.

4.1 Electrostatic reduction

As seen in section 3.1, the electric field produced by the conductors of the railway system induces a voltage in an eventual wire or in any conductive part not directly connected to the earth. Furthermore, the possible presence of electric field, due to atmospheric electrical charge present in that region of space, must be also the cause of the appearance of induced voltage in the victim conductor.

For safety reasons any mass, in principle, is directly connected to the earth, to prevent a potential that can be dangerous for people, animals and things.

Generally to allow the proper operation of the signaling systems the wires connecting the different elements are wires having a protective shield. This shield must be connected to earth in order to fix the potential to earth. If the shield is earthed, voltage $U_{shield} = 0$. For electrostatic reduction is enough to connect just an extreme of the shield to earth. In this case of connection, to guarantee the condition $U_{shield} = 0$, the shield must be earthed in more points along the line, as close as possible. In this way in the extreme not connected to earth, the voltage to earth will be low and it can be assumed $U_{shield} \approx 0$. If the section is too long there will be fewer guarantees to maintain $U_{screen} = 0$. Considering equation (15), (16) and (17) it can be seen that capacitances depend on the length of the shielding: decreasing the length decreases the capacitance, and consequently the electrostatic effect is reduced.

The shield could also be earthed in both ends. In this case it is very important to earth the shield along the line often, to reduce the magnetic coupling. Closing the shield it forms a spire and consequently it can be victim of electromagnetic coupling.

4.2 Electromagnetic reduction

Protection against electromagnetic induction is hard to achieve. As explained in section 3.2, the magnetic coupling between the railway system and the victim conductor produces induced voltages depending on the position of the different conductors. In the case of high-speed lines, the inducing circuit is composed by the loop power substation-overhead wire-train-rails. The current that is circulating varies in time with a given frequency and produces a magnetic field of the same frequency. This magnetic field induces an electromotive force in the circuit composed by the victim conductor and the earth, and consequently an induced current in the victim circuit will be generated.

To protect the signaling system from voltages induced by electromagnetic coupling in general in railways are used cables with a factor reduction. Besides of the electrostatic shields, these kinds of cables have another shield against the electromagnetic effects. Using these kind of shields it is possible to reduce the magnetic field and consequently to reduce the induced voltages and currents.

The reduction factor or factor shielding as the mitigation of electromagnetic fields produced by a power line by the presence of shields or other metal objects:

$$K = \frac{V_1}{V_2} \tag{47}$$

Where V_1 is the voltage induced in presence of these shields and V_2 is the voltage that would be induced in their absence. Thus the reduction factor will always be less than the unit.

In general there are 2 families of magnetic screen:
- Magnetic screen with magnetic material;
- Magnetic screen with conductive material.

Magnetic screen with magnetic material are based on the Snell's Law. If a magnetic material with high permeability is put in the interface between two zones (Figure 6), the magnetic flux lines change direction and reduce the magnitude according to the magnetic permeability of the material. Reducing the magnetic field, the induced voltages will be lower.

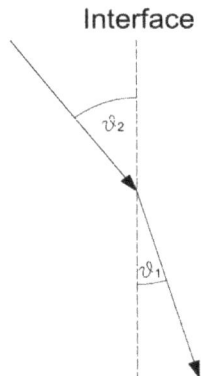

Fig. 6. Snell's Law

Magnetic screen with conductive material are the most used in railways. The principle is to create a spire in the victim conductor where inducing an electromotive force in order to create a magnetic field contrary to the one that is induced by the high speed line. In this way the two electromotive forces are summed up reducing the original effect. In order to give a more visual approach, Figure 7 shows how this kind of shielding works.

The two extremes of the shields must be earthed to create a closed loop where creating the opposite induced voltage. There are -in the case of connecting the shield to earth at both ends- the possibility of occurrence of earth loops in the event that land is not equipotential. This potential difference would generate a disturbing current on the shield. This effect can be reduced by a good earthing, improving the land and making the earthing as often as possible so that the potential difference between both ends of the cable is as close to zero as possible.

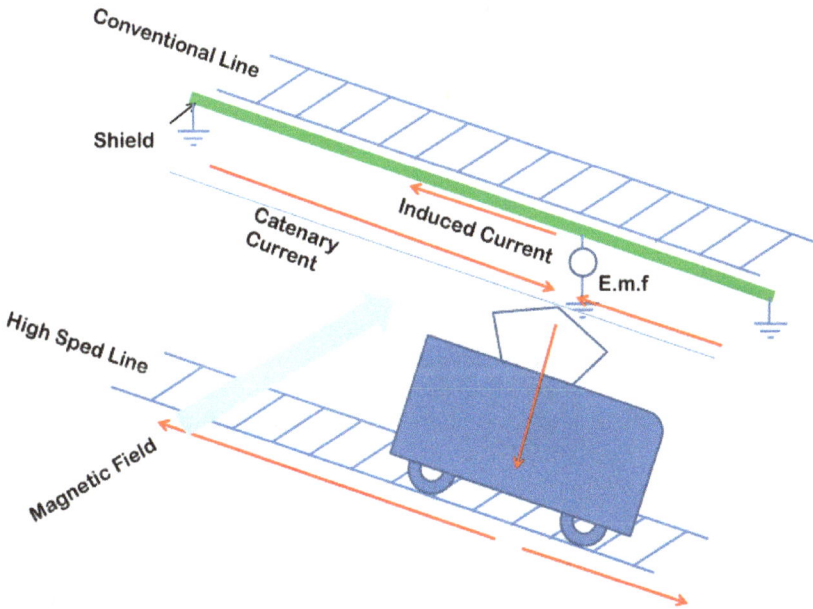

Fig. 7. Interaction between the magnetic field produced by the AC railway system and the conventional DC railway system in presence of magnetic shield

When both ends are connected to earth, the effect of stray currents or earth return currents must be considered. These currents depend on the quality of insulation of the ballast and earth conductivity. Presumably, earthing the shield at both ends of the return current will ease the conduction by the shield.

One way of addressing this problem is the connection of capacitors in an end of the screen. A capacitor acts like an open circuit for DC preventing its movement along the screen.

Moreover, the capacitor value can be calculated for a resonance frequency of 50Hz if the value of the inductance of the shield for a given field is known. The inductance of the shield is a value that depends of a nonlinear magnetic field as it is composed of a ferromagnetic material. This makes the calculation of this capacitor complex because it should be calculated for a magnetic field value undetermined. In addition, the inductance of the cable depends on the type of cable, the thickness of it, the reduction factor and the length of the span.

5. How to protect systems and equipment from electrical disturbations

Once that the problems have been described and the physical phenomena were formulated in the above sections this section will describe how electrical disturbances in already existing railway lines can be lessen or eliminated. In the next subsections a compendium of solutions is exposed.

5.1 Track circuits

Track circuits, as has been shown in this chapter, are safety elements that can be affected by disturbances. Since it has been demonstrated that this kind of devices can be altered by disturbances, the recommendation is to change them for audiofrequency track circuits in order to avoid malfunctioning. In order to determine which track circuits should be replaced simulations are performed in order to determine the induced voltage level between both rails. When this simulated voltage level is above a determined disturbance level, a decision to change them is carried out. The induced level of voltage is function of the current circulating through the overhead wire system in the high speed line, the distance between both lines, the angle between them, the distance between the rails (UIC gauge, Spanish gauge…), and the length of the track circuit section.

When direct contact disturbances might occur, usually nearby a gauge changing facility or a electrification changing section isolation rail joints might be used. In order to avoid that train units could shortcut both electrical zones two isolation joints can be placed with a separation bigger than the length of the largest train that will pass by that lines. When isolation joints cannot be deployed due to traction return problems, impedances can be used. In the Spanish example impedances are installed in order to offer a high impedance at 50 Hz and low impedance at zero frequency. This 'high impedance' circuit for 50 Hz and higher frequency is built as a low pass filter to offer maximum impedance at 50 Hz. Thus the 50 Hz current that would flow through the DC rails is dramatically reduced. According to the measures made in some gauge changing facilities the reduction of the flowing current through conventional rails can be reduced from 1/4 to 1/2 of the current that would be present if no impedance was present. These impedances are useless in transitions between electrified and non electrified lines since only isolation joints are needed to avoid the presence of 50 Hz current through the rails. The electrified rail cannot have isolation joints in order to allow the traction current to return to the feeder station. Since these impedances have a direct effect in the safety of the line they should be monitored in order to detect malfunctions. If they are not monitored, no assumption can be made regarding the reduction of the section of the line where 50 Hz track circuits should be replaced by audiofrequency ones. The way to calculate the boundary between affected and not affected zones is to determine the distance where there's a current through the rails equivalent to that of the normal operation of the track circuits. This distance can be as long as 18 km depending on the technology.

5.2 Signalling and communications wires

Wires are used in railway signalling in order to connect the interlocking with the field elements. These wires are spread in parallel along the track and some of them can have a length of several kilometres. The input signals to the interlocking are basically the state of the track circuits (occupied/unoccupied), the position of the points and the current state of the light signals, and the outputs are the orders to the points and the orders to the signals.

These orders are usually 50 Hz signals. They can be of a different range of voltages that can vary between 50 to 220 volts. So there is a chance that induced voltages can either turn off one aspect of the signal, which would only affect operation of the line, or can turn it on what can affect safety.

Communication wires are laid along the line from one station to their collaterals. Voice services are transmitted via these wires and also data. 50 Hz disturbances in these wires can affect communications, but are not likely to have an influence in safety issues.

Block wires connect collateral stations. This is a coded signal that is not likely to be affected by 50 Hz disturbances, but can be affected by harmonic disturbances caused by the rolling stock.

In order to protect these services from disturbances several actions can be carried out:

- Wires can be substituted by fiber optic that is immune to electrical disturbances. This is a good solution, especially for communications and services between stations.
- Wires can be shielded. This shield has to be earthed on both ends in order to provide effective shielding against both inductive coupling and capacitive coupling. This way of shielding is very effective against disturbances, but gives a good return for DC circulating through earth back to the feeder station.
- Wires can be shielded with both ends earthed, but in one side earthing is made through a capacitor (Koopal & Evertz, 2008). Thus no DC would return using the shield. This is an expensive way because each capacitor has to be calculated depending on the length of wire and the resistivity of the earth in order to obtain a minimum impedance at 50 Hz frequency so shielding against 50 Hz electrical disturbances is effective.

5.3 Level crossings

Railway level crossings are particular signalling systems that can be or not be connected to other signalling systems such as interlocking. These systems are in charge of protecting cars and citizens from railways. Level crossings electronic systems can be triggered by the interlocking or by track pedal that gives a train announcement to the system. These facilities have also wires that can have a length of more than two kilometres.

The solutions to immunize these systems from disturbances are the same one that those stated above for signalling and communication wires.

5.4 Accessible metallic elements

According to EN 50122-1 we can define the following terms:

- Rail potential (EN 50122-1): The voltage occurring under operating conditions when the running rails are utilised for carrying the traction return current or under fault conditions between running rails and earth.
- Accessible voltage (EN 50122-1):That part of the rail potential under operating conditions which can be bridged by persons, the conductive path being conventionally from hand to both feet through the body or from hand to hand.
- Touch voltage (EN 50122-1): Voltage under fault conditions between parts when touched simultaneously.

As was described above, usually DC rails used in traction lines are isolated from earth in order to prevent stray currents. This measure can raise rail potential. This can happen even without tractor trains in the line if there's a parallel high speed line nearby. Passenger might want to cross from one platform to other crossing the rails even if it's forbidden. This is more common in small stations with low traffic density.

There are other elements that have similar problems like the wire fences used to prevent intrusions in high speed lines. Also metal structures in big stations like metallic shelters that are not adequately earthed can have accessible voltages that have to be measured in order to check that there are not over the limits established by the standards.

In order to keep the accessible voltage or the touch voltage below the established limits, and as is reflected in the standard EN 50122-1, voltage limiting devices (VLD) can be connected between earth and rail, fence or structure. These devices keep both parts isolated from each other until the voltage between them exceeds a certain defined value, then it connects both parts so there is an electrical circulation between both systems and after that the VDL opens the circuit and checks the voltage between both parts.

Another solution that cannot be used for rails is to have a good earthing circuit. Metallic shelters or metallic fences should be adequately connected to earth in order to prevent accidents or electrocutions.

6. Measurements methodology

CITEF has performed several measurements campaigns in order to determine whether the systems and facilities used in conventional lines described above are well protected against disturbances or not. A measurement methodology was used and will be presented here. This methodology has been enhanced with the experience acquired and the analysis of the results. Measurements were carried out in the following elements:

- 50 Hz track circuits placed in the boundary section between 50 Hz track circuits unmodified and audiofrequency track circuits.
- Communication wires between stations. They are measured at each end of the electrical pairs or quads and loaded with the characteristic impedance of the line.
- Signal aspects. In this position, a lit aspect and a switched off aspect are measured in order to determine whether or not are affected by disturbances.
- Touch voltage in the metallic elements like fences, signal posts, point engines, lampposts, etc. surrounding the measurement position and of the rails according to EN 50122-1 Annex E: Measurement methods for touch voltage.

In order to obtain good accurate results a pole is disposed in a determined position of the high speed line to shortcut the electrical overhead wire system and the rail. Calculations are made in order to determine that point considering the position of the feeder stations, the impedance of the rail and the impedance of the overhead wires. 1x25 kV scheme is adopted even if electric trackside facilities are 2x25 kV. This is made so because this electrification scheme is more disturbing and return impedance is higher. This is done this way in order to obtain a steady state current as close as possible as 1200 A. This current is determined by the estimated amount of energy needed when two double train compositions accelerate at full throttle simultaneously in the same electrical section. This is an ADIF (Spanish Railway Infrastructure Manager) requirement. Sometimes because of the topology of the line, the situation of the feeder stations and the electric network, that current level cannot be obtained. In that case, lower currents are considered and measured.

In order to determine the position of the short-circuit pole, calculations are performed using a CITEF self developed application that implements these electrical calculations, but this calculations can be estimated easily knowing the impedance of the elements (rail and overhead contact line), the voltage level in the feeder station and the position of the feeder stations in the line.

The use of this shortcut steady state current is of great value in order to assure the repetitiveness of the measures. A real traction unit won't allow a steady state current to be maintained in order to measure the effects in the determined elements.

To measure touch voltage the **EN 50122-1 Annex E methodology** is used. The methodology described in the standard can be summarized in the following aspects:

- The voltage shall be measured with a voltmeter that has an internal resistance of 1 kΩ.
- Each measuring electrode, in order to simulate one feet, shall have a total area of 400 cm² and shall be pressed on the earth with a minimum force of 500 N. Alternatively, a sensor driven 20 cm into the earth may be used instead of the measuring electrode.
- To measure the touch voltages/accessible voltage in concrete surfaces or dry up soils, a wet cloth or a water film shall be placed between the foot electrodes and the earth. The foot electrodes shall be positioned at a distance not less than 1 meter from the accessible conductive part.
- A measuring electrode, usually a tip electrode, shall be used in order to simulate a hand. In this case paint coatings (but not insulations) shall be pierced.
- One clamp of the voltmeter shall be connected to the so called hand electrode and the other clamp shall be connected to the so called foot electrode.
- It is enough to carry out such measurements by random checks of an installation.

In Figure 8 it is shown the connection scheme to measure the communication wires.

In order to perform the measures a quad is used. This quad is loaded in the determined ends with the characteristic impedance of the wire that is usually a value between 500 to 600 Ω depending on the wire. These measures are carried out with the shielding of both ended connected and with one of the ends unconnected in order to determine the disturbance induced when the shielding in not properly connected.

These measures are usually done by two measurement teams each one of them placed at each one of the ends usually in a station communication facility. In order to comply with the measurement procedure a continuity of the shielding from one end to the other of the wire needs to be achieved. This was not always possible because DC stray currents could be a more important problem than preventing AC disturbances. In these cases, only prevention against electrostatic field was achieved since only one end of the shield was correctly earthed while the other end was unconnected.

The differentiation between electromagnetic shield and electrostatic shield needs to be clarified. In order to prevent electromagnetic disturbances both shields must be connected between them and connected to earth according to the standard **ITU K14 Provision of a metallic screen in plastic-sheathed cables.** The electrostatic shield is just a copper conductor. It is enough to earth just one end of the shield to prevent electrostatic disturbances.

When measures were carried out in a 50 Hz track circuit the measured values were:

- The track voltage in the receiver.
- The transformed track voltage (if transformers were used) in the relay.
- The state of the track circuit (occupied/unoccupied)

These measures should be carried out shunting the emitter and without shunting it in order to determine the disturbance in both situations. As said before, one of them –occupancy of an unoccupied track circuit- has operational implications and the other one –liberation of an occupied track circuit- has safety implications.

End A   **End B**

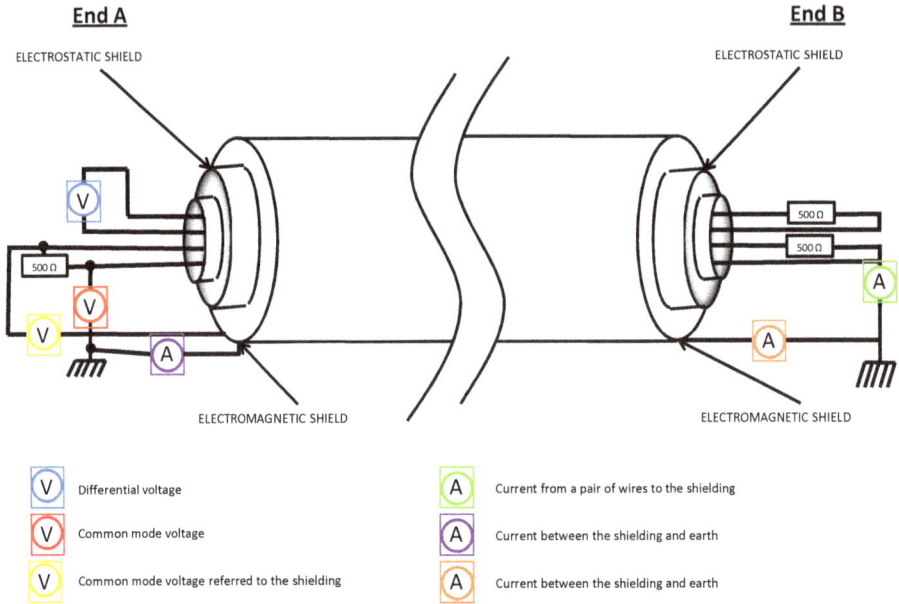

Ⓥ Differential voltage	Ⓐ Current from a pair of wires to the shielding
Ⓥ Common mode voltage	Ⓐ Current between the shielding and earth
Ⓥ Common mode voltage referred to the shielding	Ⓐ Current between the shielding and earth

Fig. 8. Measurement scheme for communication wires used in the measurement campaigns by CITEF

7. Results

In this section, some results from the measurement campaigns will be showed to enhance the understanding of the disturbances. These figures are extracted from real measurements. In order to publish them, location information is not included as well as any other information regarding the date or line where they were acquired.

Figure 9 shows the registered current in the feeder station during one measurement interval. The peak current is 1689 A that means an rms value of 1194 Arms. This is quite close of the above referred value of 1200 Arms.

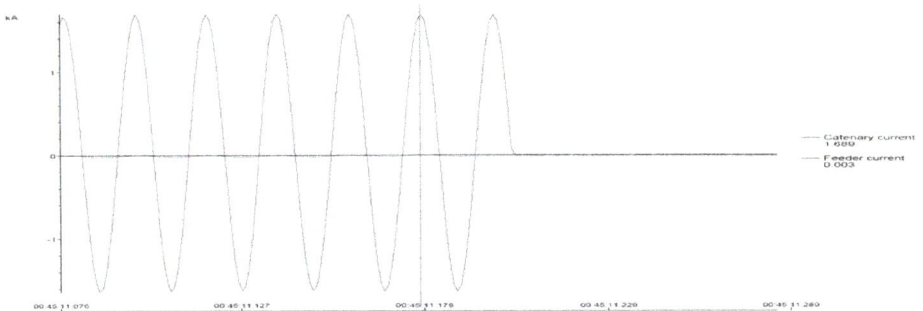

Fig. 9. Shortcut current as recorded in feeder station facility

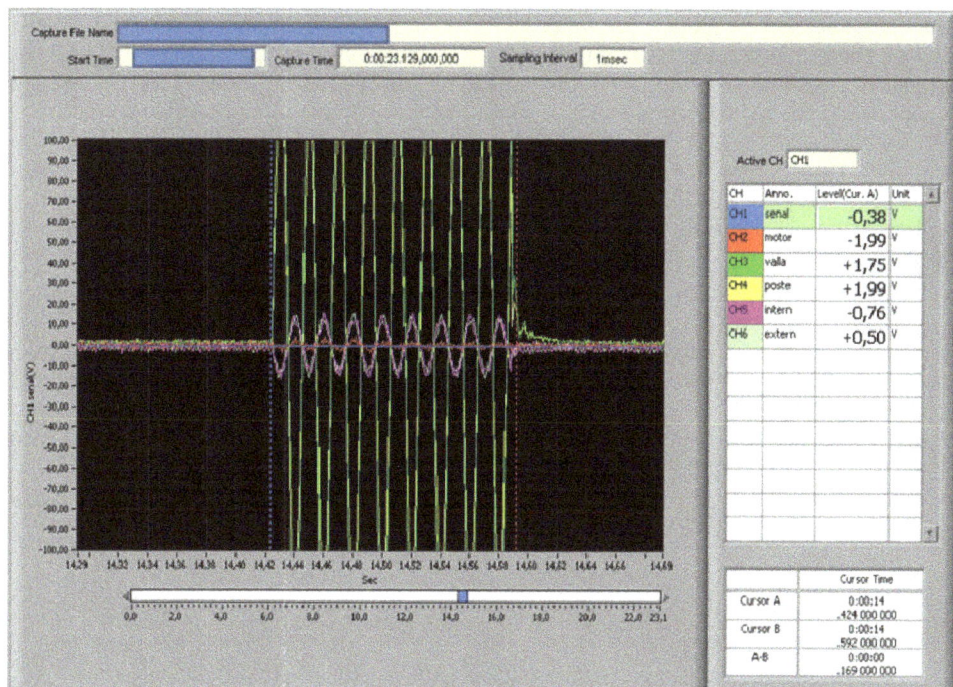

Fig. 10. Graphical representations of the measurements around one signal

For Figure 9 current values, Figure 10 offers the registered values for a series of measurement items nearby a signal the measured items were:

- Signal (blue)
- Point engine (red)
- Fence (green)
- Pole (yellow)
- Farther rail (purple)
- Nearer rail (pale green)

And the measured rms values were:

- Signal: negligible (0V)
- Point engine: 1,1 Vrms
- Fence: 114,9 Vrms
- Pole: 85,4 Vrms
- Farther rail: 46,4 Vrms
- Nearer rail: 46,6 Vrms

Some of the items measured had a 10/1 reduction probe. In this case both rails have this relation. This example shows in a very visual way that the effects described in the above sections do occur. These items were measured according to the EN 50122-1 described methodology and some of the obtained values overpass the limits described in the standard. This example shows in a very clear way that the effect does exist and that measures have to be adopted in order to prevent their effects. That figure shows the effect in elements that have to be protected against direct contact.

8. Conclusion

Electrical disturbances are close related to power quality, and service quality, and even safety, can be seriously compromised when two linear systems with different electrical feeder schemes are deployed one next to other and have even direct electrical contact between them. Electrical systems in DC and AC railway systems are designed separately, especially considering earthing, and this leads to new problems that weren't faced when old DC lines were design, deployed and operated.

Disturbances are a common issue that an engineer has to face while designing a new facility or system. Disturbances can also appear when the system is already in operation. These systems need to be protected from the effects of the disturbances in the most effective ways. These ways are usually the replacement of equipments (or at least part of them) and the shielding against those electrical disturbances.

In the above sections it has been explained the effects that disturbances from high speed lines can cause on conventional DC lines. A measurement methodology has been proposed. This methodology was used in on site measuring campaigns and their results have been used as a validation of the protecting works carried out before the high speed line comes into operation. The determination of the boundaries between what needs to be changed and what does not need to be changed is a complex problem that must be solved in a cost effective way, but without reducing safety issues.

The different ways of disturbance have been presented and explained so a global vision of the problem can be easily achieved.

9. References

Arturi, C. M. (2008). Campi elettrici, magnetici e di conduzione – introduzione ai metodi computazionali, Maggioli Editore, ISBN: 978-88387-4248-4. Milano, Italy.

Carson, J. R. (1926). Wave propagation in overhead wires with earth return, *Bell systems technical journal*, Vol. 5, No. 4, (Oct. 1926), pp 539 – 554.

Dommel, H. W. et al (1992). Electro – Magnetic Transient Program (EMTP). Theory Book.

González Fernández, F. J. & Fuentes Losa. (2010) (2nd edition). J. Ingeniería Ferroviaria (2nd edition), UNED Editorial, ISBN: 978-84-362-6074-8. Madrid, Spain.

Koopal, R. & Evertz, E. P. (2008). 50 Hz Track Circuits Parallel to a 25 kV 50 Hz Railway Line, *proceedings of World Congress on Railway Research*, ISBN, Seoul, May 2008.

Perticaroli, F. (2001). Sistemi elettrici per i transporti. Trazione Elettrica (2nd edition), Casa Editrice Ambrosiana, ISBN: 88-408-1035-8. Milano, Italy.

7

Power Quality and Voltage Sag Indices in Electrical Power Systems

Alexis Polycarpou
Frederick University
Cyprus

1. Introduction

In modern electrical power systems, electricity is produced at generating stations, transmitted through a high voltage network, and finally distributed to consumers. Due to the rapid increase in power demand, electric power systems have developed extensively during the 20th century, resulting in today's power industry probably being the largest and most complex industry in the world. Electricity is one of the key elements of any economy, industrialized society or country. A modern power system should provide reliable and uninterrupted services to its customers at a rated voltage and frequency within constrained variation limits. If the supply quality suffers a reduction and is outside those constrained limits, sensitive equipment might trip, and any motors connected on the system might stall.

The electrical system should not only be able to provide cheap, safe and secure energy to the consumer, but also to compensate for the continually changing load demand. During that process the quality of power could be distorted by faults on the system, or by the switching of heavy loads within the customers facilities. In the early days of power systems, distortion did not impose severe problems towards end-users or utilities. Engineers first raised the issue in the late 1980s when they discovered that the majority of total equipment interruptions were due to power quality disturbances. Highly interconnected transmission and distribution lines have highlighted the previously small issues in power quality due to the wide propagation of power quality disturbances in the system. The reliability of power systems has improved due to the growth of interconnections between utilities.

In the modern industrial world, many electronic and electrical control devices are part of automated processes in order to increase energy efficiency and productivity. However, these control devices are characterized by extreme sensitivity in power quality variations, which has led to growing concern over the quality of the power supplied to the customer.

According to the IEEE defined standard (IEEE Std. 1100, 1999), power quality is "The concept of powering and grounding electronic equipment in a manner suitable to the operation of that equipment and compatible with the premise wiring system and other connected equipment". Some authors use the term 'voltage quality' and others use 'quality of supply' to refer to the same issue of power quality. Others use the term 'clean power' to refer to an intolerable disturbance free supply. Power quality is defined and documented in established standards as reliability, steady state voltage controls and harmonics. Voltage sag is defined as a short reduction in voltage magnitude for a duration of time, and is considered to be the most common power quality issue. The economic impact of power quality on a utility is of great importance. Living in a world where making money is the major objective of electricity companies, quality is often overlooked. Thus a need for specific, easy to assess and well-defined performance criteria for use worldwide encouraged the definition and establishment of Power Quality indices.

In the first section of this chapter Power quality disturbances, such as Voltage sag, Interruption, transient overvoltage, swell, Harmonic issues and voltage imbalance, are Introduced. Statistical voltage sag Indices used for characterization and assessment of power quality are then presented. These indices classify within the following categories: Single event, Site indices and System indices. Furthermore the development and verification of Mathematical voltage sag indices is presented, applicable for power quality improvement through optimization techniques. The impact on the voltage profile of heavy induction motor load switching is predicted, and the possibility to mitigate potential power quality violations before they occur is created. Finally the chapter conclusions are presented, highlighting the importance of the statistical indices and how the mathematical indices could further enhance the power quality of an electric power system.

2. Power quality disturbances

There is a wide variety of power quality disturbances which affect the performance of customer equipment. The most common of these are briefly described in this section of the chapter.

2.1 Voltage sags

Voltage Sag is defined as a short reduction in voltage magnitude for a duration of time, and is the most important and commonly occurring power quality issue. The definitions to characterise voltage sag in terms of duration and magnitude vary according to the authority. According to the IEEE defined standard (IEEE Std. 1159, 1995),voltage sag is defined as a decrease of rms voltage from 0.1 to 0.9 per unit (pu), for a duration of 0.5 cycle to 1 minute. Voltage sag is caused by faults on the system, transformer energizing, or heavy load switching.

2.2 Interruptions

Interruption is defined as a 0.9 pu reduction in voltage magnitude for a period less than one minute. An interruption is characterized by the duration as the magnitude is more or less constant. An interruption might follow a voltage sag if the sag is caused by a fault on the source system. During the time required for the protection system to operate, the system sees the effect of the fault as a sag. Following circuit breaker operation, the system gets isolated and interruption occurs. As the Auto-reclosure scheme operates, introduced delay can cause a momentary interruption.

2.3 Transient overvoltages, swells

Overvoltage is an increase of Root Mean Square (RMS) voltage magnitude for longer than one minute. Typically the voltage magnitude is 1-1.2 pu and is caused by switching off a large load from the system, energizing a capacitor bank, poor tap settings on the transformer and inadequate voltage regulation. Overvoltages can cause equipment damage and failure. Overvoltages with duration of 0.5 cycle to 1 min are called voltage swells. A swell is typically of a magnitude between 1.1 and 1.8 pu and is usually associated with single line to ground faults where voltages of non-faulted phases rise.

2.4 Harmonic issues

The increasing application of power electronic devices like adjustable speed drives, uninterruptible power supplies and inverters, raises increasing concerns about harmonic distortion in the power system. These devices can not only cause harmonics in the system but are also very sensitive to voltage-distorted signals. The presence of harmonics in the system could also cause several unwanted effects in the system including excessive transformer heating or overloading and failure of power factor correcting capacitors.

The maximum total harmonic distortion which is acceptable on the utility system is 5.0% at 2.3-69kV, 2.5% at 69-138kV and 1.5% at higher than 138kV voltage levels (IEEE Std. 1250, 1995).

2.5 Voltage imbalance

This type of power quality disturbance is caused by unequal distribution of loads amongst the three phases. At three-phase distribution level, unsymmetrical loads at industrial units and untransposed lines can result in voltage imbalance. Voltage imbalance is of extreme importance for three-phase equipment such as transformers, motors and rectifiers, for which it results in overheating due to a high negative sequence current flowing into the equipment. The asymmetry can also have an adverse effect on the performance of converters, as it results in the production of harmonic.

3. Voltage sag statistical indices

For many years, electricity companies have used sustained interruption indices as indicators describing the quality and reliability of the services they provide. In order to compare power quality in different networks, regulators need to have common, standardised quality indices. The number of these indices should be kept at a minimum, easy to assess, and be representative of the disturbance they characterise. This section briefly discusses various voltage sag indices proposed by electrical association organisations and indices suggested by recent researchers. These indices are used to characterise any voltage sag, according to the individual index point of view. The procedure to evaluate the quality of supply, reference to non-rectangular events and equipment compatibility issues are also discussed.

3.1 Types of indices

Any available voltage sag index can be classified within the following three categories (Bollen, 2000).

a. Single-event index: a parameter indicating the severity of a voltage or current event, or otherwise describing the event. Each type of event has a specific single-event index.

b. Single-site index: a parameter indicating the voltage or current quality or a certain aspect of voltage or current quality at a specific site.

c. System index: a parameter indicating the voltage or current quality or a certain aspect of voltage or current quality for a whole or part of a power system.

The procedure to evaluate the power systems performance regarding voltage sag is as follows (Bollen, 2000).

a. Step 1- Obtain sampled voltages with a certain sampling rate and resolution.
b. Step 2- Calculate event characteristics as a function of time, from the sampled voltages.
c. Step 3- Calculate single event indices from the event characteristics.
d. Step 4- Calculate site indices from the single-event indices of all events measured during a certain period of time.
e. Step 5- Calculate system indices from the site indices for all sites within a certain power system.

Each step mentioned above is discussed in the following parts of this chapter, with more detail given on the steps involving the various index types. Step 1 is not discussed at all since it simply represents the procedure to obtain voltage samples for every event with a certain sampling rate and resolution.

3.2 Event characteristics

From the sampled voltages, the characteristic voltage magnitude as a function of time can be obtained. Methods to accomplish this, using three phase measurements are (Bollen & Styvaktakis, 2000):

a. Method of Symmetrical components: From the voltage magnitude and phase angle, a sag type is obtained along with the characteristic voltage, and Zero sequence voltage. The characteristic magnitude (the absolute value of the characteristic complex voltage) can be used to characterize three-phase unbalanced dips without loss of essential information. Using the characteristic magnitude and duration for three-phase unbalanced dips, corresponds to the existing classification (through magnitude and duration) for single-phase equipment.
b. Method based on six rms voltages: The procedure used in this method is to calculate the zero sequence component of the voltage and remove it from the phase voltages. The new phase to phase voltages can then be calculated. From the phase to phase voltages and the three phase voltages, the rms values can be calculated. The characteristic magnitude would then be the lowest of the six rms voltages.

3.3 Single event indices

From the event characteristics as a function of time, a number of indices are determined that describe the event. For some applications the phase angle at which the sag begins is important and called 'point on wave of dip initiation'. Using the concept of point on wave the exact beginning of the voltage dip can be identified (Bollen, 2001). In addition the point on wave of voltage recovery allows precise calculation of the sag duration. The maximum phase shift can be obtained from the voltage characteristic versus time and can be used for accurate phase-angle jump calculation. The maximum slew rate, and zero sequence voltage are mentioned as potential single event indices. The Canadian electrical association uses two approved quality indices: RMS Overvoltage (RMSO) and Undervoltage (RMSU) (Bergeron,R., 1998). The indices are assessed over intervals equal to the time needed for equipment to reach its steady-state temperature. The heating time constant, which varies with the size and nature of the equipment, has been divided into three classes:

a. Highly sensitive electronic systems with a heating time constant of less than 600 ms.
b. Varistors and power electronics with a 2-min heating time constant.

c.   Electrical apparatus with a heating time constant exceeding 24 min.

Canadian utilities have retained 2-min heating time constant to assess the factor related to overvoltages and undervoltages. This factor is measured over 10 min intervals (5x2 min).

The most-commonly used single event indices for voltage dips are "retained voltage" and "duration". It is recommended to only use the rms voltage as a function of time, or the magnitude of the characteristic voltage for three-phase measurements, to calculate the duration. Using the characteristic voltage magnitude versus time, the retained voltage and duration can be obtained as follows.

The basic measurement of a voltage dip and swell is U_{RMS} (1/2) on each measurement channel. Where U_{RMS} (1/2) is defined as the value of the RMS voltage measured over one cycle and refreshed each half cycle (Polycarpou et al., 2004).

3.3.1 For single-phase measurements

A voltage sag begins when the U_{RMS} (1/2) voltage falls below the dip threshold, and ends when the U_{RMS} (1/2) voltage is equal to or above the dip threshold plus the hysterisis voltage(Polycarpou, A. et al., 2004). The retained voltage is the smallest U_{RMS} (1/2) value measured during the dip.

The duration of a voltage dip is the time difference between the beginning and the end. Voltage sags may not be rectangular. Thus, for a given voltage sag, the duration is dependent on a predefined threshold sag value.

The user can define the sag threshold value either as a percentage of the nominal, rated voltage, or as a percentage of pre-event voltage. For measurements close to equipment terminals and at distribution voltage levels, it is recommended to use the nominal value (Bollen, 2001). At transmission voltages, the pre-event voltage may be used as a reference.

The choice of threshold obviously affects the retained voltage in per cent or per unit. The choice of threshold may also affect the measurement of the duration for voltage dips with a slow recovery. These events occur due to motor starting, transformer energizing, post-fault motor recovery, and post-fault transformer saturation.

3.3.2 For multi-channel three phase measurement

The voltage sag starts when the RMS voltage U_{RMS} (1/2) , drops below the threshold in at least one of the channels, and ends when the RMS voltage recovers above the threshold in all channels. The retained voltage for a multi-channel measurement is the lowest RMS voltage in any of the channels.

Several methods have been investigated leading to a single index for each event. Although this leads to higher loss of information, it simplifies the comparison of events, sites and systems. The general drawback of any single-index method is that the result no longer directly relates to equipment behaviour. Single indices are briefly described below.

Loss of Voltage:

The loss of voltage "LV" is defined as the integral of the voltage drop during the event.

$$L_V = \int \{1 - v(t)\} dt \tag{1}$$

Loss of Energy:

The loss of energy "LE" is defined as the integral of the drop in energy during the event:

$$L_E = \int \{1 - v(t)^2\} dt \tag{2}$$

Method proposed by R. S. Thalam, 2000, defines the energy of voltage sag as:

$$E_{Vs} = (1 - V_{pu})^2 \times t \tag{3}$$

Where t is the sag duration.

Method proposed by Thallam & Heydt, 2000: The concept of "lost energy in a sag event" is introduced, in such a way that the lost energy for events on the Computer Business Equipment Manufacturers Association (CBEMA) curve is constant for three-phase measurements. The lost energy is added for the three phases:

$$W_a = \{1 - \frac{V_a}{V_a \text{nominal}}\}^{3.14} \times t \tag{4}$$

To include non-rectangular events an integral expression may again be used. The event severity index 'S_e' is calculated from the event magnitude (in pu) and the event duration. Also essential for the method is the definition of a reference Curve.

$$S_e = \frac{1 - V}{1 - V_{ref}(d)} \tag{5}$$

Where $V_{ref}(d)$ is the event magnitude value of the reference curve for the same event duration. This method is illustrated by the use of the CBEMA and Information Technology Industry Council (ITIC) curves as reference curves. However the method is equally applicable with other curves.

3.4 Site indices
Usually the site indices have as inputs the retained voltage and duration of all sags recorded at a site over a given period. Available RMS Variation Indices for Single Sites are described below.

3.4.1 SAFRI related index and curves
The **SAFRI** index (System Average RMS Variation Frequency Index) relates how often the magnitude of a voltage sag is below a specified threshold. It is a power quality index which provides a rate of incidents, in this case voltage sags, for a system (Sabin, 2000).

SARFI-X corresponds to a count or rate of voltage sags, swell and interruptions below a voltage threshold. It is used to assess short duration rms variation events only.

SARFI-Curve corresponds to a rate of voltage sags below an equipment compatibility curve. For example **SARFI-CBEMA** considers voltage sags and interruptions that are not within the compatible region of the CBEMA curve.

Since curves like CBEMA do not limit the duration of a RMS variation event to 60 seconds, the SARFI-CBEMA curve is valid for events with a duration greater than ½ cycle. To demonstrate the use of this method the following table is assumed for a given site (Polycarpou et al., 2004).

Event Number	Magnitude (pu)	Duration (s)
1	0.694	0.25
2	0.459	0.1
3	0.772	0.033
4	0.47	0.133
5	0.545	0.483
6	0.831	0.067
7	0.828	0.05
8	0.891	0.067
9	0.008	0.067
10	0.721	0.067
11	0.684	0.033
12	0.763	0.033

Table 1. Voltage sag event characteristics

The values of the table above are used with various types of scatter plot to illustrate the use of known curves in equipment compatibility studies.

a. CBEMA curve Scatter Plot

The CBEMA chart presents a scatter plot of the voltage magnitude and event duration for each RMS variation. The CBEMA group created the chart as a means to predict equipment mis-operation due to rms variations. An RMS variation event with a magnitude and duration that lies within the upper and lower limit of the CBEMA curve, has a high probability to cause mis-operation of the equipment connected to the monitored source. Observing Figure 1, the number of events which are below the lower limit of the CBEMA curve is seven, giving a SAFRI-CBEMA of seven events.

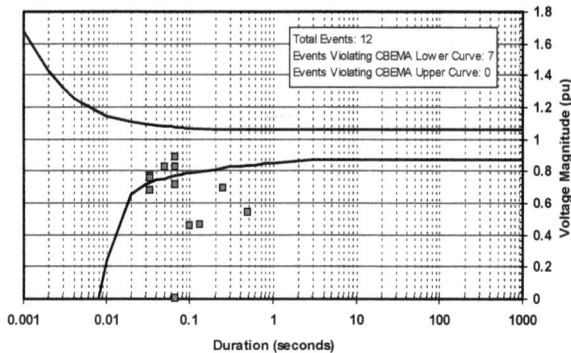

Fig. 1. The CBEMA curve scatter plot

b. ITIC Curve Scatter Plot

The ITIC Curve describes an AC input voltage boundary that typically can be tolerated (Sabin, 2000). Events above the upper curve or below the lower curve are presumed to cause the mis-operation of information technology equipment. The curve is not intended to serve as a design specification for products or ac distribution systems. In this case, the number of events, which are below the lower limit of the ITIC curve, in Figure 2, is six, giving a SAFRI-ITIC of six events.

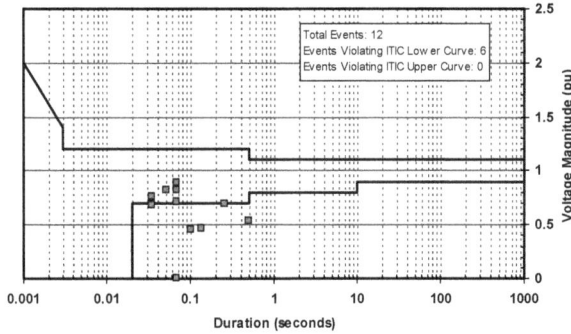

Fig. 2. The ITIC curve scatter plot

c. SEMI Curve Scatter Plot

In 1998, the Semiconductor Equipment and Materials International (SEMI) group, power quality and Equipment Ride Through Task force, recommended the SEMI Standard F-47 Curve to predict voltage sag problems for semiconductor manufacturing equipment. Figure 3 shows the application of the data of Table 1on the SEMI curve.

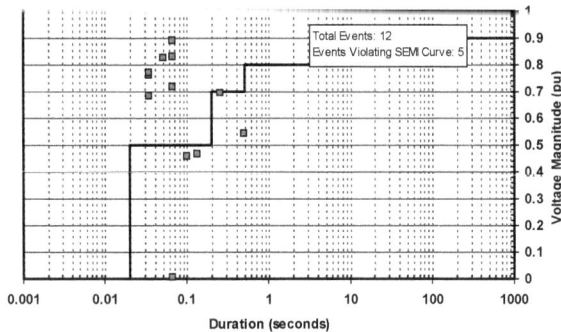

Fig. 3. SEMI Curve scatter plot

The results obtained from each combination of curves can be presented with the use of tables, such as UNIDEPE DISDIP or ESKOM Voltage sag Table, (Sabin, 2000).

d. Voltage sag co-ordination chart-IEEEStd.493 and 1346

The chart contains the supply performance for a given site through a given period, and the tolerance of one or more devices. It illustrates the number of events as a function of event severity. Observing the graph shown in Figure 4, there are 5 events per year where the voltage drops below 40% of nominal Voltage for 0.1 s or longer. Equally there are 5 events per year where the voltage drops below 70% magnitude and 250 ms duration.

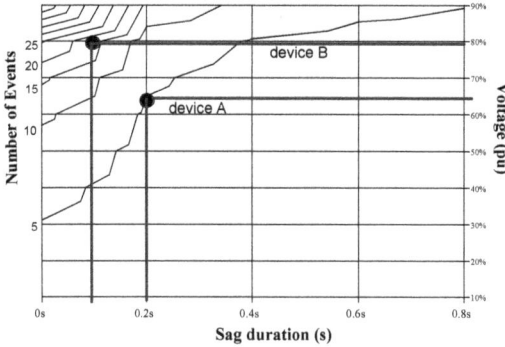

Fig. 4. Voltage sag co-ordination chart

The advantage of this method is that equipment behavior can be directly compared with system performance, for a wide range of equipment. The disadvantage of the method is that a two-dimensional function is needed to describe the site. For comparison of different sites a smaller number of indices would be preferred.

3.4.2 Calculation methods

a. Method used by Detroit Edison

The method calculates a "sag score" from the voltage magnitudes in the three phases (Sabin, 2000).

$$S = 1 - \frac{V_a + V_b + V_c}{3} \tag{6}$$

This sag score is equal to the average voltage drop in the three phases. The larger the sag score, the more severe the event is considered to be.

b. Method proposed by Thallam

A number of site indices can be calculated from the "voltage sag energy" (Thallam, 2000). The "Voltage Sag Energy Index" (VSEI) is the sum of the voltage sag energies for all events measured at a given site during a given period:

$$VSEI = \sum_i E_{VS\_i} \tag{7}$$

The "Average Voltage Sag Energy Index" (AVSEI) is the average of the voltage sag energies for all events measured at a given site during a given period:

$$AVSEI = \frac{1}{N} \sum_{i=1}^{N} E_{VS\_i} \tag{8}$$

A sensitive setting will result in a large number of shallow events (with a low voltage sag energy) and this in a lower value for AVSEI.

The sag event frequency index at a particular location and period is suggested as the number of qualified sag events at a location and period (Thallam & Koellner, 2003).

The System sag count index is the total number of qualified voltage sag events over the number of monitor locations. By the expression qualifying events, it implies a voltage less than 90%, with event duration limited to 15 cycles and energy greater or equal to 100.

3.4.3 Non-rectangular events

Non-rectangular events are events in which the voltage magnitude varies significantly during the event. A method to include non-rectangular events in the voltage-sag coordination chart is also applicable according to the IEEE defined standard (IEEE Std.493, 1997). Alternatively, the function value can be defined as the number of times per year that the RMS voltage is less than the given magnitude for longer than the given duration.

EPRI-Electrotek mentions that each phase of each ms variation measurement may contain multiple components (Thallam, 2000). Consequently, these phase rectangular voltage sag measurements are easily characterized with respect to magnitude and duration. Approximately 10% of the events are non-rectangular. These events are much more difficult to characterize because no single magnitude-duration pair completely represent the phase measurement.

The method suggested for calculating the indices used by EPRI-Electrotek is called the "Specified Voltage" method. This method designates the duration as the period of time that the rms voltage exceeds a specified threshold voltage level used to characterize the disturbance.' The consequence of this method is that an event may have a different duration when being assessed at different voltage thresholds as shown in Figure 5.

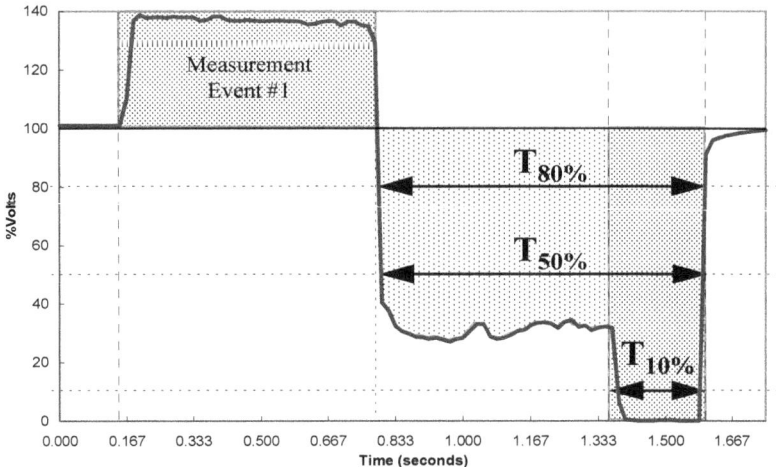

Fig. 5. Illustration of "specified voltage" characterization

Most of the single site indices relate the magnitude and duration of the sag and the number of events. These events can be grouped in order to make their counting easier and more practical. Power quality surveys in the past have just referred to the number of voltage sags per year for a given site. This value could include minor events, which do not affect any equipment.

The Canadian Electrical Association recommends tracking 4 indices for sag magnitudes (referring to the remaining voltage), of 85%, 70%, 40% and 1%. The latter refers to interruptions rather than sags.

ESKOM (South African Utility), groups voltage sags into five classes (Sabin, 2000):

class Y: 80% – 90% magnitude, 20 ms - 3 sec duration

class X: 40% - 80% magnitude, 20 ms - 150 ms duration

class S: 40% - 80% magnitude, 150 ms - 600 ms duration

class T: 0 - 40% magnitude, 20 ms - 600 ms duration

class Z: 0 – 80% magnitude, 600 ms - 3 sec duration

EPRI -Electrotek suggests the following five magnitudes and three duration ranges to characterize voltage thresholds:

a. **RMS variation** Frequency for voltage threshold X: with X=90%, 80%, 70%, 50%, 10%: the number of events per year with magnitude below X, and duration between 0.5 cycle and 60 sec.

b. **Instantaneous RMS variation** Frequency for voltage threshold X: with X=90%, 80%, 70%, 50%: the number of events per year with magnitude below X, and duration between 0.5 cycle and 0.5 sec.

c. **Momentary RMS variation** Frequency for voltage threshold X: with X=90%, 80%, 70%, 50%: the number of events per year with magnitude below X, and duration between 0.5 sec and 3 sec.

d. **Momentary RMS variation** Frequency for voltage threshold 10%: the number of events per year with a magnitude below 10%, and a duration between 0.5 cycle and 3 sec.

e. **Temporary RMS variation** Frequency for voltage threshold X: with X=90%, 8070, 70%, 50%, 10%: the number of events per year with magnitude below X, and duration between 3 sec. and 60 sec.

The duration ranges are based on the definition of instantaneous, momentary and temporary, as specified by IEEE (IEEE Std. 1159, 1995).

3.5 System indices

System Indices are typically a weighted average of the single-site indices obtained for all or a number of sites within the system. The difficulty lies in the determination of the weighting factors. In order to assess any indices for the system, first monitoring of the quality of supply must take place. When the Electric Power Research Institute (EPRI)-Distribution Power Quality (DPQ) program placed monitoring equipment on one hundred feeders, these feeders needed to adequately represent the range of characteristics seen on distribution systems. This required the researchers to use a controlled selection process to ensure that both common and uncommon characteristics of the national distribution systems were well represented in the study sample. Thus a level of randomness is required. Many devices are susceptible to only the magnitude of the variation. Others are susceptible to the combination of magnitude and durationOne consideration in establishing a voltage sag index is that the less expensive a measuring device is, the more likely it will be applied at many locations, more completely representing the voltage quality electricity users are experiencing.

With this consideration in mind, sag monitoring devices are generally classified into *less expensive devices* that can monitor the gross limits of the voltage sag, and *more expensive devices* that can sample finer detail such as the voltage-time area and other features that more fully characterize the sag.

The sag limit device senses the depth, of the voltage sag. The sag area device can sample the sag in sufficient detail to plot the time profile of the sag. With this detail it could give a much more accurate picture of the total sag area, in volt-seconds, as well as the gross limits; the retained voltage, V_r, is also shown.

The developed RMS variation indices proposed by EPRI-Electrotek, are designed to aid in the assessment of service quality for a specified circuit area. The indices are defined such that they may be applied to systems of varying size (Bollen, 2001).Values can be calculated for various parts of the distribution system and compared to values calculated for the entire system.

Accordingly, the four indices presented assess RMS variation magnitude and the combination of magnitude and duration.

a. System Average RMS (Variation) Frequency Index$_{voltage}$ (SARFI$_x$)

SARFI$_x$ represents the average number of specified rms variation measurement events that occurred over the assessment period per customer served, where the specified disturbances are those with a magnitude less than x for sags or a magnitude greater than x for swells. Notice that SARFI is defined with respect to the voltage threshold 'x' (Sabin, 2000).

$$SARFI_x = \frac{\sum N_i}{N_T}$$

(9)

where

x = percentage of nominal rms voltage threshold; possible values - 140, 120, 110, 90, 80, 70, 50, and 10

N_i = number of customers experiencing short-duration voltage deviations with magnitudes above x% for x >100 or below x% for x <100 due to measurement event i

N_T =number of customers served from the section of the system to be assessed

b. System Instantaneous Average RMS (Variation) frequency Index$_{voltage}$(SIARFI$_x$)

SIARFI$_x$ represents the average number of specified instantaneous rms variation measurement events that occurred over the assessment period per customer served. The specified disturbances are those with a magnitude less than x for sags or a magnitude greater than x for swells and duration in the range of 0.5 - 30 cycles.

$$SIARFI_x = \frac{\sum NI_i}{N_T}$$

(10)

Where:

x = percentage of nominal rms voltage threshold; possible values - 140, 120, 110, 90, 80, 70, and 50

NI_i = number of customers experiencing instantaneous voltage deviations with magnitudes above x% For x>100 or below x% for x <100 due to measurement event i

Notice that SIARFI$_x$ is not defined for a threshold value of x = 10%. This is because IEEE Std. 1159, 1995, does not define an instantaneous duration category for interruptions.

c. System Momentary Average RMS (Variation) Frequency Index$_{vortage}$ (SMARFIx)

In the same way that SIARFIx is defined for instantaneous variations, SMARFIx is defined for variations having a duration in the range of 30 cycles to 3 seconds for sags and swells, and in the range of 0.5 cycles to 3 seconds for interruptions.

$$SMARFI_x = \frac{\sum NM_i}{N_T} \tag{11}$$

x = percentage of nominal rms voltage threshold; possible values - 140, 120, 110, 90, 80, 70, 50, and 10

NM =number of customers experiencing momentary voltage deviations with magnitudes above X% for X >100 or below X% for X <100 due to measurement event i.

d. System Temporary Average RMS (Variation) Frequency Index$_{vortage}$ (STARFI$_x$)

STARFI$_x$ is defined for temporary variations, which have a duration in the range of 3 - 60 seconds.

$$STARFI_x = \frac{\sum NT_i}{N_T} \tag{12}$$

x = percentage of nominal rms voltage threshold; possible values - 140, 120, 110, 90, 80, 70, 50, and 10.

NT_i = number of customers experiencing temporary voltage deviations with magnitudes above x% for x >100 or below x% for x <100 due to measurement event i.

As power networks become more interconnected and complex to analyse, the need for power quality indices to be easily assessable, and representative of the disturbance they characterise with minimum parameters, arises. This section has presented the various Voltage sag indices available in literature. Most of these indices are characterized through the sag duration and magnitude. To demonstrate the theory of equipment compatibility, with the use of the System Average RMS Variation Frequency Index, various power acceptability curves were used.

Electricity distribution companies need to assess the quality of service provided to customers. Hence, a common index terminology for discussion and contracting is useful. Future voltage sag indices need to be adjustable and adaptable to incorporate future changes in technology and system parameters. This would enable implementation of indices into the next generation of power system planning software.

4. Voltage sag mathematical indices

In this section of the chapter, the mathematical formulation of two voltage sag indices (ξ and $\zeta_{1,2}$) is introduced as well as the results of the investigation towards their accuracy establishment. The Mathematical equations describing the development of a Combined Voltage Index (CVI) are also presented as well as the results obtained by the verification process. The index supervises the power quality of a system, through characterising voltage sags. The voltage sags are caused by an increase in reactive demand due to induction motor starting.

A feeder can be modeled by an equivalent two-port network, as shown in Figure 6.

The sending end voltage and current of the system can be represented by equations 13 and 14.

$$U_s \angle \delta_s = AU_r \angle \delta_r + BI_r \tag{13}$$

$$I_s = CU_r \angle \delta_r + DI_r \tag{14}$$

Where U_s is the sending end voltage, I_s the sending end current, U_r the receiving end voltage, I_r the receiving end current, δ_s the sending end voltage angle, δ_r the receiving end voltage angle, and A,B,C, D are the two port network constants. For a short length line, corresponding to distribution network, the two port network parameters can be approximated as: A=D=1, B= $Z \angle \theta$, C=0.Where Z is the transmission line impedance vector magnitude, and θ the transmission line impedance vector angle.

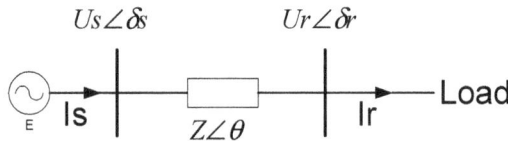

Fig. 6. The equivalent two port network model.

The line power flow, for the active power at the sending and receiving end of the line, can be described by (15) and (16).

$$P_s = \frac{U_s^2}{Z}Cos(\theta) - \frac{U_sU_r}{Z}Cos(\theta + \delta_s - \delta_r) \tag{15}$$

$$P_r = \frac{U_sU_r}{Z}Cos(\theta + \delta r - \delta s) - \frac{U_r^2}{Z}Cos(\theta) \tag{16}$$

4.1 'ζ' index
If the index ζ signifies the voltage magnitude during the sag as a per unit function of the sending voltage ($U_r = \zeta U_s$), and is substituted in equation 16, equation 17 yields (Polycarpou & Nouri, 2005).

$$\frac{\zeta U_s^2}{Z}Cos(\theta + \delta_r - \delta_s) - \frac{(U_s\zeta)^2}{Z}Cos(\theta) - P_r = 0 \tag{17}$$

Thus the solution of the second order equation, resulting from (17), can be calculated using equation (18).

$$\zeta_{1,2} = \frac{Cos(\theta + \delta_r - \delta_s) \pm \left(Cos^2(\theta + \delta_r - \delta_s) - \frac{4ZP_rCos\theta}{U_s^2} \right)^{\frac{1}{2}}}{2Cos\theta} \tag{18}$$

Equation 18 provides a tool to calculate the voltage sag ,as a per unit value of the sending end voltage, through angles and power demand.

However, since the equation is obtained through a quadratic equation, it has two solutions. ζ_1 will be valid for a specific range of parameters. In the same way ζ_2 will be valid for a different range of parameters. The validity of the two solutions, ζ_1 and ζ_2, with the use of various line X/R ratios is investigated in (Nouri et al., 2006). X/R ratio varies from Distribution to Transmission according to the cables used for the corresponding voltages. Typical values of X/R ratio are: for a 33kV overhead line -1.4, for a 132kV overhead line -2.4, for a 275kV overhead line -8.5, for a 400kV overhead line -15. A distribution line example is the IEEE34, 24.9kV overhead line with X/R ratio of 0.441.

According to the parameters either ζ_1 or ζ_2 will be the correct answer which should match the receiving end voltage.

The point of intersection of U_r with ζ_1 and ζ_2, occurs when $(\zeta_1 - \zeta_2)\cos\theta$ is equal to zero, when both solutions are identical. However, in practice a gap develops when both solutions approach the U_r axis, where none of the two solutions accurately represent the receiving voltage U_r, as seen in Figure 7.

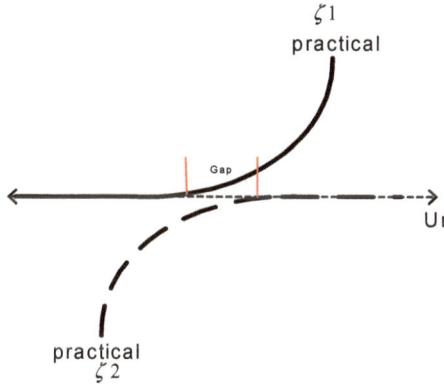

Fig. 7. Developed gap of inaccuracy

The distance between the two curves at the point of the gap can be defined by equation 19.

$$\zeta_1 - \zeta_2 = \frac{\left(Cos^2(\theta + \delta_r - \delta_s) - \dfrac{4ZP_rCos\theta}{U_s^2} \right)^{\frac{1}{2}}}{Cos\theta} \tag{19}$$

In order to fully investigate the range of accuracy of the two solutions, X/R ratio values of 1 to 15 are used for the line impedance. Since the receiving end power varies according to the load, five loading conditions are used in the investigation. Each loading consists of induction motors. The loads are switched in the system one by one to create the effect of supplying minimum load (one motor) and maximum load (five motors).

Using MathCad, the value of $(\zeta_1 - \zeta_2)\cos\theta$ is calculated for five different loadings and each X/R ratio, starting from one up to fifteen in steps of one. The results can be seen in Figure 8 (Nouri et al., 2006).

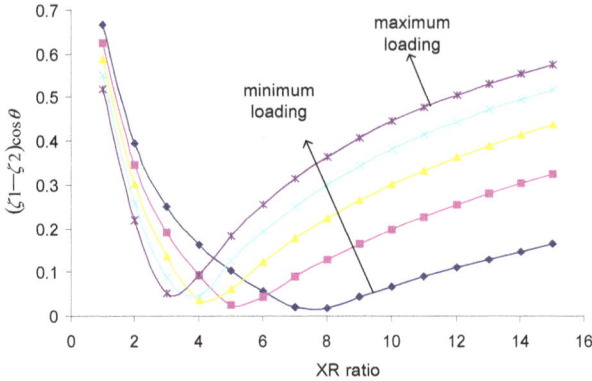

Fig. 8. Mathematical results obtained for $(\zeta_1 - \zeta_2)\cos\theta$

It can be observed from Figure 8, that the minimum values of $(\zeta_1 - \zeta_2)\cos\theta$ occur within X/R ratio values of 3 to 8, for all test cases. Therefore during those points, the gap of inaccuracy for the index can be expected for the two solutions. Taking under consideration Figure 7, solution ζ_1 should cover the ranges less than three and solution ζ_2 should cover X/R greater than eight. Between those X/R values the gap position varies according to the loading and the X/R ratio of the line, thus it cannot be generalized. The accuracy of the defined location of the gap is and verified through application on a two-bus system within Power system Computer Aided Design software. The resulting data for a test system of X/R ratio equal to five, shown in Figure 9, verifies the mathematical theory concerning the gap.

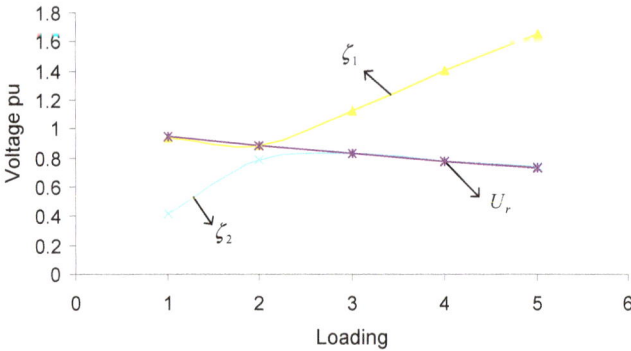

Fig. 9. ζ_1, ζ_2 and U_r for line X/R ratio of five

As shown in Figure 9, the plot of ζ_1 has a negative slope until loading two, and then it becomes positive. Whereas the plot of ζ_2 has a positive slope for the initial loadings and becomes negative when the third load is switched in.

Throughout the investigation of various X/R ratios a pattern was established regarding the slope of ζ_1 and ζ_2. When the slope of ζ_1 is negative it is the accurate solution. When $(\zeta_1 - \zeta_2)\cos\theta$ reaches minimum, ζ_1 deviates and ζ_2 becomes the correct answer with negative slope. Thus their slope is directly related to the minimum value of $(\zeta_1 - \zeta_2)\cos\theta$ and to the accuracy of each solution. The relationship between the slope of ζ_1 and ζ_2 with the index accuracy and choice of solution is described by equation 20. The value of 'i' is 1for ζ_1 or 2 for ζ_2.

$$\forall \frac{\partial(\zeta_i)}{dLoading} < 0 \Rightarrow \zeta i = valid \tag{20}$$

4.2 Combined voltage index
If ξ signifies the voltage magnitude during the sag as a per unit function of the sending voltage ($U_r = \xi U_s$), and is substituted in equation 15, equation 21 yields.

$$\xi = \frac{Cos\theta - \frac{ZP_s}{U_s^2}}{Cos(\theta + \delta_s - \delta_r)} \tag{21}$$

ξ and $\zeta_{1,2}$ signify the voltage magnitude during the sag as a per unit function of the sending voltage. When the two equations are combined, the resulting Combined Voltage Index (CVI), described by equation 22 features improved accuracy (Polycarpou & Nouri,2005). The value of CVI is the value of the receiving end voltage of the system power line.

$$\frac{a\xi + \zeta}{a + 1} = CVI \tag{22}$$

Where 'a' is the value of the scaling factor (Polycarpou & Nouri, 2009) and is defined as shown in equation 23.

$$a = \frac{1}{n} \sum_{l=1}^{n} \left[\frac{1 + \sqrt{1 - hl} - \left[wl + \sqrt{wl^2 - hl} \right]}{2kl} \right] \tag{23}$$

Where:

$w = \cos(\theta + \delta_r - \delta_s)$

$j = \cos(\theta + \delta_s - \delta_r)$

$k = \cos\theta$

$h = \frac{4ZP_rK}{U_s^2}$

and n= Number of loads supplied.

For simplicity, the value of scaling factor setting is 1.6 for the entire range of line X/R ratios investigated in the next sections of the chapter. Equations 18, 21 and 22 provide a tool to calculate the load voltage, as a per unit value of the sending end voltage. The equations are functions of receiving end variables such as the the receiving end voltage angle, δ_r, and the receiving end power P_r. The receiving end power can be described by $P_r = P_s - I^2 Z \cos\theta$. The angle δ_r of the receiving end voltage can be represented by sending end quantities through equation 24 (Nouri & Polycarpou, 2005).

$$\delta_r = a\tan\left[\frac{U_s \sin(\delta_s) - ZI \sin(\theta + i)}{U_s \cos(\delta_s) - ZI \cos(\theta + i)}\right] \tag{24}$$

Assuming the presence of an infinite bus at the sending end, equation 24 can be reduced to equation 25.

$$\delta_r = a\tan\left[\frac{ZI \sin(\theta + i)}{ZI \cos(\theta + i) - 1}\right] \tag{25}$$

4.2.1 Combined voltage index accuracy investigation

Most distribution power system loads have a power factor of 0.9 to 1. Industrial companies have to keep their power factor within limits defined by the regulatory authorities, or apply power factor correction techniques, or suffer financial penalties. In order to cover a wider area of investigation it is decided to simulate loads of power factor 0.8 to 0.99.

The relationship between the power factor and the X/R ratio of a load is: X/R ratio $= \tan\theta$, where $\theta = \cos^{-1} pf$.

In order to achieve the load X/R ratio variation the circuit model of the double cage induction motor, used within the PSCAD environment, is considered. The Sqc100 Motor circuit diagram is shown in Figure 10 (Polycarpou &Nouri, 2002).

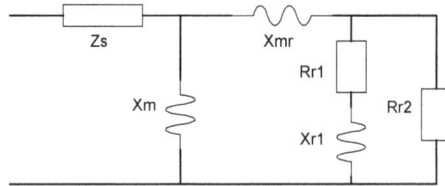

Fig. 10. The Double cage Induction motor model circuit diagram

The motor circuit parameters are:
Slip: 0.02, Stator resistance(Rs) 2.079 pu, First cage resistance(Rr1) 0.009 pu, Second cage resistance(Rr2)0.012 pu, Stator reactance (Xs)0.009 pu, Magnetizing reactance (Xm) 3.86 pu Rotor mutual reactance (Xmr) 0.19 pu, First cage reactance (Xr1)0.09 pu

The resistance of the stator winding is varied in order to achieve the required power factor and X/R ratio. Load X/R ratios of 0.1 to 0.75 are investigated. Two distribution line X/R ratios are used in the investigation in order to observe the accuracy of the index while varying both load as well as line X/R ratio for distribution system lines. The line X/R ratios are 0.12087, and 1. The amount of loading is varied through introducing five identical motors for each investigated case. The results of this investigation are presented in the following subsections.

a. Distribution Line X/R ratio is 0.120817

The per unit receiving voltage, obtained with variation of the load X/R ratio while line X/R ratio is 0.120817, can be seen in Figure 11. M_1 Signifies the minimum loading with the first motor being switched in. As any switched in motor reaches rated speed, the next load is switched in the system. M_5 corresponds to the Maximum loading with the fifth motor being switched in while the previous four are in steady state operation. The Combined Voltage Index (CVI) deviation corresponding to this scenario can be seen in Figure 12.

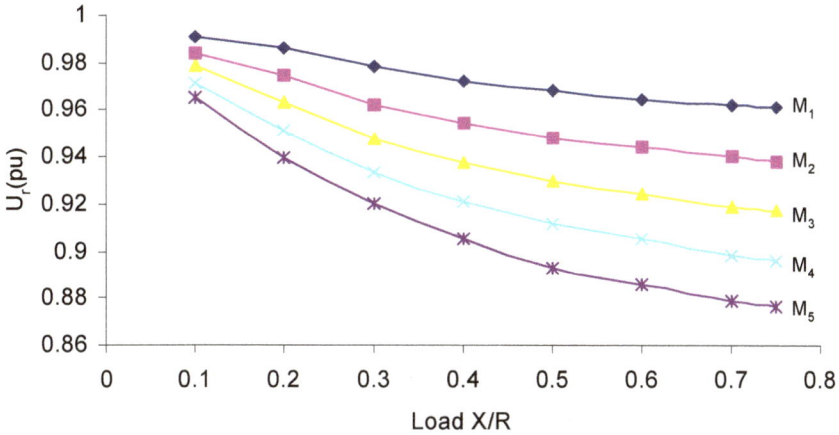

Fig. 11. U_r for various Load X/R ratios and loadings whilst Line X/R is 0.120817

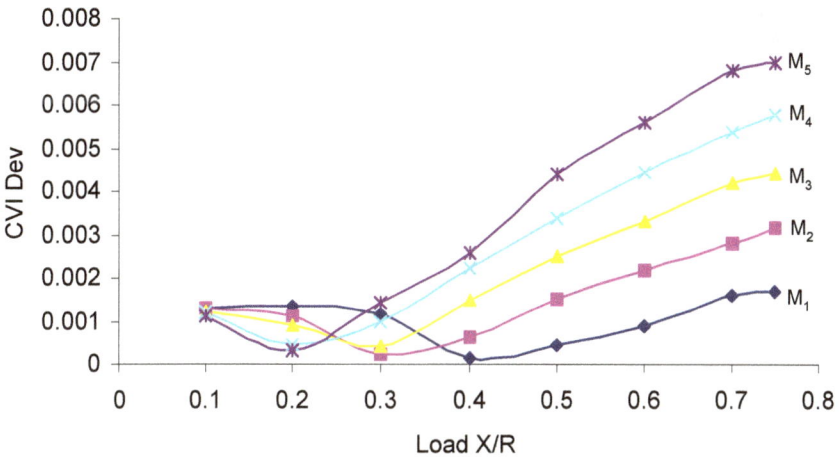

Fig. 12. CVI deviation for load X/R variation whilst line X/R=0.120817

Observing the figures above it can be concluded that as loading levels (M_1 to M_5) and X/R ratio increases, the receiving end voltage naturally decreases. This is due to the voltage drop occurring on the impedance of the transmission line. The accuracy of the proposed index is well within acceptable limits(0.7% in the worst case). Thus for line X/R ratio of 0.120817 the index is capable of calculating the receiving end voltage for variation of load X/R ratio.

b. Distribution Line X/R ratio is 1

The resulting receiving end voltage, obtained through variation of the load X/R ratio for the specific line X/R ratio can be seen in Figure 13.

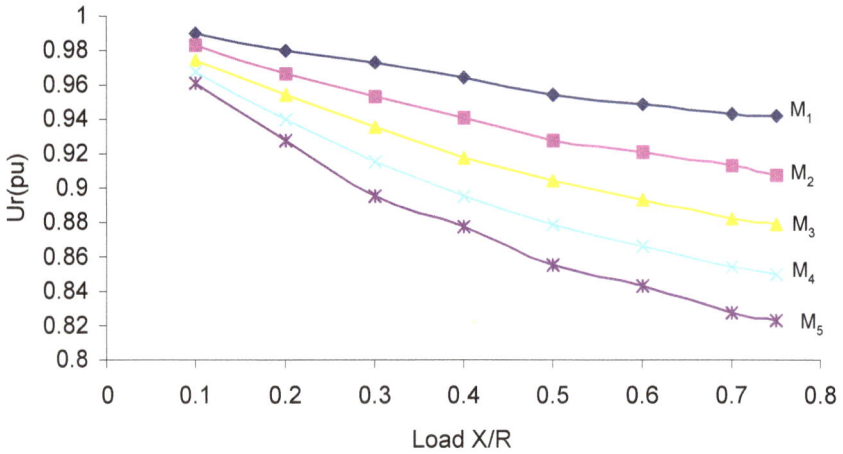

Fig. 13. U_r for load X/R variation whilst line X/R=1

Comparing Figure 11 to Figure 13, it is concluded that as the X/R ratio of the line increases, any load increment has more severe impact on the receiving end voltage due to the impedance magnitude of the line. The CVI deviation corresponding to this investigation case can be seen in Figure 14.

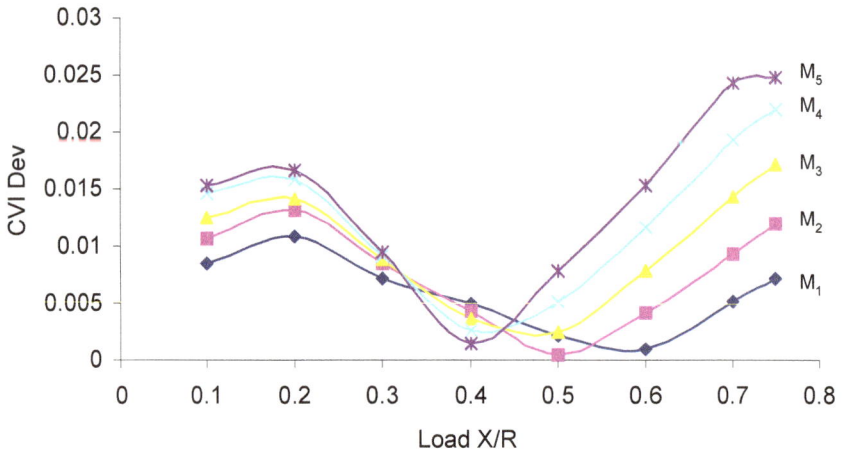

Fig. 14. CVI deviation for load X/R variation whilst line X/R=1

As the X/R ratio of the line increases the index accuracy did not decrease homogeneously. Between the values of 0.3 and 0.5 for load X/R an area of decreased inaccuracy can be observed. This is due to the scaling factor setting being 1.6 for the entire range of line X/R ratios investigated. The accuracy of the index is within acceptable margins as the largest deviation is within 2.5% (Polycarpou & Nouri, 2009). The index is within acceptable accuracy limits.

5. Conclusion

The first part of this chapter presents various statistical voltage sag indices proposed by electrical association organisations and indices suggested by recent researchers. These indices are used to characterise any voltage sag, according to the individual index point of view. The procedure to evaluate the quality of supply, reference to non-rectangular events and equipment compatibility issues are also presented. To demonstrate the theory of equipment compatibility, with the use of System Average RMS Variation Frequency Index, various power acceptability curves were used.

Furthermore the formulation defining a set of Mathematical voltage sag indices, leading to the Combined Voltage Index, is presented. Various motor power factors and loading levels are used in order to establish the behavior of the index for a wide range of loads. Mathematical description of voltage angle characteristics, relating to line X/R ratio variation is also illustrated. A relationship is established between the slope of ζ and the range of accuracy for each solution of the quadratic index. The CVI has proven to be easy to assess, accurate and representative of the disturbance it characterizes at distribution level. If better accuracy is required for distribution system applications, the scaling factor can be varied to achieve it. The described CVI index can be used in conjunction with optimization techniques for power quality improvement as well as power system operation optimization. The index is adaptable to incorporate future changes in technology and system parameters. This enables its implementation into the next generation of power system planning software.

6. Acknowledgment

The author would like to express his appreciation to the University of the West of England, UK, and to Prof. Hassan Nouri, for the opportunity to carry out significant portion of the research work presented in this chapter at their establishments, as a member of the Power Systems and Electronics Research Group.

7. References

Bergeron,R. (1998). Canadian electrical association approved quality indices. *IEEE Power summer meeting*

Bollen, M. (2000). Voltage sag indices-Draft 2. *Working document for IEEE P1564and CIGRE WG 36-07*

Bollen, M. (2001).Voltage Sags in Three-Phase Systems. *IEEE Power Eng. Review*, pp. 8-15.

Bollen, M.&Styvaktakis,S. (2000). Characterization of three phase unbalanced sags, as easy as one, two, three. *IEEE Power summer meeting.*

IEEE Std. 1159 (1995). Recommended practice for monitoring electric power quality.

IEEE Std. 1250 (1995) IEEE Guide for Service to Equipment Sensitive to Momentary Voltage Disturbances –Description, Corrected Edition Second Printing

IEEE Std. 493 (1997). Gold book, *IEEE recommended practice for the design of reliable industrial and commercial power systems*

IEEE Std. 1100, (1999). IEEE Recommended Practice for Powering and Grounding Electronic Equipment

Nouri, H., & Polycarpou, A. (2005). Load Angle Characteristic Analysis For A Radial System Using Various Line X/R Ratios During Motor Load Increment, Paper presented at the *Universities Power Engineering Conference*, Cork, Ireland

Nouri H, Polycarpou A. and Li z.(2006). Mathematical Development, Investigation and Simulation of a New Quadratic Voltage Index, *IEEE 41st Int. Universities Power Engineering Conference*, Newcastle-Upon-Tyne, UK, pp 1-6

Polycarpou, A., &Nouri,H. (2002). Analysis and Simulation of Bus Loading Conditions on Voltage Sag in an Interconnected Network, Paper presented at the *Universities Power Engineering Conference*, Staffordshire, UK

Polycarpou, A., Nouri,H., Davies,T., &Ciric, R. (2004). An Overview Of Voltage Sag Theory, Effects and Equipment Compatibility, Paper presented at the *Universities Power Engineering Conference*, Bristol, UK

Polycarpou, A., &Nouri,H. (2005). A New Index for On Line Critical Voltage Calculation of Heavily Loaded Feeders, Paper presented at the *Power Tech Conference*, St. Petersburg, Russia

Polycarpou A. and Nouri H.(2009). Investigation into the accuracy limits of a proposed Voltage Sag Index, *IEEE 44th Int. Universities Power Engineering Conference*, Glasgow, Scotland, UK, pp 1-5

Sabin, D. (2000). Indices used to assess RMS Voltage variations. *IEEE P1564. summer meeting.*

Thallam, R. (2000). Comments on voltage sag indices. IEEE P1564 internal document.

Thallam, R.&Heydt, G. T.(2000). Power acceptability and voltage sag indices in the three phase sense. *IEEE Power summer meeting*

Thallam, R. & Koellner, K.(2003). SRP Voltage sag index methodology-Experience and FAQs. *IEEE PES Meeting*

Part 3

Software Tools

8

A Power Quality Monitoring System Via the Ethernet Network Based on the Embedded System

Krisda Yingkayun[1] and Suttichai Premrudeepreechacharn[2]
[1]Rajamangala University of Technology Lanna,
[2]Chiang Mai University
Thailand

1. Introduction

The problems about power quality have increasingly caused a failure or a malfunction of the end user equipment for the past few years up to now. The problems have concerned with either voltage or current frequency deviation. To have the power quality monitoring done flowingly and completely, the measurement takes an important role on voltage, current, frequency, harmonic distortion and waveforms. Many researchers have used methods of power quality measurement (Dugan et al., 2002; Baggini, 2008) while other researchers have used various protocols to control the system (Auler & d'Amore, 2002). Others have presented the data acquisition based on PC (Batista et al., 2003) or Power Line Communication (Hong et al., 2005) or TMS320CV5416 DSP Processor (Rahim bin Abdullah & Zuri bin Sha'ameri, 2005). Another researcher has applied ARM and DSP processor (Yang & Wen, 2006) or has only applied DSP processor (Salem et al., 2006) to monitoring power quality in real time. In the meantime, the detecting fault signals of power fluctuation in real time and a power quality monitoring for real-time fault detection using real-time operating system (RTOS) are proposed (Yingkayun & Premrudeepreechacharn, 2008,2009) and the low cost power quality monitoring system is suggested (So et al., 2000; Auler & d'Amore, 2009), for example.

This chapter has developed the idea of power quality monitoring system via the Ethernet network based on the embedded system with the two selected ARM7 microcontrollers: ADUC7024 and LPC2368. On account of ADUC7024, it has a function of sampling waveforms and of writing the sampling signals to the external memory. Meanwhile, LPC2368 can execute the main tasks: detecting the fault signals; storing fault data in SD-CARD up to 2 GB; and communicating with PC or laptop via the Ethernet network. The power quality monitoring on the embedded system suggested can acquire the voltage, the status and the frequency. It can send them via network at real time, can operate as stand alone equipment and can display the fault signals in real time of power fluctuation. But anyhow, when being absent, we can download the fault data from the site place, depending on the program configuration. In this case the fault signals can be displayed on the screen of the PC or laptop at real time or can be done after as desired. Moreover, there can be a single phase or a 3-phase voltage measurement supported by the power quality monitoring hardware. When working in different places, we can establish the network in various sites and connect via the Ethernet network from a single PC or laptop. The network has the capability to send from the board site to the PC or laptop with high speed up to 100 megabit per second (Mb/s). Nevertheless, it is easy to monitor the power quality monitoring system via the Ethernet network from PC or laptop.

This chapter is organized as follow: Section 2 is the architecture of the power quality monitoring being composed of 2 sets of ARM7 microcontroller boards, 3 signal conditioning modules, external memory board and energy measurement board. Section 3 is the embedded software design with the details of sampling concept, power quality monitoring concept, data frames, configuration data and Ethernet packet structures. Section 4 is the application software to interface the power quality monitoring hardware. Section 5 is the experimental results which are displaying the fault signals from AC lines while being on operation or being done after. Section 6 is the conclusion from the research with the future work suggestion.

2. Architecture of the power quality monitoring system

The architecture of the power quality monitoring system is planned for specific purposes which are to detect, to store, to download and to display the fault signals while being on operation or being done later on. The main structures are divided into 5 circuit boards. The block diagram of the architecture of the power quality monitoring system is shown in Fig. 1.

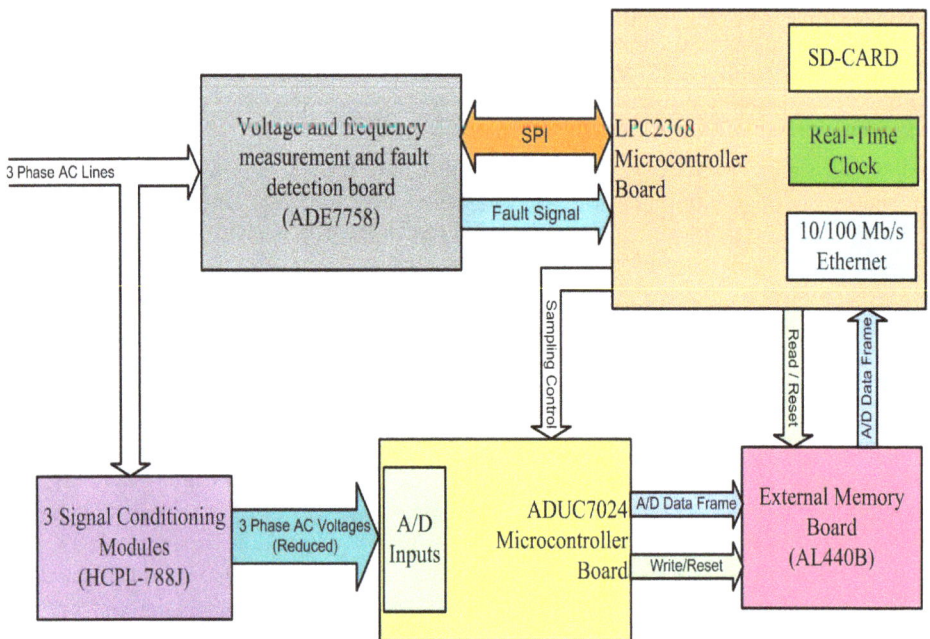

Fig. 1. Block diagram of the architecture of the power quality monitoring system

2.1 Signal conditioning modules (HCPL-788J)

There are actually 3 signal conditioning modules with 3 HCPL-788J integrated circuits for measuring the voltage of the three-phase AC lines in order to send the attenuated AC voltage to the waveform sampling board which the ADUC7024 microcontroller is embedded in. The 3 signal conditioning modules have used the same electronic circuit. The signal conditioning circuit is shown in Fig. 2.

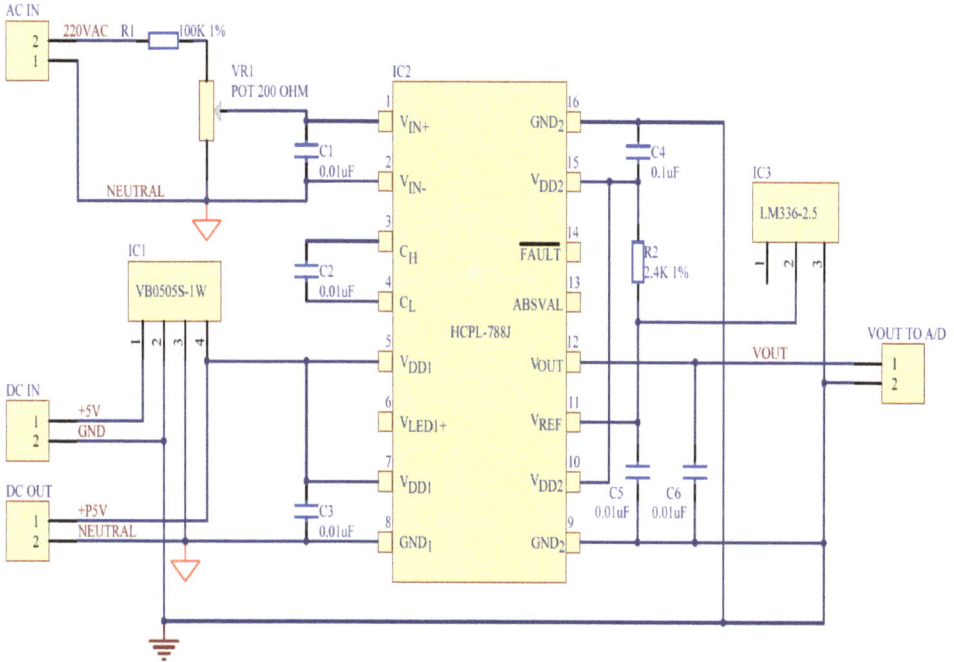

Fig. 2. The circuit of the signal conditioning module

2.2 The voltage and frequency measurement and fault detection board (ADE7758)

This board is the key part of monitoring system of fault detection which refers to sags and overvoltages. In this chapter, ADE7758 integrated circuit has been chosen to operate because of its suitable qualification to detect the fault signals in time. The circuit of the board is shown in Fig. 3.

2.3 Microcontroller boards (ADUC7024 and LPC2368)

As it is known that the embedded system with the two selected ARM7 microcontrollers which are ADUC7024 and LPC2368 are the developed microcontroller boards, using in this chapter. The first board is used for sampling waveforms and for writing the sampling signals to the external memory and the latter is used for various purposes: (1) reading voltage, frequency and fault detection from ADE7758; (2) storing fault data in SD-CARD up to 2 GB, for communicating with PC or laptop via the Ethernet Network and; (3) controlling the sampling process of the ADUC7024 board. The picture of the two microcontroller boards are shown in Fig. 4 and Fig. 5.

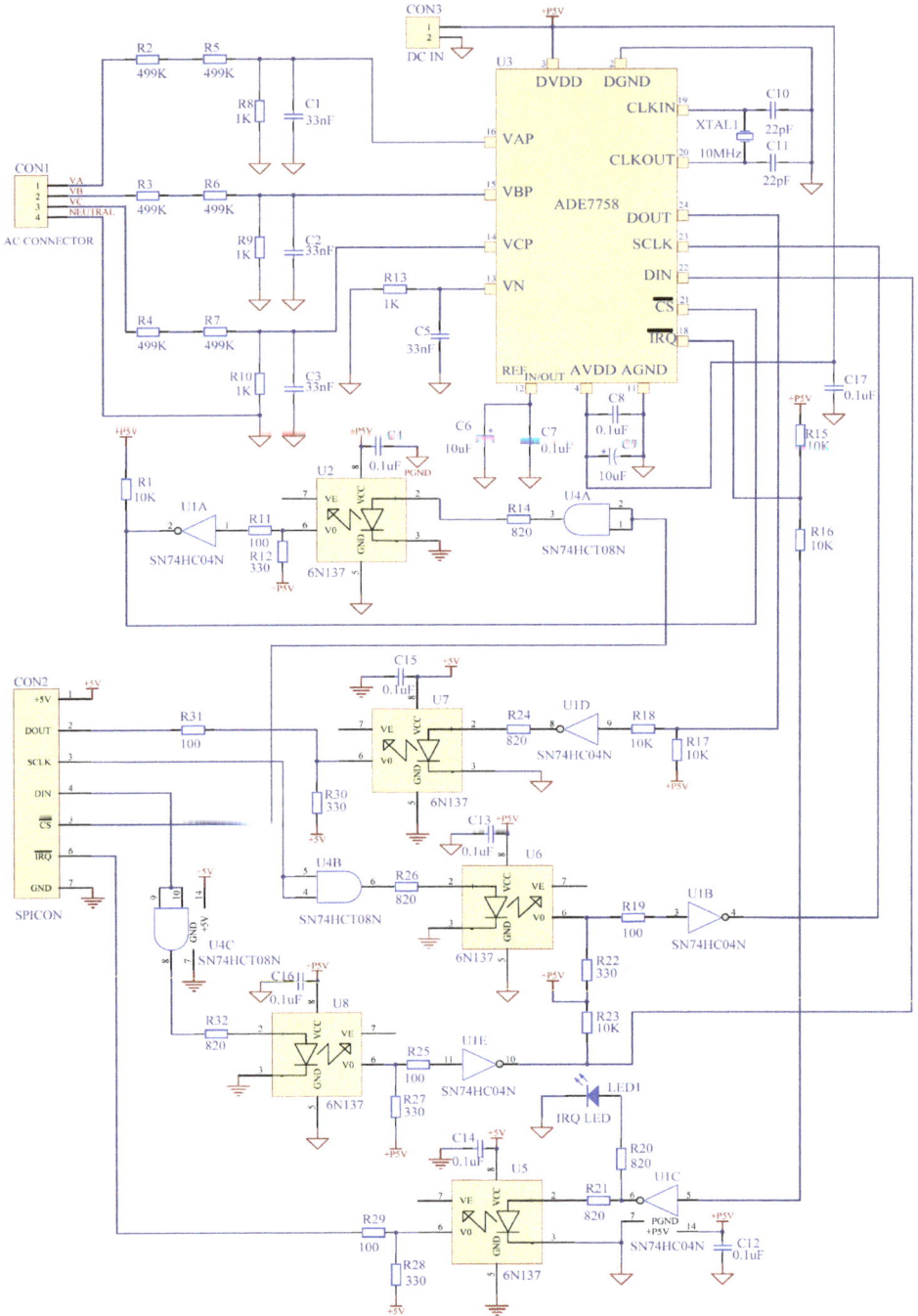

Fig. 3. The circuit of the voltage and frequency measurement and fault detection board

Fig. 4. The development board of ADUC7024 microcontroller

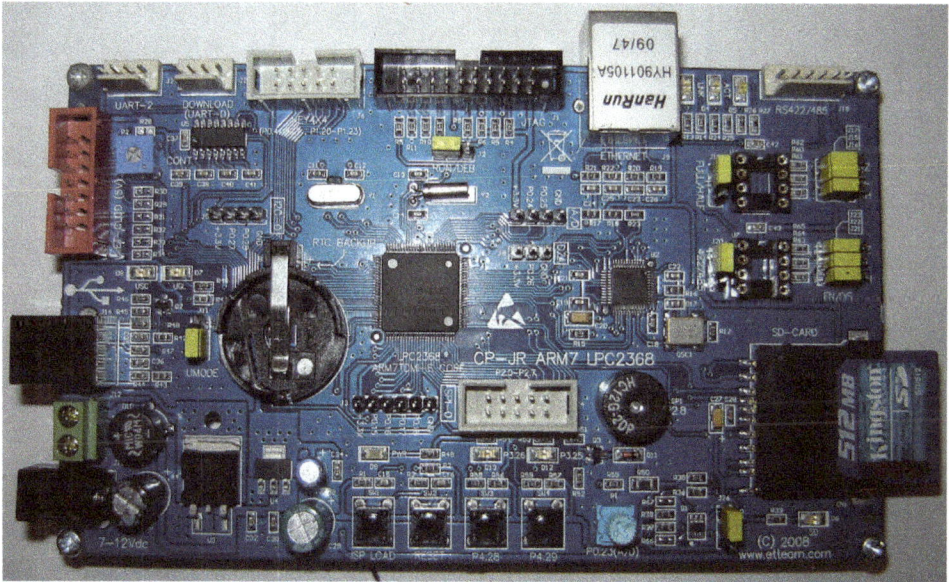

Fig. 5. The development board of LPC2368 microcontroller

2.4 External memory board (AL440B)

This external memory board is one device of this power quality monitoring architecture to collect the fault data from ADUC7024 microcontroller into its memory in series. The data will be read by the LPC2368 microcontroller board and will be stored in SD-CARD for recalling or downloading later via the Ethernet network when of necessity needed. The circuit of the external memory board is shown in Fig. 6.

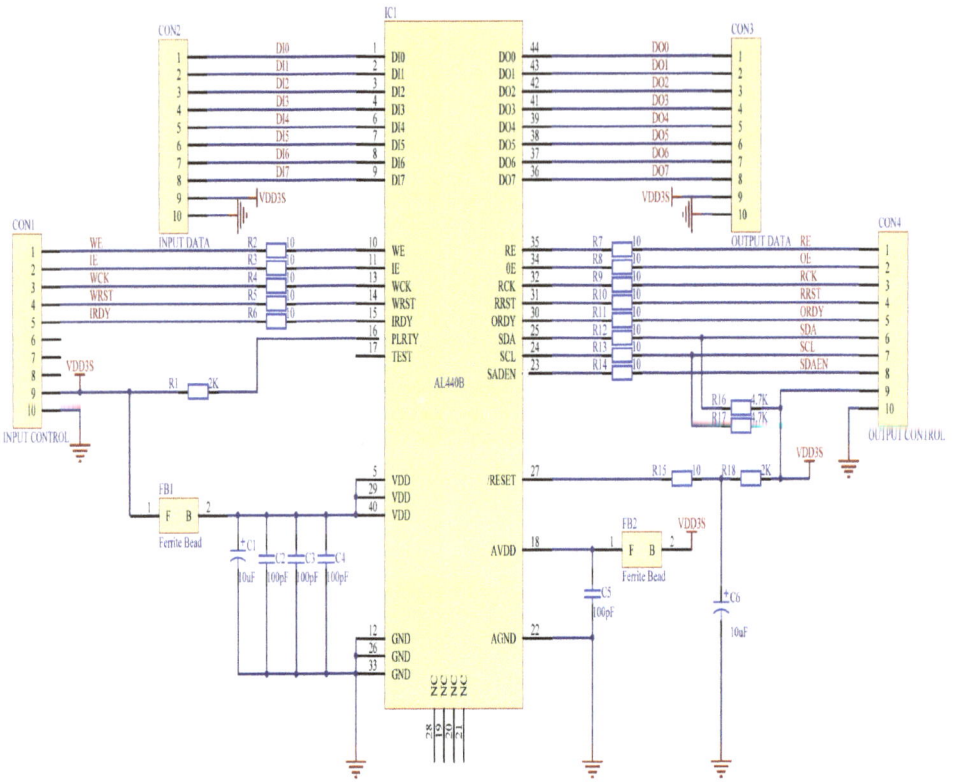

Fig. 6. The circuit of the external memory board

2.5 The process of the power quality monitoring system

The picture of the power quality monitoring system hardware is shown in Fig. 7, consisting of five parts as the followings:

1. The voltage and frequency measurement and fault detection board (ADE7758)
2. Signal conditioning module board (HCPL-788J)
3. ADUC7024 microcontroller board
4. LPC2368 microcontroller board
5. External memory board (AL440B)

From the architecture planned, the 3-phase AC lines connect to ADE7758 and to HCPL-788J for being measured of their voltages and frequencies by the ADE7758 board which is interfaced with the LPC2368 microcontroller board in order to acquire the voltage and frequency values and also to initialize the ADE7758 for detecting fault signal at one time. Then the HCPL-788J attenuates the voltage inputs and isolates the output signals. After that, the output signal is sent to become the signal inputs to the ADUC7024 microcontroller board for sampling waveforms and writing the sampling signals to the external memory board. In the ordinary state, the ADUC7024 microcontroller board will do the sampling and will write the data frames continuously, and in the same time, the LPC2368 microcontroller board will attain the voltage and frequency data from ADE7758, as well.

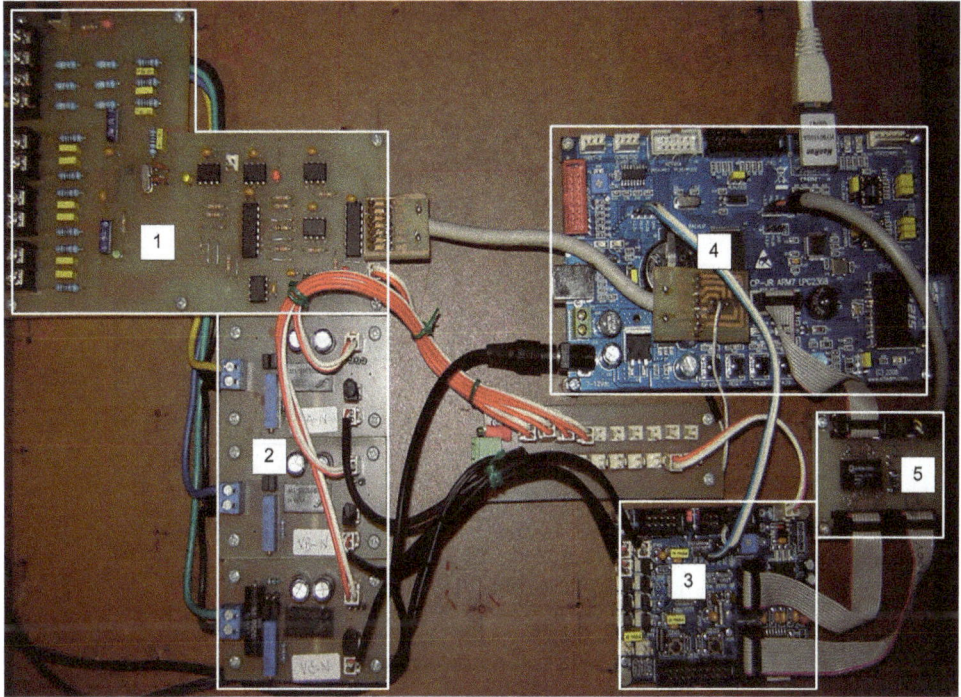

Fig. 7. The power quality monitoring system hardware

If the fault signal which is detected , in case, is an uncommon state, LPC2368 microcontroller board will control the ADUC7024 board to stop sampling and writing process temporarily, will read the sampling signal from the external memory board and will store the signal into SD-CARD. When it comes to the normal condition, both the microcontroller board will go back to their usual tasks as before.

In case of the network that has the connection with Transmission Control Protocol/Internet Protocol (TCP/IP) and User Datagram Protocol (UDP) is established between the LPC2368 and the PC or laptop, we can monitor the status of power quality monitoring system hardware via the network in real time. In addition, the fault data can be downloaded from the SD-CARD and displayed later when it is necessarily needed.

3. Embedded software design

The embedded software design is a flowchart of the two selected microcontrollers: ADUC7024 and LPC2368 used in this chapter. It is to explain the process of the power quality monitoring system based on the embedded system. For this embedded software design, there are the sampling concept, the power quality monitoring concept, data frame and the Ethernet packet structures.

3.1 Sampling concept

The flowchart of the sampling concept shown below is to illustrate how the concept direction works according to the objectives of this chapter.

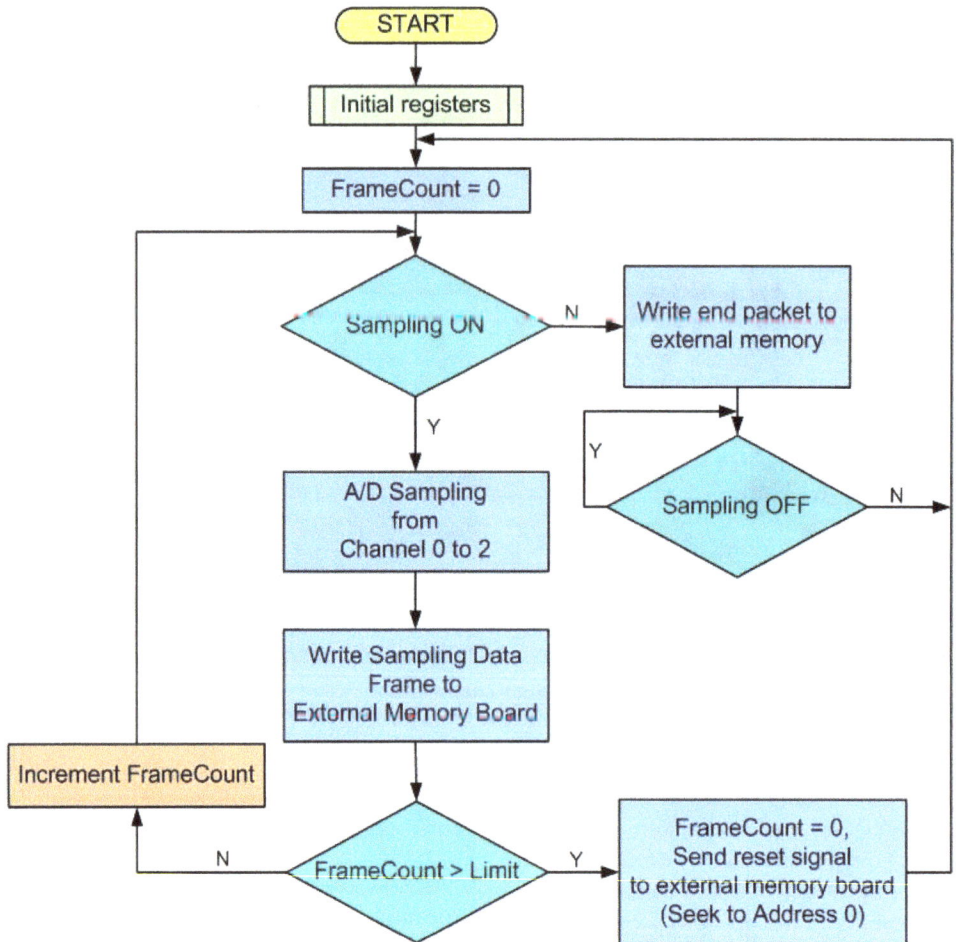

Fig. 8. Flowchart of the sampling concept

To explain the process of the sampling concept flowchart in Fig. 8., it is supposed to follow as the picture shown. After the used registers and utilized variables have been initialized, the procedures will work according to the conditions. That is, if the sampling condition is on operation, the procedure will do the sampling from the analog signals to the digital ones of the 3 channels. And the next procedure is to write the digital sampling data which are packed in the form of data frames to the external memory board. But if the counted data packet from the frame count value is excessive over the maximum limit, it will seek to the beginning of the first address in the external memory. Meanwhile, the data packet which is counted will be cleared. And if the counted data packet does not reach to the limit value, it will return to the sampling condition for rechecking.

In case, the sampling condition is off-operation, it will change the content of the data frame, will write it to the external memory and will return to the sampling condition to be rechecked. The procedures with the two sampling conditions mentioned above are done repeatedly time after time.

3.2 Power quality monitoring concept

It is generally a conceptual method to monitor power quality used with the embedded system of LPC2368 microcontroller board as shown in the flowchart of Fig. 9.

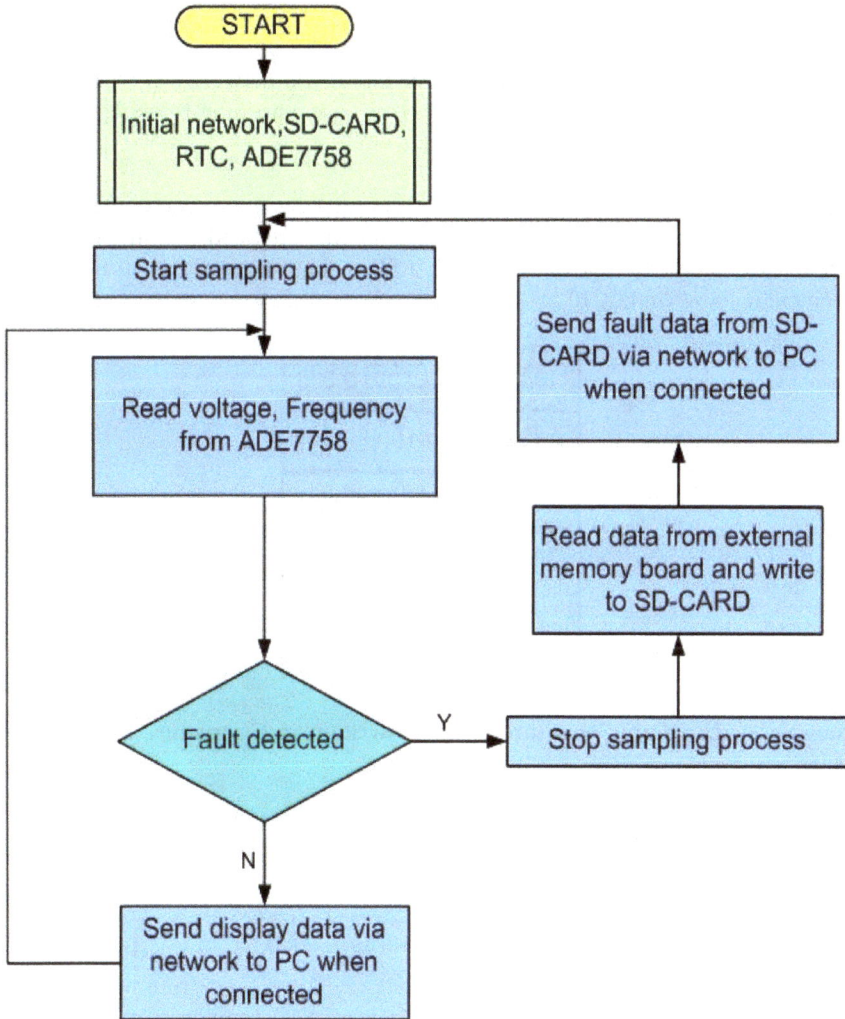

Fig. 9. Flowchart of the power quality monitoring concept

To run the flowchart in order, it must initialize the network chip on LPC2368 microcontroller board, SD-CARD, real-time clock, the RTOS tasks that are modified from A Power Quality Monitoring System for Real-Time Detection of Power Fluctuations (Yingkayun & Premrudeepreechacharn, 2008) and ADE7758 board to get ready for operating its functions. Next, LPC2368 microcontroller sends a control signal to ADUC7024 microcontroller board in order to start doing the sampling process. It reads the voltage and frequency data from ADE7758 board. In case, there is no any fault signals, which are read by LPC2368 microcontroller, the display information of the following values: the data of voltage; frequency; date; time; status and so on, will be sent via network to PC or laptop when it is connected. The LPC2368 microcontroller will take turn to operate its function repeatedly from the start once more. But if ADE7758 board detects the fault signals, it will send the fault signal to LPC2368 microcontroller which stops the sampling process, then, reads the fault data from the external memory board and writes the fault data in SD-CARD for storing. The fault data will be sent to PC or laptop via network when it is connected. LPC2368 microcontroller will start doing the process once more after receiving the next fault signals.

3.3 Data frame

The data frame, sending to the external memory board, is defined with the head byte, the samples of the 3-phase voltages with 12-bits A/D resolution and the tail byte. The data frame content is shown in Fig. 10.

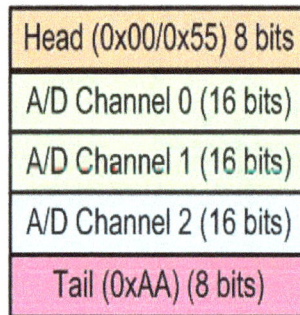

Head (0x00/0x55) 8 bits

A/D Channel 0 (16 bits)

A/D Channel 1 (16 bits)

A/D Channel 2 (16 bits)

Tail (0xAA) (8 bits)

Fig. 10. Data frame

The example of data frame structure which is written in C structure format is shown below:

```
struct  adc_info {  /* Frame structure */
    unsigned char head; /* defined as 0x00 when sampling ON and 0x55
                           when sampling OFF */
    unsigned int samples[3];/* 12 bits Analog to Digital Signal from
                           AC lines */
    unsigned char tail;  /* defined as 0xAA */
};
```

To calculate the floating point voltages from a raw A/D sample, it can be expressed by

$$V_n = \frac{X_n}{(2^{12}-1)} - Offset_n \tag{1}$$

Where V is the floating point voltage, X is the 12-bits raw data from an A/D sample, n is each A/D channel number which varies from 0 to 2, and *Offset* is the zero crossing voltage offset.

3.4 Configuration data

To initialize the power quality monitoring system hardware, board configuration file stored on SD-CARD is needed. The board configuration file is a text file named "board.cfg" and the contents of the configuration file are shown below:

```
ip = 192.168.0.200      //Hardware IP
id = 1                  //Hardware Number
port= 2000              // UDP Port
offset0 = 1.246         //Channel 0 Offset
offset1 = 1.242         //Channel 1 Offset
offset2 = 1.251         //Channel 2 Offset
phase = 3               //Select 1 or 3 Phase System
```

The advantage of using the configuration file from the SD-CARD is convenient to apply by setting its contents when more power quality monitoring boards are used at each node.

3.5 Ethernet packet stream structure

To define the Ethernet packet stream structure sending over the Ethernet network, it is consisted of fault data and electrical measurement information, which is in the place of the TCP packet and will be packed to the UDP packets, then it is transmitted to PC or laptop. The structure of Ethernet packet stream is shown below:

```
struct  display_info { /* Display structure  type*/
    char type;              /* Type of information defined as 0 */
    char line1 [42];        /* Data buffer for the 3-phase voltages */
    char line2 [42];    /* Data buffer for the status of the AC Lines */
    char line3 [42];        /*Data buffer for the frequency values for the
                            3-phase system. */
    char line4 [42];        /* Data buffer for current date and time */
};
struct fault_info {
    char type;              /* Type of information defined as 1 */
    char buffer[200];       /* Data buffer for fault data */
    char status;            /* Status of the data buffer */
};
```

The LPC2368 microcontroller board uses UDP protocol packet to transmit data to a remote computer. UDP protocol is one of protocols in the TCP/IP protocol suite that is used in the place of TCP when a reliable delivery is not required. There is less processing of UDP packets than the one of TCP. UDP protocol is widely used for streaming audio and video, voice over IP (VoIP) and video conferences, because there is no time to retransmit erroneous or dropped packets. The UDP packet which is in Ethernet frame is shown in Fig. 11.

Ethernet Frame

Ethernet Header	IP Header	UDP Header	Data (Ethernet Stream Structure)	Ethernet Trailer

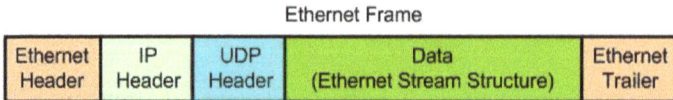

Fig. 11. UDP within Ethernet frame

The structures of the Ethernet stream can be illustrated in C programming language are as followings:

```
struct Frame_Header
{
  char Dest_MAC[6];
  char Src_MAC[6];
  char Ethernet_Type[2];
};
struct IP_Header
{
   char IP_Version;
   char TypeOfService;
   unsigned int DataSize;
   unsigned int Identification;
   unsigned int Flag;
   char TimeToLive;
   char Protocol;  //UDP=0x11
   unsigned int CheckSum;
   char SrcAddr[4];
   char DestAddr[4];
};
struct UDP_Header
{
   unsigned int SrcPort;
   unsigned int DestPort;
   unsigned int DataGramLength;
   unsigned int CheckSum;
};
struct Ethernet_Stream0
{
   struct Frame_Header frame;
   struct IP_Header ip;
   struct UDP_Header udp;
   struct display_info dp;
   char ZeroByte;
};
struct Ethernet_Stream1
{
   struct Frame_Header frame;
   struct IP_Header ip;
   struct UDP_Header udp;
   struct fault_info fault;
   char ZeroByte;
};
```

4. Application software

Application software is developed and written in Delphi 7 in order to control and to receive informational data from power quality monitoring system hardware and to save the data to the computer. Actually, this application gets the display information from the hardware mentioned above and then displays on the PC or laptop screen via the Ethernet network. The application can set date and time and can also receive the fault data from the referred hardware via the Ethernet network. The ordinary state and the uncommon one of the power lines are displayed by application software on PC or laptop shown in Fig. 12 and Fig. 13 respectively.

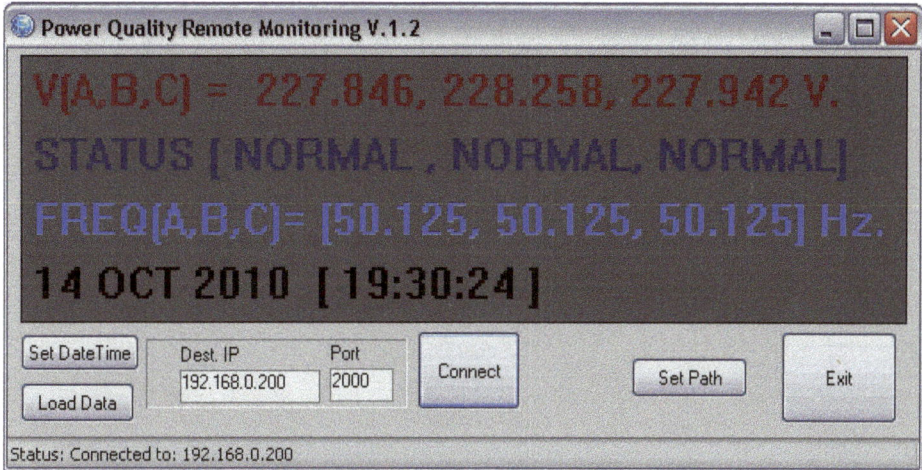

Fig. 12. Display window in ordinary state

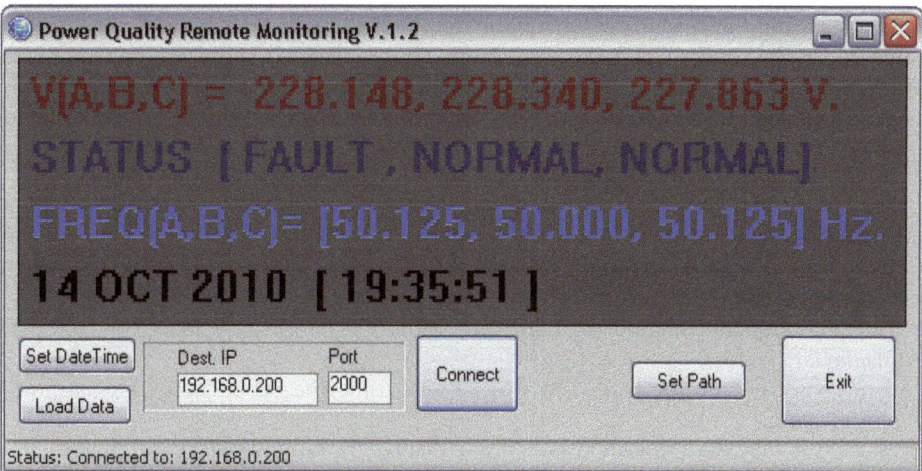

Fig. 13. Display window in uncommon state

And the researchers have developed the application software for displaying the fault signal which is sent from the monitoring hardware to PC or laptop in order to illustrate the fault data. This developed application software can either save in picture file or print out to take the data under the considerable analysis of the cause in the faults at later time. The developed application software is shown in Fig. 14.

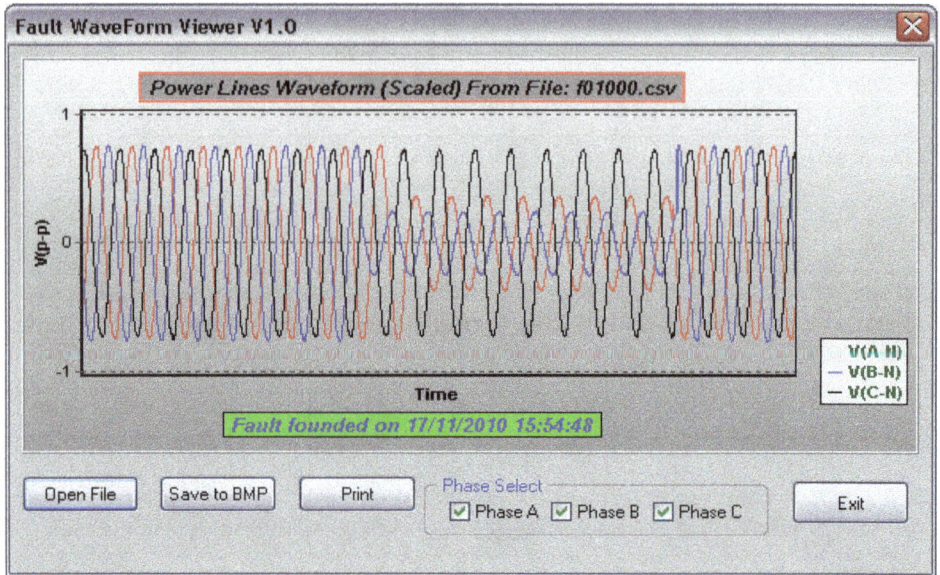

Fig. 14. Application software for waveform display

5. Experimental results

In order to communicate to the power quality monitoring hardware, application software is created to display and to plot the waveform. The application software connects to the monitoring hardware via the Ethernet network. It both gets the display information and saves the fault data to PC or laptop. It also shows the fault waveforms on the PC or laptop screen. The communication procedure connecting between a monitoring hardware to PC or laptop is to get the necessary information which is executed in the following steps:

1. A user sends a request signal with the application program to the power quality monitoring hardware.
2. The power quality monitoring hardware receives the request signal and establishes the connection to PC or laptop.
3. The power quality monitoring hardware sends the information to PC or laptop for displaying.
4. If any fault occurs, the fault data will be saved into the PC or laptop storage in the specific path.
5. A user can take the application software to display the fault waveform for investigating the problem in power lines from the saving path above.

When the connection procedures have been already established between the hardware and the computer in consequence, the user can monitor the voltage, the frequency and the fault in power lines on the PC or laptop screen from the remote site. Additionally, the user can also open the data file which is saved in the computer.

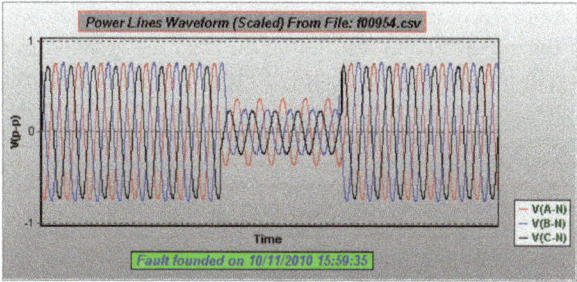

(a)

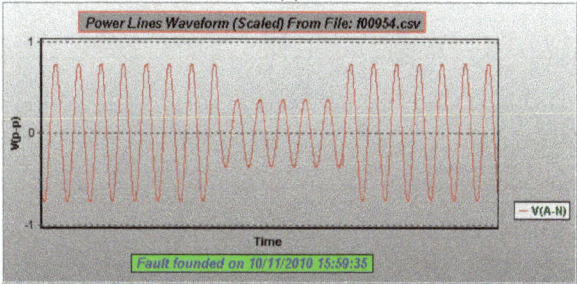

(b)

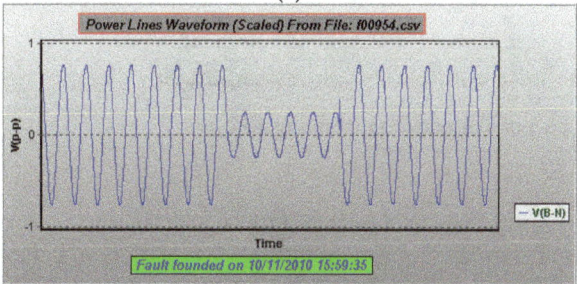

(c)

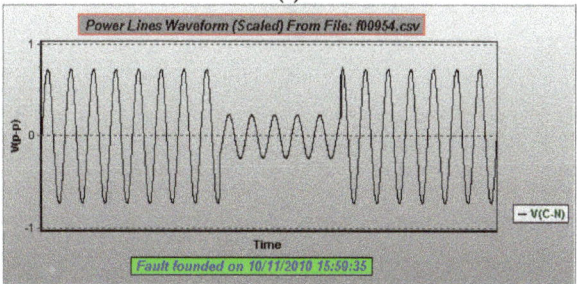

(d)

Fig. 15. (a) Voltage sags on phase A,B and C, (b) Voltage sag on phase A, (c) voltage sag on phase B and (d) voltage sag on phase C

For more advantage, the communication of the power quality monitoring hardware and PC or laptop is not only limited with the only one hardware but also connected to other hardwares by executing more developed application programs that are shown in Fig. 12 and Fig. 13. The examples of the experimental result, that the data file is saved in the form of bitmap file (.BMP format) by the application program of Fig. 14, is illustrated in Fig. 15, Fig. 16 and Fig. 17.

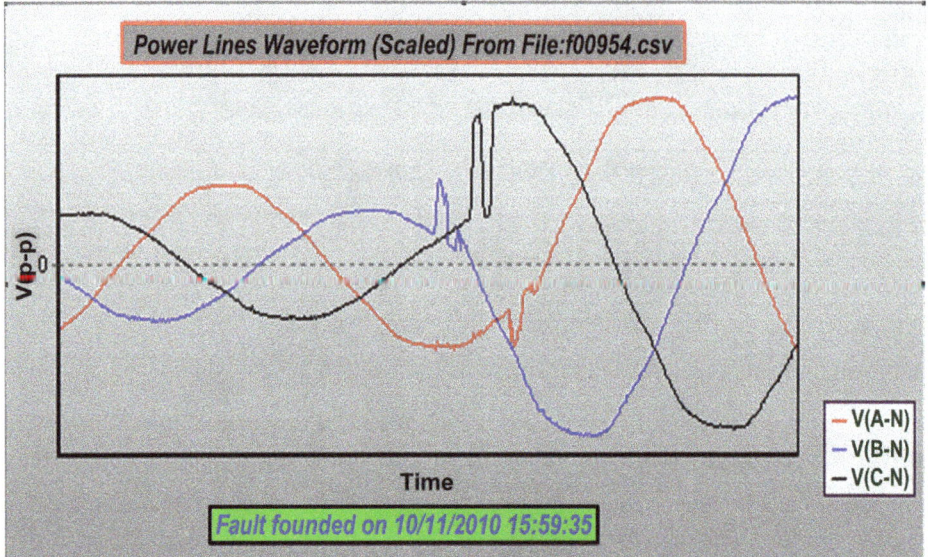

Fig. 16. Zoomed signals of voltage sags from Fig.15(a)

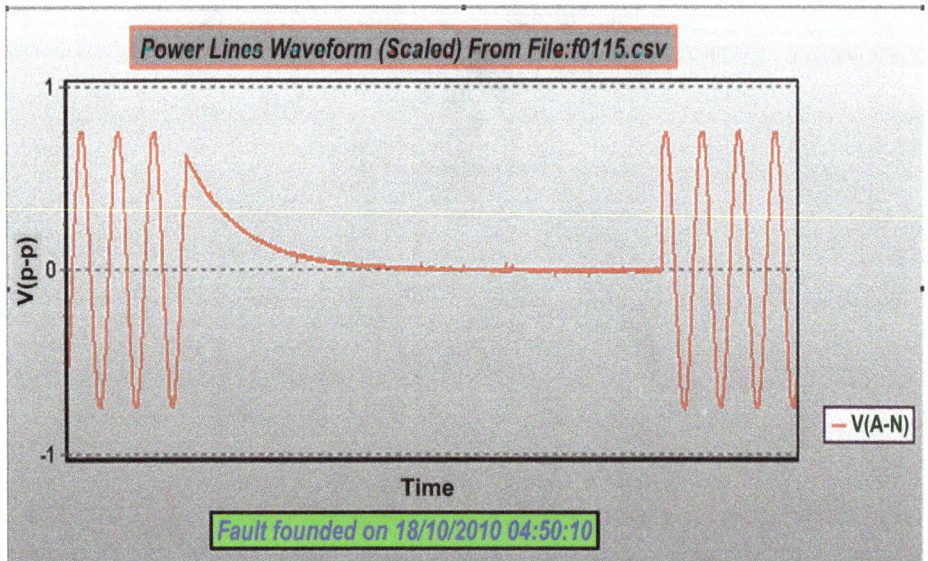

Fig. 17. Interruption on phase A (measured on a single phase system)

Fig. 15(a) is the picture of sags for all 3 phases. The experimental results appeared in Fig. 15(b), (c) and (d) are the examples of sag in phase A, B and C chronically from Fig. 15(a). It is to separate the signal for testing each one in each phase that is easily studying.

Fig.16 is the zoomed picture from Fig. 15(a) to show the detail characteristics of the fault signals in each phase.

Another experiment of this chapter is applied to detect the fault on a single phase system. From Fig. 17 shown above is an example of the interruption for a short time.

6. Conclusion and future work

A power quality monitoring system via the Ethernet network based on the embedded system has been proposed in this chapter in order to monitor the power quality in case of faults detection and also to measure voltage and frequency in power lines. ADUC7024 and LPC2368 of ARM7 microcontroller are selected to apply in the power quality monitoring system for not only detecting the fault signals that cause any problems in either the system or the end user equipment but also reading and writing them in real time of power fluctuation. Moreover, the fault signal data can be sent and stored in SD-CARD to display later on the screen of PC or laptop at the site place. However, the users can download and analyze the fault signal data which have already sent and stored in SD-CARD via the Ethernet network using TCP/IP and UPD protocol at some other time when of necessity needed.

For future work, the researchers tend to substitute ARM7 with ARM9 in order to monitor power quality and to detect the transient in power lines. In any case, the researchers have always concerned with the same primitive ideas and objectives.

7. Acknowledgements

The authors gratefully acknowledge to National Science and Technology Development Agency (NSTDA), Ministry of Science and Technology of Thailand, Thailand Research Fund (TRF), and Provincial Electricity Authority (PEA) for supports.

8. References

Auler, L.F. & d'Amore, R. (2003). Power Quality Monitoring and Control using Ethernet Networks, *Proceedings of 10th International Conference on Harmonics and Quality of Power*, pp. 208-213, ISBN 0-7803-7671-4, Rio de Janeiro, Brazil, October 6-9, 2002

Auler, L.F. & d' Amore, R. (2009). Power Quality Monitoring Controlled Through Low-Cost Modules, *IEEE Transactions on Instrumentation and Measurement*, Vol.58, No.3, (March 2009), pp. 557-562, ISSN 0018-9456

Baggini, A. B. (2008). *Handbook of POWER QUALITY*, WILEY, ISBN 978-0-470-06561-7, Wiltshire, Great Britain

Batista, J.; Alfonso, J.L. & Martins, J.S. (2004). Low-Cost Power Quality Monitor based on a PC, *Proceeding of ISIE'03 IEEE International Symposium on Industrial Electronics*, pp. 323-328, ISBN 0-7803-7912-8, Rio de Janeiro, Brazil, June 9-11, 2003

Dugan, R.C.; McGranaha, M.F.; Santoso, S. & Beaty, H. W. (2002). *Electrical Power Systems Quality*, McGraw-Hill, ISBN 0-07-138622-X, New York, USA

Hong, D.; Lee J. & Choi, J. (2006). Power Quality Monitoring System using Power Line Communication, *Proceeding of ICICS 2005 Fifth International Conference on Information, Communications and Signal Processing*, pp. 931-935, ISBN 0-7803-9283-3, Bangkok, Thailand, December 6-9, 2005

Rahim bin Abdullah, A. & Zuri bin Sha'ameri, A. (2005). Real-Time Power Quality Monitoring System Based on TMS320CV5416 DSP Processor, *Proceeding of PEDS 2005 International Conference on Power Electronics and Drives Systems*, pp. 1668-1672, ISBN 0-7803-9296-5, Kuala Lumpur, Malaysia, November 28 – December 1, 2005

Salem, M.E.; Mohamed, A.; Samad, S.A. & Mohamed, R. (2006). Development of a DSP-Based Power Quality Monitoring Instrument for Real-Time Detection of Power Disturbances, *Proceedings of PEDS 2005 International Conference on Power Electronics and Drives Systems*, pp. 304-307, ISBN 0-7803-9296-5, Kuala Lumpur, Malaysia, November 28 – December 1, 2005

So A.; Tse, N.; Chan W.L. & Lai, L.L. (2000). A Low-Cost Power Quality Meter for Utility and Consumer Assessments, *Proceeding of IEEE International Conference on Electric Utility Deregulation and Restructuring and Power Technologies*, pp. 96-100, ISBN 0-7803-5902-X, City University London, UK, April 4-7, 2000

Yang, G.H. & Wen, B.Y. (2006). A Device for Power Quality Monitoring Based on ARM and DSP, *Proceedings of IEIEA 2006 The 1st IEEE Conference on Industrial Electronics and Applications*, pp. 1-5, ISBN 0-7803-9513-1, Marina Mandarin Hotel, Singapore, May 24-26, 2006

Yingkayun, K. & Premrudeepreechacharn S. (2009). A Power Quality Monitoring System for Real-Time Detection of Power Fluctuations, *Proceeding of NAPS'08 The 40th North American Power Symposium*, pp. 1-5, ISBN 978-1-4244-4283-6, Calgary, Canada, September 28-30, 2008

Yingkayun, K.; Premrudeepreechacharn S. & Oranpiroj, K. (2009). A Power Quality Monitoring for Real-Time Fault Detection, *Proceedings of ISIE 2009 IEEE International Symposium on Industrial Electronics*, pp. 1846-1851, ISBN 978-1-4244-4347-5, Seoul, Korea, July 5-8, 2009

9

Understanding Power Quality Based FACTS Using Interactive Educational GUI Matlab Package

Belkacem Mahdad and K. Srairi
Department of Electrical Engineering, Biskra University
Algeria

1. Introduction

The electricity is invisible and the complexity of mathematical models deviate the graduate students attention from well understanding the underlying main concepts. Interactive educational power system software has become a fundamental teaching tool because it helps in particular the undergraduate students to assimilate theoretical issues and complex models analysis through flexible graphic visualization of data inputs and the results (Abur et al., 2000), (Milano, F., 2005). From the educational point of view software developed for educational purposes should be flexible and interactive, easy to use and reliable. In particular, software for power system education should contain a user interface not only to allow graduate student to analyse and understand the physical phenomena, but also to improve the existing models and algorithms (Mahdad, B., 2010).

Flexible AC Transmission Systems (FACTS) philosophy was first introduced by Hingorani (Hingorani N. G., and Gyugyi L, 1999) from the Electric power research institute (EPRI) in the USA in 1988, although the power electronic controlled devices had been used in the transmission network for many years before that. The objective of FACTS devices is to bring a system under control and to transmit power as ordered by the control centers, it also allows increasing the usable transmission capacity to its thermal limits. With FACTS devices we can control the phase angle, the voltage magnitude at chosen buses and/or line impedances.

The avantages of the graphical user interface tool proposed lie in the quick and the dynamic interpretation of the results and the interactive visual communication between users and computer solution processes. The physical and technical phenomena and data of the power flow, and the impact of different FACTS devices installed in a practical network can be easily understood if the results are displayed in the graphic windows rather than numerical tabular forms (Mahdad, 2010).

The application programs in this tool include power flow calculation based Newton-Raphson algorithm, integration and control of different FACTS devices, the economic dispatch based conventional methods and global optimization methods like Parallel Genetic Algorithm (PGA), and Particle Swarm Optimization (PSO). In the literature many educational Graphical tools for power system study and analysis developed for the purpose of the power system education and training (Milano et al., 2005).

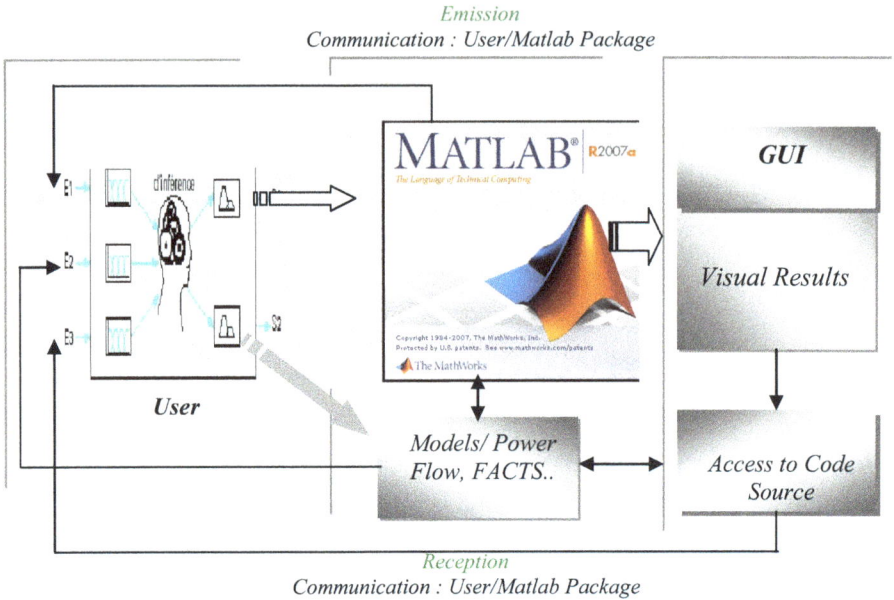

Fig. 1. Strategy for understanding power quality based FACTS technology

To carry out comprehensive studies on FACTS devices, to understand the basic principle of FACTS devices, and to determine the role that FACTS technology may play in improving power quality, it is mandatory to have an interactive educational tool using graphic user interface based Matlab, this is the main object of this chapter. This chapter is limited to show how the simplified software package developed works by showing the effects of the introduction of different FACTS devices like shunt Controllers (SVC, STATCOM), series Controllers (TCSC, SSSC) and the hybrid Controllers (UPFC) on a practical network under normal and abnormal situation. Fig.1 shows the strategy for understanding power quality based FACTS technology using an interactive graphical user interface (GUI).

2. Basic principles of power flow control

To facilitate the understanding of the basic principle of power flow control and to introduce the basic ideas behind the different type of FACTS controllers, the simple model shown in Fig. 2 is used (Mahdad, B., 2010). The sending and receiving end voltages are assumed to be fixed. The sending and receiving ends are connected by an equivalent reactance, assuming that the resistance of high voltage transmission lines is very small. The receiving end is modeled as an infinite bus with a fixed angle of 0°.

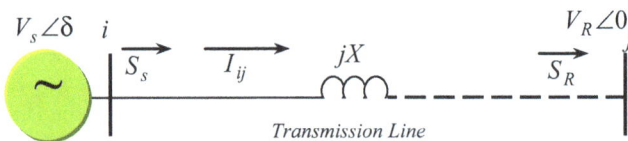

Fig. 2. Model for calculation of real and reactive power flow control

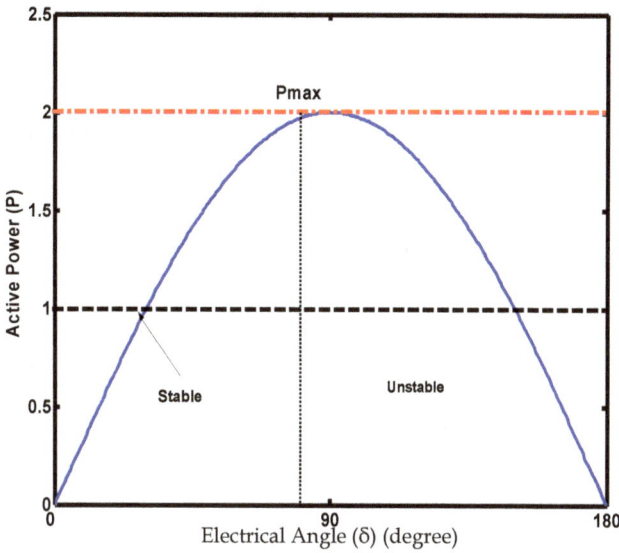

Fig. 3. Power angle curve

Complex, active and reactive power flows in this transmission system are defined, respectively, as follows:

$$S_R = P_R + jQ_R = V_R I^* \qquad (1)$$

$$P_R = \frac{V_S V_R}{X} \sin \delta = P_{max} \sin \delta \qquad (2)$$

$$Q_R = \frac{V_S V_R \cos \delta - V_R^2}{X} \qquad (3)$$

Similarly, for the sending end:

$$P_S = \frac{V_S V_R}{X} \sin \delta = P_{max} \sin \delta \qquad (4)$$

$$Q_S = \frac{V_S^2 - V_S V_R \cos \delta}{X} \qquad (5)$$

Where V_S and V_R are the magnitudes of sending and receiving end voltages, respectively, while δ is the phase-shift between sending and receiving end voltages. Fig. 3 shows the evolution of the active power delivered.

It's clear from the demonstrated equations, that the active and reactive power in a transmission line depend on the voltage magnitudes and phase angles at the sending and receiving ends as well as line impedance.

2.1 Example of power flow control

The concepts behind FACTS controller is to enable the control of three parameters which are:

1. Voltage magnitude (V)
2. Phase angle (δ)
3. And transmission line reactance (X) in real-time and, thus vary the transmitted power according to system condition.

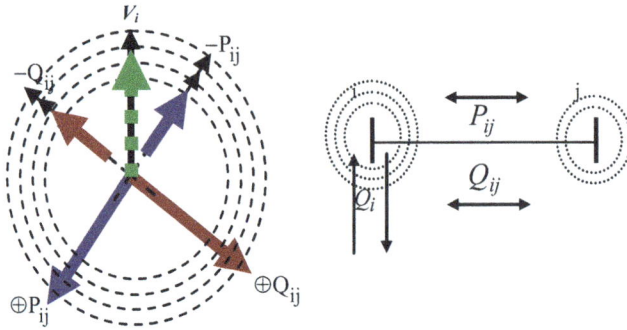

Fig. 4. Three vector control structure (Voltage control -Active power control - Reactive power control) based FACTS technology

The ability to control power rapidly, within appropriately defined boundaries, can increase transient and dynamic stability as well as the damping of the system.

The following section illustrate the basic principle of the FACTS Controllers designed to be integrated in a practical network. Fif. 4 shows the three mode control related to FACTS compensators.

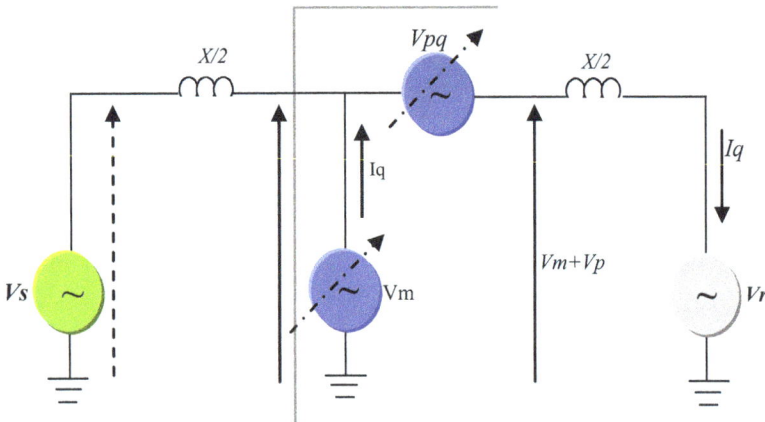

Fig. 5. Generalized schematic of power flow controller

The simplified genralized power flow controller consists of two controllable elements, a voltage source (V_{pq}) inserted in series with the line, and a current source (I_q), connected in shunt with the line at the midpoint. The four classical cases of power transmission are considered:

1. Without line compensation,
2. With series compensation,
3. With shunt compensation,
4. and with phase angle control.

The different operation mode can be obtained by appropriately specifying V_{pq} and I_q in the generalized schematic power flow controller is shown in Fig. 5.

Case 1 Power flow controller is off. Then the power transmitted between the sending and receiving end generators can be expressed by:

$$P_1 = \frac{V^2}{X_l}\sin(\delta) \tag{6}$$

Where δ is the angle between the sending and receiving-end voltage phasors.

Case 2 Assume that $I_q = 0$ and $V_{pq} = -jkXI$, the voltage source acts at the fundamental frequency precisely as a series compensating capacitor. The degree of series compensation is defined by coefficient k ($0 \le k \le 1$), the relationship between P and δ becomes:

$$P_2 = \frac{V^2}{X(1-k)}\sin(\delta) \tag{7}$$

Case 3 The reactive current source acts like an ideal shunt compensator which segments the transmission lines into independent parts, each with an impedance of $X/2$, by generating the reactive power necessary to keep the mid-point voltage constant, independently of angle δ, for this case the relationship between P and δ becomes:

$$P_3 = \frac{2V^2}{X}\sin(\frac{\delta}{2}) \tag{8}$$

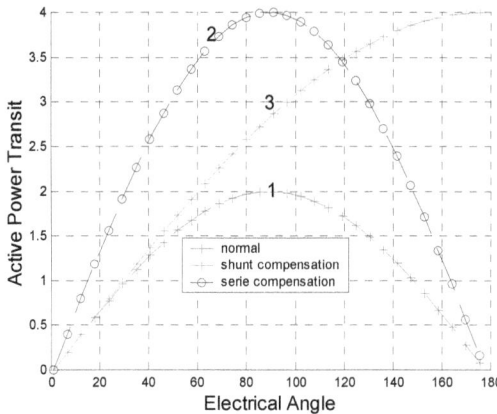

Fig. 6. Active power transit with different compensation types

Case 4 The basic idea behind the phase shifter is to keep the transmitted power at a desired level independently of angle δ in a predetermined operating range. Thus for example, the power can be kept at its peak value after angle δ exceeds π/2, by controlling the amplitude of quadrature voltage V_{pq} . Fig. 6 shows the evolution of active power transit based different compensation types.

$$P_4 = \frac{V^2}{X}\sin(\delta + \alpha) \tag{9}$$

2.2 Role of FACTS devices in power system operation and control

To further understand the strategy of FACTS devices in power system operation and control, consider a very simplified case in which generators at two different regions are sending power to a load centre through a network consisting of three lines.

Fig. 7 shows the topology of simple electrical network, suppose the lines 1-2, 1-3 and 2-3 have continuous ratings of 1000MW, 2000MW, and 1250MW, respectively, and have emergency ratings of twice those numbers for a sufficient length of time to allow rescheduling of power in case of loss of one of these lines (Hingorani, N. G., and Gyugyi L, 1999).

For the impedances shown, the maximum power flow for the three lines are 600, 1600, and 1400, respectively, as shown in Fig. 7, such a situation would overload line 2-3 (loaded 1600 MW for its continuous rating of 1250 MW), and there for generation would have to be decreased at unit 2, and increased at unit 1, in order to meet the load without overloading line 2-3. The following simplified studies cases demonstrate the main objective of integration of FACTS technology in a practical power system to enhance power system security.

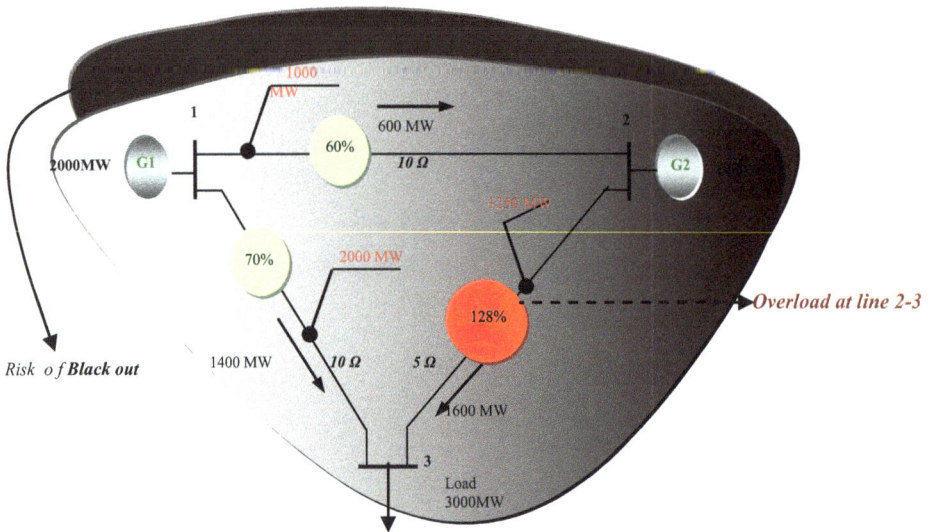

Fig. 7. Topology of the electrical network 3-bus with technical characteristics without dynamic compensators

Case 1: Capacitive Series Compensation at line 1-3

If the dynamic series FACTS Controller (type capacitive)installed at line 1-3 adjusted to deliver a capacitive reactance, it decreases the line's impedance from 10Ω to 4.9919Ω, so that power flows through the lines 1-2, 1-3, and 2-3 will be 250 MW, and 1750 MW, respectively. Fig. 8 illustrates the per cent loading of lines. It is clear that if the series capacitor is adjustable, then other power flow levels may be realized in accordance with the ownership, contract, thermal limitations, transmission losses, and wide range of load and generation schedules. Fig. 8 shows clearly the effect of series capacitive compensation to control the active power flow with another degree of compensation ($X_C = 6\,\Omega$).

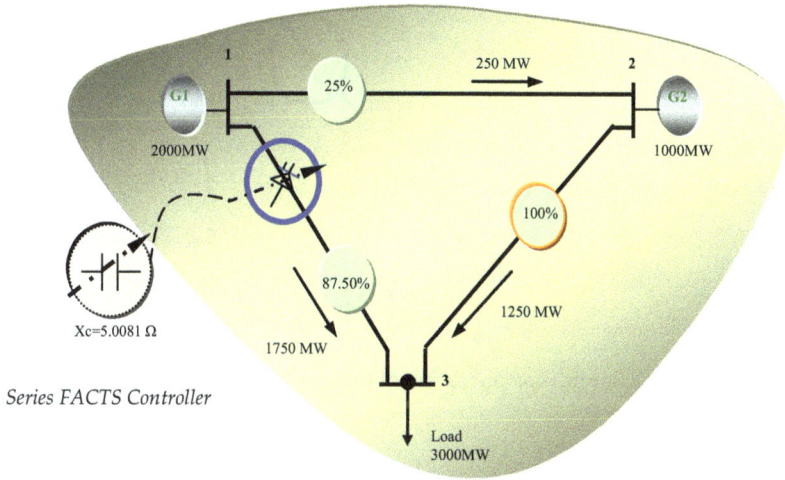

Fig. 8. Load flow solution with consideration of dynamic compensators: Case1

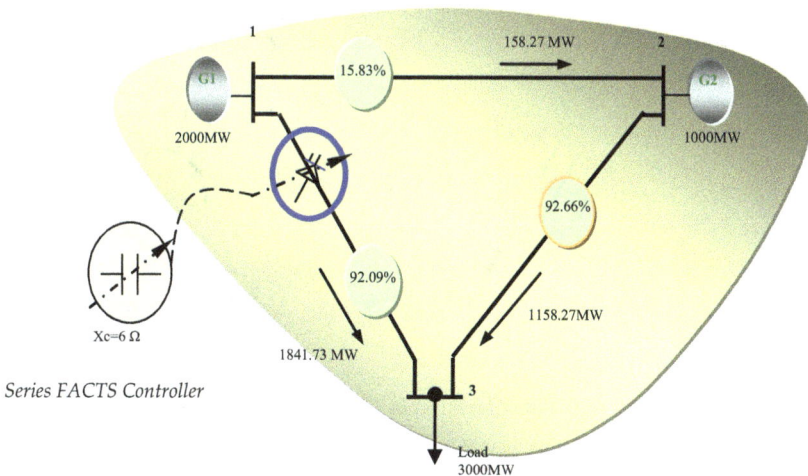

Fig. 9. Load flow solution with consideration of dynamic compensators: Case1

Case 2: Inductive Series Compensation at line 2-3

If the dynamic series FACTS Controller (type inductive) installed at line 2-3 adjusted dynamically to deliver an inductive reactance, it increase the line's impedance from 5 Ω to 12.1Ω, so that power flows through the lines 1-2, 1-3, and 2-3 will be 248.22 MW, 1751.78 MW and 1248.22 MW, respectively.

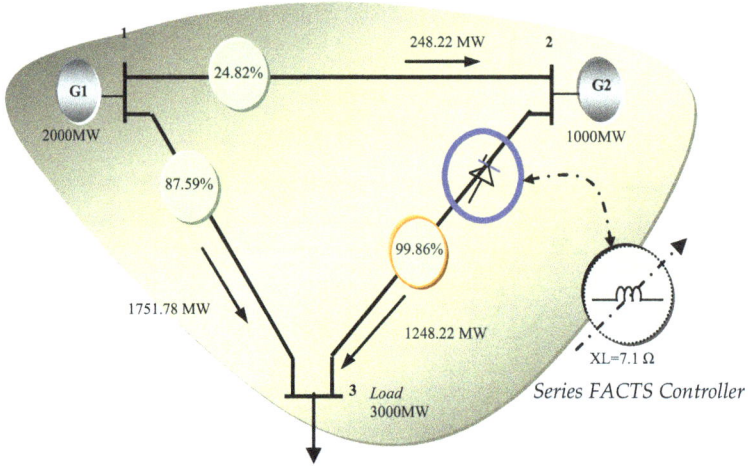

Fig. 10. Load flow solution with consideration of dynamic compensators: Case2

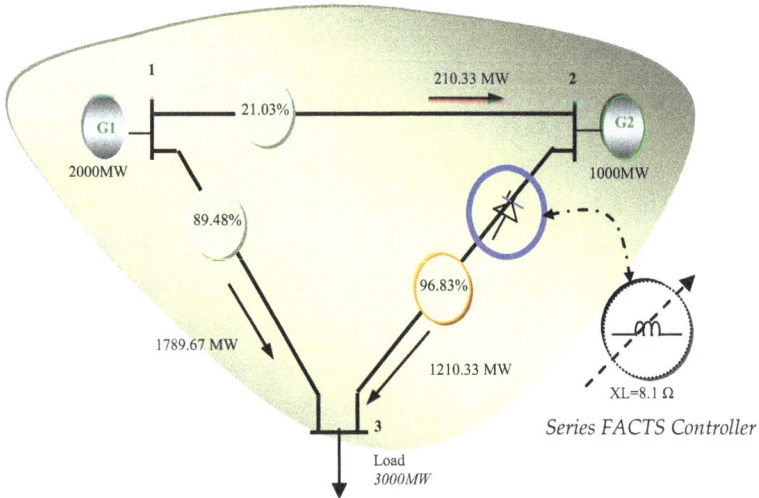

Fig. 11. Load flow solution with consideration of dynamic compensators: Case2

It is clear from Fig. 9 and Fig. 10, that if the series inductance is adjustable, then other power flow levels may be realized in accordance with the ownership, contract, thermal limitations, transmission losses, and wide range of load and generation schedules.

As we can see from simulation results depicted in different Figures; the location of series FACTS devices affect significtly the perfermances of power system in term of lines loading and total power losses.

2.3 Basic types of FACTS controllers

In general, FACTS Controllers can be classified into three categories (Hingorani, NG., and Gyugyi L, 1999) :

* Series Controllers
* Shunt Controllers
* Combined series-shunt Controllers

a. Series Controllers

In Fig. 12 the series controllers could be variable impedance, such as capacitor, reactor, etc., in principle; all series controllers inject voltage in series with the line. Even variable impedance multiplied by the current flow through it, represents an injected series voltage in the line. As long as the voltage is in phase quadrature with the line current, the series Controller only supplies or consumes variable reactive.

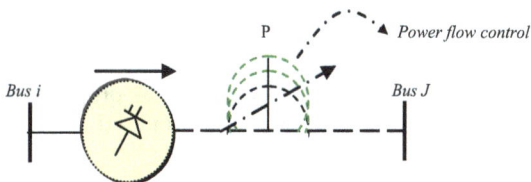

Fig. 12. Series Controller

b. Shunt Controllers

In Fig. 13 as in the case of series Controllers, the shunt controllers may be variable impedance, variable source, or a combinaison of these.

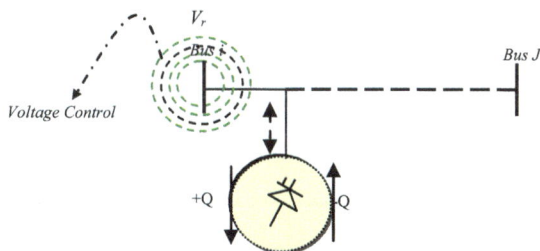

Fig. 13. Shunt Controller

In principle, all shunt controllers inject current into the system at the point of connection. Even a variable of shunt impedance connected to the line voltage causes a variable current flow and hence represents injection of current into the line (Mahdad, 2010).

c. Hybrid Controllers (Combined series-shunt)

This could be a combination of separate shunt and series compensators, which are controlled in coordinated manner, or a unified power flow with series and shunt elements.

In Fig. 14 combined shunt and series controllers inject current into the system with the shunt part of the controller and voltage in series in the line with the series part of the controller. However, when shunt and series controllers are unified, there can be a real power exchange between the series and shunt controllers via the power link (Achat et al., 2004).

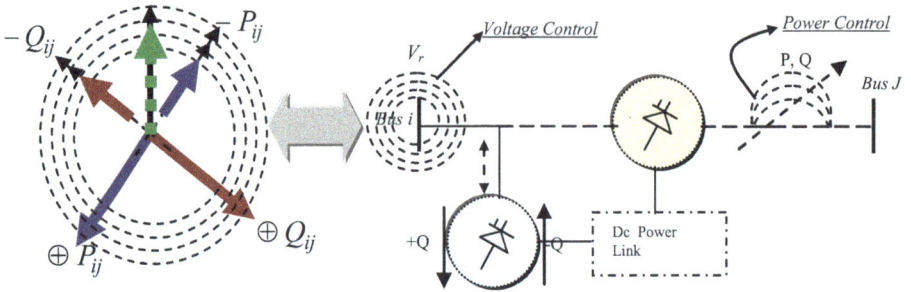

Fig. 14. Unified series-shunt Controller

3. FACTS modeling

Since their apparition, many models of FACTS devices are proposed by researchers to improve the power quality delivered to consumer, the proposed models are integrated in the standard power flow problem, and to the optimal power flow problem. The objective of this section is to investigate the integration of many types of FACTS controllers (shunt, series, and hybrid Controllers) in a practical electrical network to enhance the power quality.

3.1 Static VAR Compensator (SVC)

The steady-state model proposed by Acha et al. (Achat et al., 2004) is used here to incorporate the SVC on the standard power flow problems based Newton Raphson. This

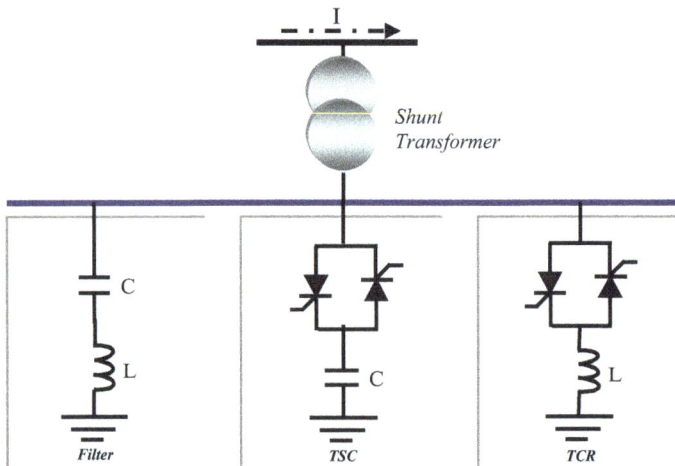

Fig. 15. Static var Compensator (SVC)

model is based on representing the controller as a variable impedance, assuming an SVC configuration with a fixed capacitor (FC) and Thyristor-controlled reactor (TCR) as depicted in Fig. 15, the controlling element is the Thyristor valve. The thyristors are fired symmetrically, in an angle control range of 90 to 180 with respect to the capacitor (inductor) voltage. Fig. 16 shows the two SVC models basic representation.

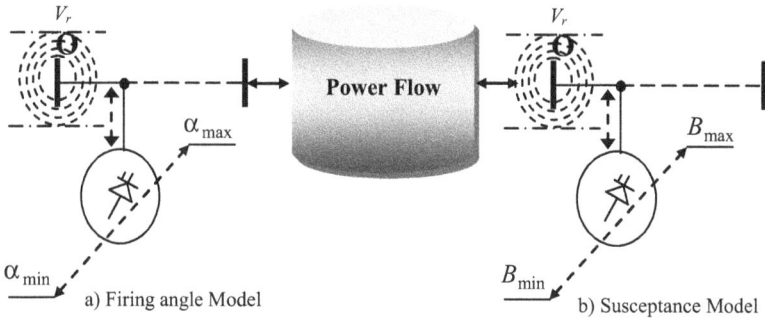

Fig. 16. Two SVC models representation

3.2 Unified Power Flow Controller (UPFC)

An equivalent circuit of the UPFC as shown in Fig. 17 can be derived based on the operation principle of the UPFC (Achat et al., 2004) , (Mahdad et al., 2005).

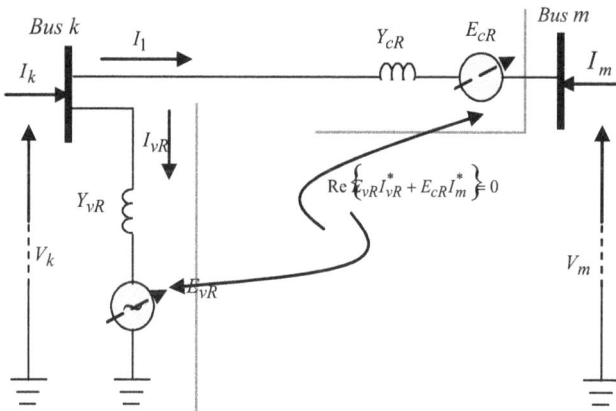

Fig. 17. Equivalent circuit based on solid state voltages sources

The UPFC equivalent circuit described in Fig. 17 is represented by the following voltage sources:

$$E_{vR} = V_{vR}\left(\cos\left(\delta_{vR}\right) + j\sin\left(\delta_{vR}\right)\right) \tag{10}$$

$$E_{cR} = V_{cR}\left(\cos\left(\delta_{cR}\right) + j\sin\left(\delta_{cR}\right)\right) \tag{11}$$

Where V_{vR} and V_{cR} are the controllable magnitude,

$V_{vR}^{\min} \le V_{vR} \le V_{vR}^{\max}$, and phase angle,

$0 \le \delta_{vR} \le 2\pi$ of the voltage source representing the shunt converter. The magnitude V_{cR} and phase angle δ_{cR} of the voltage source representing the series converter are controlled between limits:ij $V_{cR}^{\min} \le V_{cR} \le V_{cR}^{\max}$, and $0 \le \delta_{cR} \le 2\pi$.

3.3 Thyristor Controlled Reactor (TCSC)

The TCSC power flow model presented in this section is based on the simple concept of a variable series reactance, the value of which is adjusted automatically to constrain the power flow across the branch to a desired value.

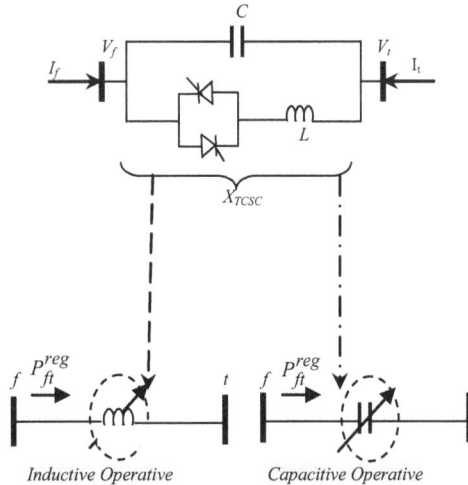

Fig. 18. Principle of thyristor controlled series capacitor (TCSC)

The amount of reactance is determined efficiently using Newton's method. The changing reactance shown in Fig. 18 represents the equivalent reactance of all the series connected modules making up the TCSC, when operating in either the inductive and capacitive region. The equivalent reactance of line X_{ij} is defined as:

$$X_{ij} = X_{line} + X_{TCSC} \tag{12}$$

Where, X_{line} is the transmission line reactance, and X_{TCSC} is the TCSC reactance. The level of the applied compensation of the practical TCSC usually between 20% inductive and 80% capacitive.

4. Understanding power quality based FACTS controllers using FACTS Simulator (SimFACTS Power Flow package)

The advantages of the proposed graphical user interface tool lie in the quick and the dynamic interpretation of the results and the interactive visual communication between users and computer solution processes. The physical and technical phenomena and data of the power flow, and the impact of different FACTS devices installed in a practical network can be easily understood if the results are displayed in the graphic windows rather than numerical tabular forms. Fig. 19 illustrates the components of the proposed strategy based FACTS technology. The SimFACTS tool includes the following application programs:

- Power flow calculation based Newton-Raphson algorithm
- Understanding power quality based FACTS devices
- Voltage Stability based continuation power flow (CPF)
- Economic dispatch based conventional methods and global optimization methods like Genetic Algorithm (GA), and Particle Swarm Optimization (PSO).

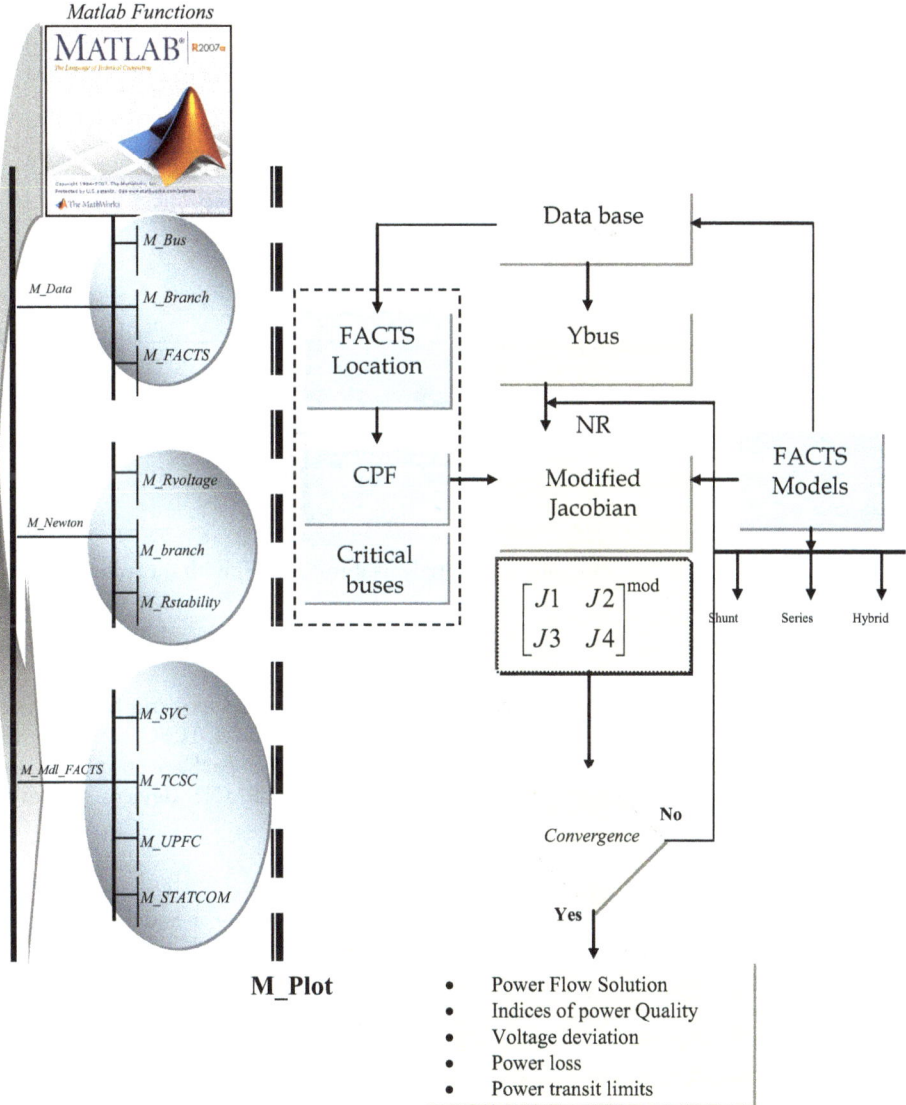

Fig. 19. Flowchart of the proposed basic SimFACTS

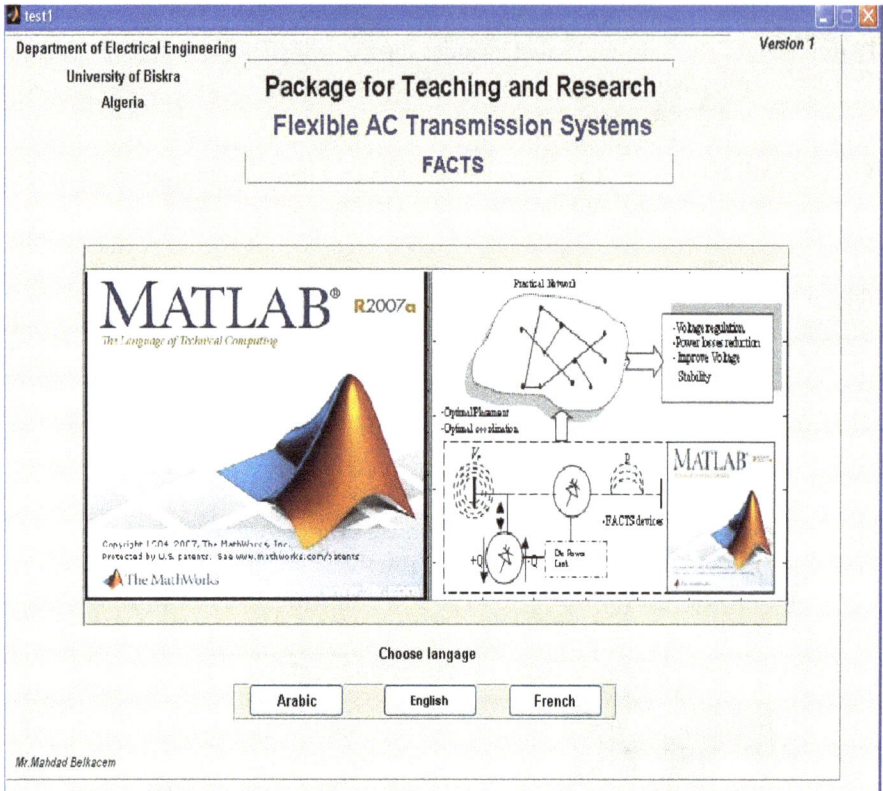

Fig. 20. The package for FACTS modelling and analysis (SimFACTS) with three languages: Arabic, English and French

In the literature many educational Graphical tools for power system study and analysis developed for the purpose of the power system education and training ().
This section reveals how the software package developed works by showing the effects of the introduction of different FACTS devices like the SVC, STATCOM, TCSC, SSSC and the UPFC Controllers. Fig. 20 shows the global functionality of the package graphical user interface based Matlab as a tool to demonstrate the impact of FACTS devices on power system operation and control.

4.1 Structure of SimFACTS
A working main screen appears as shown in Fig. 20, first the user asked to choose the working language: Arabic, English or French.
The functions of the menus are:
- **File:** To do standard file storage or retrieve files for operations.
- **Network Configuration**: To display data of the network test system, make changes, display the topology.
- **Power Flow**: This is the standard application calculation part: the Newton Raphson algorithm included to calculate the power flow; user has to click to 'Power Flow' after

data entry. The submenu 'FACTS Controller' designed to enter in details the data base of the different FACTS Controllers.

- **Optimal Power Flow**: In this version (ver.1.0): the user can calculate the OPF using three methods:
  - Basic economic dispatch based Lagrange method
  - Simple Genetic Algorithm (SGA)
  - Particle swarm Optimization (PSO)
- **Reactive Power Control**: In this version (ver.1): the user can choose the FACTS controllers to control the reactive power at a specified location.
- **Voltage Stability Analysis based Continuation Power Flow:** this section alows user to test the impact of FACTS devices in voltage stability and system loadability using continuation power flow analysis.
- **Results**: This is an option provided for the user to view all results.
- **Help**: The objectives, scope and functions of each of the components are briefly given in this option.

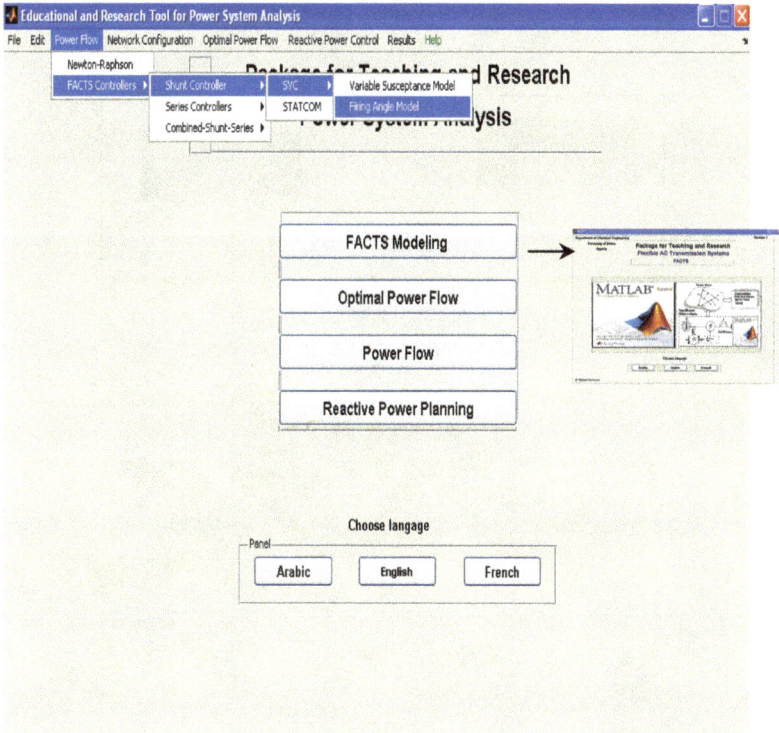

Fig. 21. Structure of the developed simulator incorporating FACTS devices

The Newton-Raphson algorithm modified based on the FACTS models and used to calculate all the necessary electrical values involved on the power flow study. The simple software proposed is capable of doing simulations for several models of FACTS controllers at different power system situation.

4.2 Graphic User Interface tool (GUI)

The MATLAB graphical user interface development environment, provides a set of tools for creating graphical user interfaces (GUIs). These tools greatly simplify the process of designing and building GUIs. We can use the GUIDE tools to:

Lay out the GUI. Using the GUIDE Layout Editor, the user can lay out a GUI easily by clicking and dragging GUI components such as panels, buttons, text fields, sliders, menus, and so on into the layout area. GUIDE stores the GUI layout in a FIG-file.

4.3 Program the GUI

GUIDE automatically generates an M-file that controls how the GUI operates. The M-file initializes the GUI and contains a framework for the most commonly used callbacks for each component the commands that execute when a user clicks a GUI component. Using the M-file editor, we can add code to the callbacks to perform the functions the user want. Fig. 22 shows the structure of the object, Fig. 23 shows the different Object contained in a GUI.

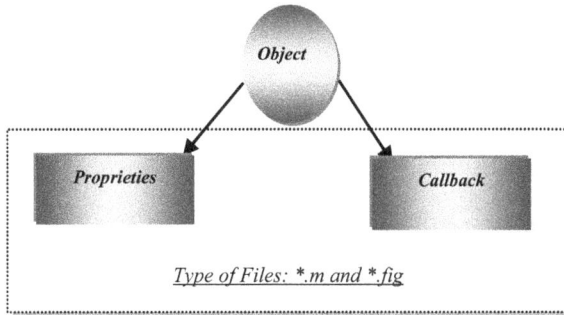

Fig. 22. Object structure

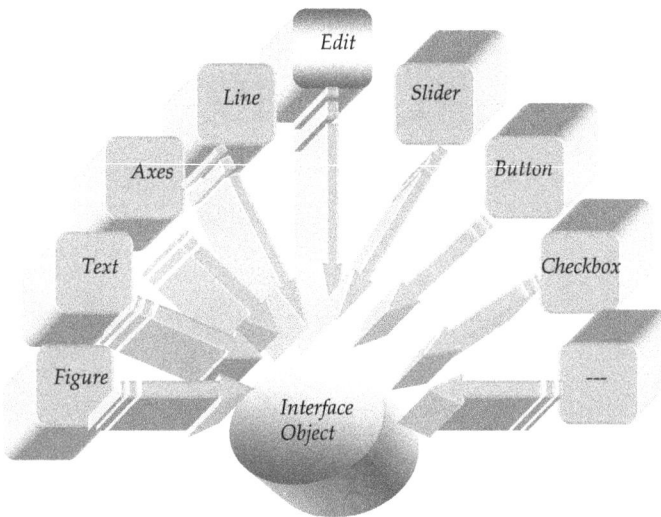

Fig. 23. Different Object of the GUIDE

4.4 Power flow program based FACTS (SimFACTS)

The interactive FACTS program proposed is implemented in a MATLAB environment, using the GUI tool; the Simulator is capable of doing simulations for several models of FACTS Controllers. First the program allows users to choose the working language (French, Arabic or English) Fig. 21 shows the global interface to simulate the integration of different FACTS Controllers in a practical network. The user may either retrieve an existing pre-saved system data or start a new system formulation; due to the limited pages we only present an example of UPFC interface description. Fig. 24 displaying the UPFC parameters to be entered and adjusted by user. For example The UPFC Controller data base parameters are:

- The insertion point
- The power flow: determines the direction of the flowing power
- Inductive reactance of Series impedance
- Inductive reactance of Shunt impedance
- Active Power Control status
- Reactive power control status
- Active power desired
- Reactive power desired
- Series voltage amplitude
- Shunt voltage amplitude
- Operational limits of the series and shunt voltage

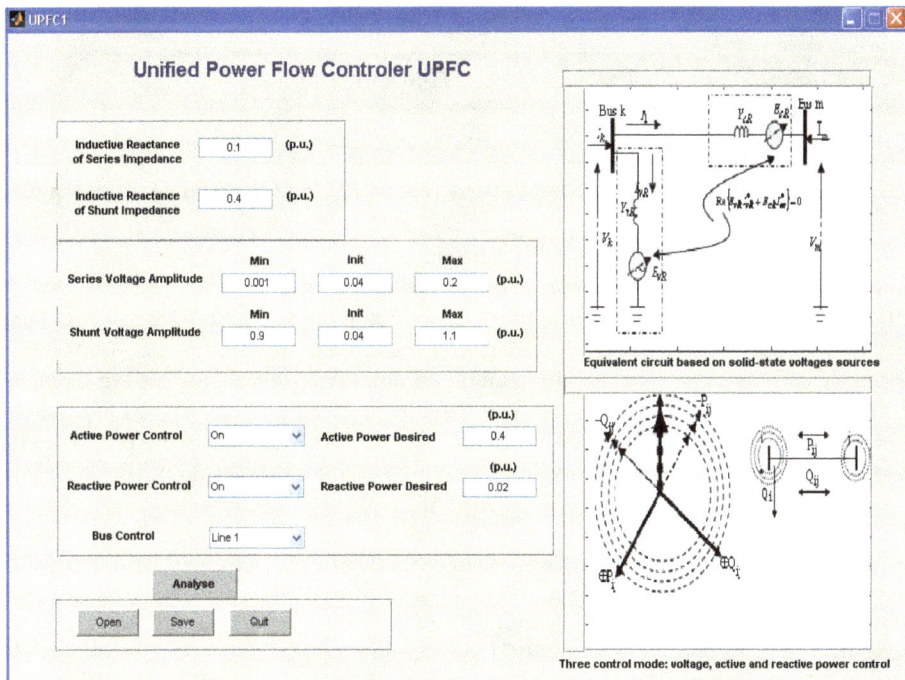

Fig. 24. UPFC parameters input data window

5. Simulation test and results using SimFACTS

The FACTS models integrated in the proposed educational power system control are those proposed by (Achat et al., 2004), and by Canizares (Canizares, C. A, at al.,). The Newton-Raphson algorithm is used to calculate all the electrical values involved in power system.. The FACTS simulation package is capable of doing interactive simulations for several types of controllers as shown in Table 1.

Using this simplified and interactive program, user can easely understand the basic concept of this new technolgy based FACTS devices introduced to power system operation and control. During simulation user can access directly to the code source program of any desired function (Ybus, Newton-Raphson, SVC Model, STATCOM model, TCSC Model, UPFC Model, Graphic functions,) , user also can modify the content of existant models, and test the efficiency of the modiefd models.

Topology	Model Identification
Shunt	*STATCOM Model*
	SVC with Variable Susceptance Model
	SVC with Firing Angle Model
Series	*SSSC Model*
	TCSC with Variable Reactance Model
	TCSC with Variable Firing Angle Model
Combined (Hybrid)	*UPFC Model*

Table. 1. List of FACTS models used in the first version (V1.0) of the SimFACTS package

5.1 Demostration example using SVC controller

The two SVC models based susceptance values and firing angle are included in the FACTS Simulator; the two models can be applied to a different practical power systems (smal, medium and large test systems). To understand the real contribution of the shunt FACTS controller (SVC) to enhance the power quality, the shunt controller integrated in a practical modified electrical network, IEEE 30-Bus. Voltage deviation (ΔV) power loss (P_{loss}), active power branch flow (P_{ij}), and system loadability (λ), are the indices of power quality considred to demonstrate the improvement of power quality, to validate the flexibilite, and the simplicity of the proposed educational SimFACTS package based Matlab.

Fig. 25 shows the improvement of voltages profiles using multi SVC controllers installed at 8 buses. Fig. 26 shows the evolution of voltage profiles at all buses based continuation power flow without SVC integration, the loading factor is 2.9449 p.u. By integration SVC Controllers at 8 critical buses, the loading factor improved to 3.1418 p.u. Fig. 27 shows clearly the contribution of shunt FACTS controllers to improve the power system loadability. Details results related to the integration of series controllers (TCSC, SSSC) and hybrid controllers (UPFC) will be given in the next contribution.

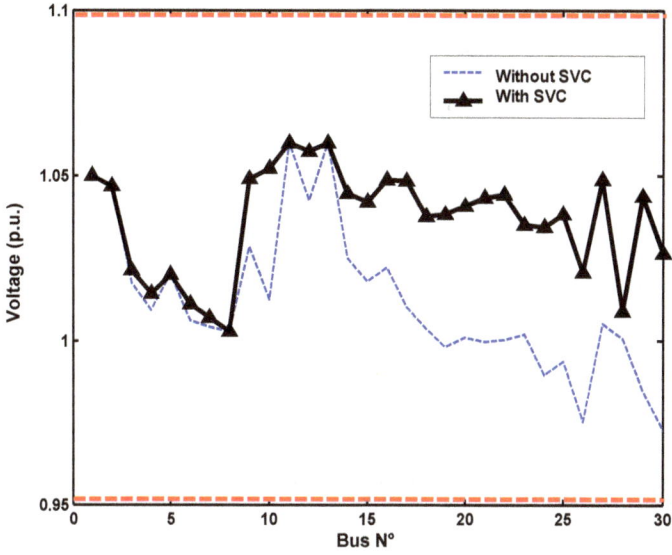

Fig. 25. Voltage profiles normal condition: case: with and without SVC installation.
NSVC=8: (10-17-19-21-22-24-27-29)

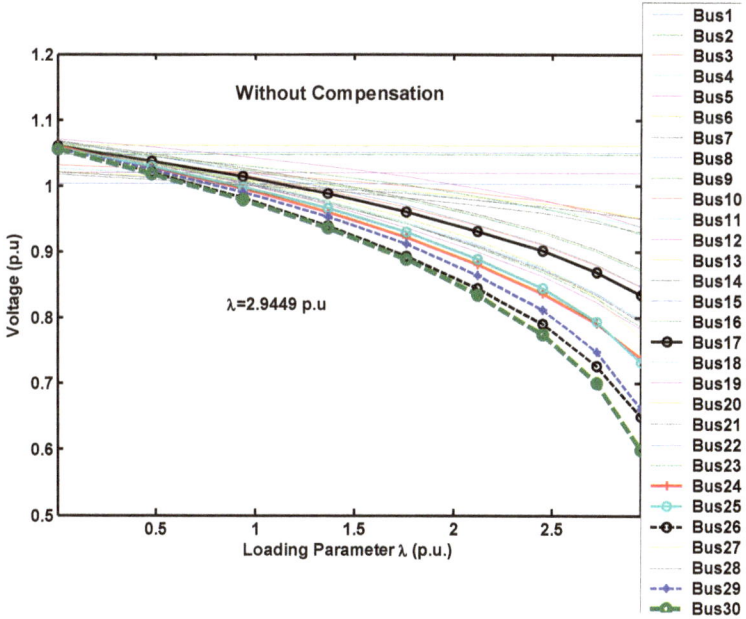

Fig. 26. Voltage profiles with continuation power flow: case: Without SVC

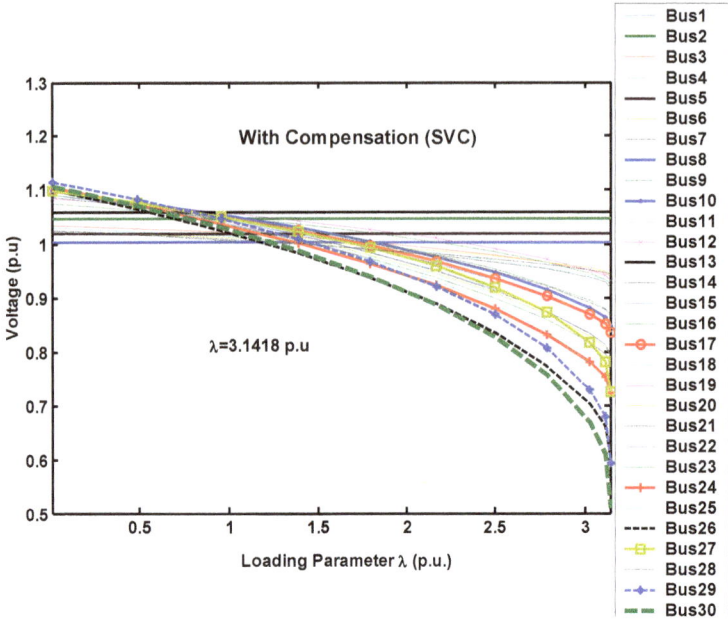

Fig. 27. Voltage profiles with continuation power flow: case: With SVC installation: NSVC=8: (10-17-19-21-22-24-27-29)

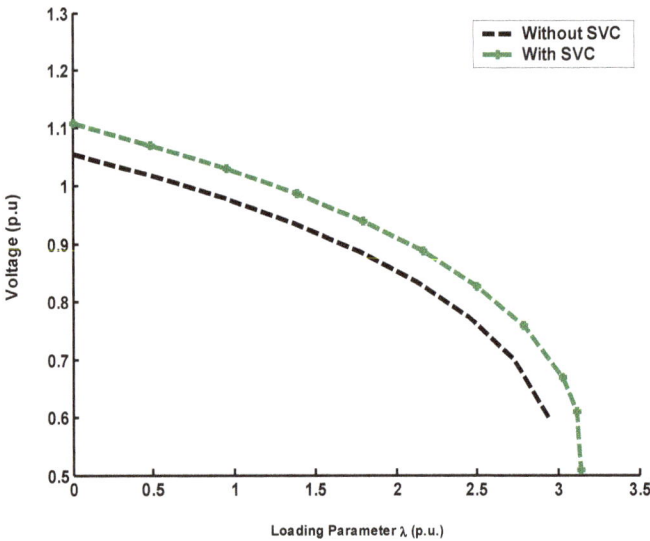

Fig. 28. Voltage profiles improvement at critical bus (bus 30) with continuation power flow: case: With and without SVC installation: NSVC=8: (10-17-19-21-22-24-27-29)

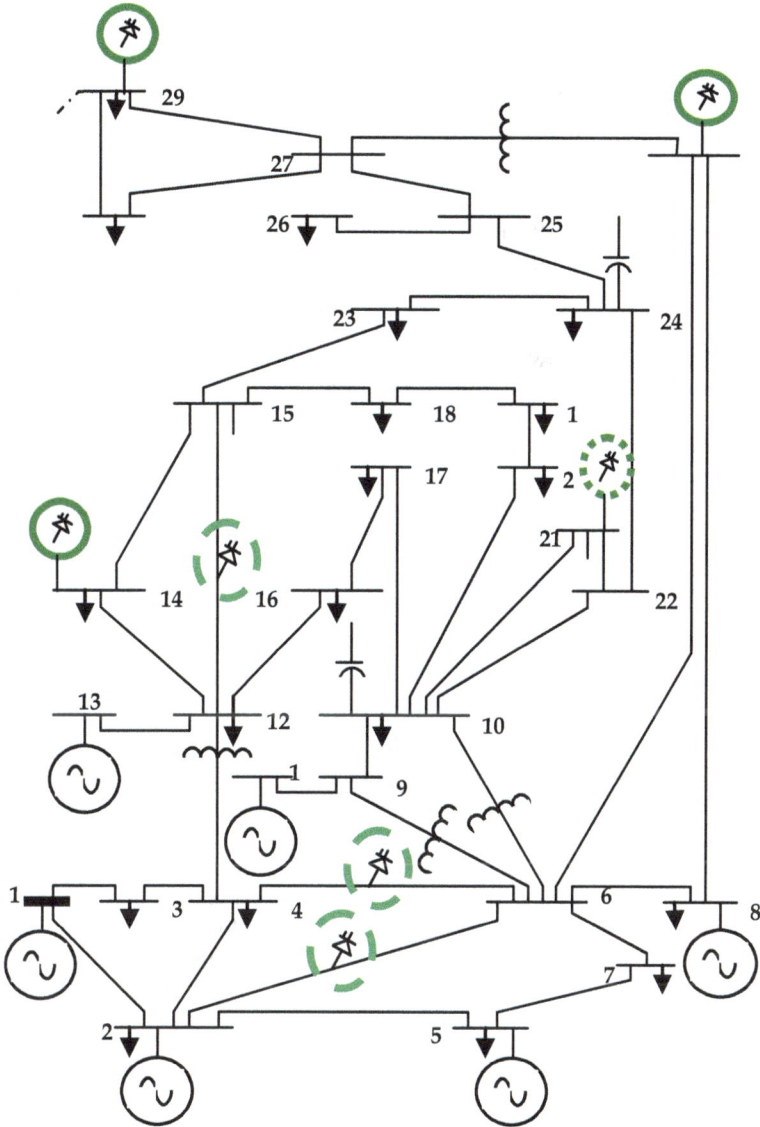

Fig. 29. Single line diagram for the modified IEEE 30-Bus test system (with FACTS devices)

6. Results discussions

The effeciency of the integrated of multi SVC controllers at different location is tested at normal condition and at critical situations.

1. Voltage magnitude is one of the important indices of power quality. For a secure operation of the power system, it is important to maintain required level of security margin.

2. One might think that the larger number of FACTS devices integrated in a practical power systems, the greater increase in the system loadability, based in experience, this supposition is not always true, there is a maximum increase on load margin with respect to the compensation level (number and size of FACTS devices).

3. System loadability analysis: To guide the decision making of the expert engineers, the power flow solution with consideration of FACTS devices should take in consideration the critical situation due to severe loading conditions and fault in power system, so it is important to maintain the voltage magnitudes within admissible values at consumer bus under abnormal situation (load increase and contingency).

4. Power loss analysis: Power loss is also an important indice used by expert engineers in power system operation and planning. Based on experience and simulation results, it is not always true that the larger number of FACTS devices integrated in a practical power systems, the greater decrease in power loss. Optimal location and coordination between multi types of FACTS devices is an important research axes.

5. Optimal location of FACTS devices is not introduced in this first version of SimFACTS, user can choose, number and location of FACTS devices based on his personal experience, for example in this case study we can get the same power quality indices (power loss, voltage deviation, and system loadability) using only three SVC Conrollers installed at critical buses.

7. Conclusion

This chapter discusses the development of an educational simulator for the FACTS devices. The motivation of this first version of simulator is to provide the undergraduate engineers students with a simple and flexible tool about the principle of FACTS modelling and the contribution of FACTS devices to enhance power quality. The simulator has been developed under the simple graphic user interface (GUI) from MATLAB program. In this first version the user can edit, modify and save the FACTS parameters proposed for each type of Controllers in a specified file (Data) and choose location of different FACTS based on the results given by power flow and his personnel practical experience.

Power quality analysis based series FACTS devices (TCSC Model), and Hybrid devices (UPFC Model) can be demostrated using the same strategy, due to the limited chapter length, new results related to these devices will be given in details with the next new chapter.

8. References

Abur, A., F. H. Magnago and Y. Lu, Educational toolbox for power system analysis, *IEEE Computer Application in Power*, vol. 13, no. 4, Oct. 2000, pp. 31-35.

Acha, E., Fuerte-Esquivel C, Ambiz-Perez (2004), FACTS Modelling and Simulation in Power Networks. John Wiley & Sons.

Canizares, C. A., Power flow and transient stability models of FACTS controllers for voltage and angle stability studies, *IEEE Proceeding* , 2000

Coelho, L. S., R. C. Thom Souza, and V. Cocco mariani, (2009) Improved differential evoluation approach based on clutural algorithm and diversity measure applied to solve economic load dispatch problems, *Journal of Mathemtics and Computers in Simulation*.

Dhaoyun, G., and T. S. Chung, Optimal active power flow incorporating FACTS devices with power flow constraints, *Electrical Power & Energy Systems*, vol. 20, no. 5, pp. 321-326, 1998.

Feurt-Esquivel, C. R., E. Acha, Tan SG, JJ. Rico, Efficient object oriented power systems software for the analysis of large-scale networks containing FACTS controlled branches, *IEEE Trans. Power Systems*, vol. 13, no. 2, pp. 464-472, May 1998.

Feurt-Esquivel, C.R., E.Acha, Unified power flow controller: A critical comparison of Newton-Raphson UPFC algorithms in power flow studies, *IEE Proc-Gener. Transm. Distrib*, vol. 144, no. 5, pp. 437-444, September 1997.

Hingorani, N.G., High Power Electronics and Flexible AC Transmission System, *IEEE Power Engineering review*, july 1988.

Hingorani, NG., Gyugyi, L, Understanding FACTS: Concepts and Technology of Flexible AC Transmission Systems. *IEEE Computer Society Press*, 1999.

Mahdad, B., (2010), Optimal Power Flow with Consideration of FACTS devices Using Genetic Algorithm: Application to the Algerian Network, Doctorat Thesis, Biskra University Algeria, 2010.

Mahdad, B., K. Srairi, T. Bouktir, M. E. H. Benbouzid, (2007) Modeling of FACTS devices with efficient location to improve voltage stability, *Association for the Advancements of Modeling and Simulation Techniques in Enterprises 'AMSE Journals'*. vol. 62, n°3, 2007.

Mahdad, B., T. Bouktir and K. Srairi , Strategy for location and controlof FACTS devices for enhancing power quality, *IEEE MELECON*, pp.1068-1072, May 2006.

Mahdad, B., T. Bouktir, K. Srairi , (2005) Dynamic Compensation of the Reactive Energy using a Fuzzy Controller, *Leonardo Electronic Journal of Practices and Technologies*, Issue 7, July-December 2005, pp 1-16, ISSN 1583-1078, Academic

Mahdad, B., T. Bouktir, K. Srairi, (2006) The Impact of Unified Power Flow Controller in Power Flow Regulation, *Journal of Electrical Engineering (JEE)*, Volume 6(1) / 2006, pages: 03-09, 2006, Romaine.

Milano, F., An open source power system analysis toolbox, *IEEE Trans. Power Systems*, vol. 20, no. 3, pp. 1199-1206, Aug 2005.

Milano, F., L. Vanfretti, and J. C. Morataya, An open source power system virtual laboratory: the PSAT case and experience, *IEEE Trans. Education*, vol. 51, no. 1, pp. 17-23, February 2008.

Nikman,T., (2010) A new fuzzy adaptive hybrid particle swarm optimization algorithm for non-linear, non-smooth and non-convex economic dispatch, *Journal of Applied Energy*, vol. 87, pp. 327-339.

Shao, W., and V. Vittal, LP-based OPF for corrective FACTS control to relieve overloads and voltage violations, *IEEE Trans. Power Systems*, vol. 21, no. 4, pp. 1832-1839, November 2006.

Simon, D., Biogeography-based optimization, (2008) *IEEE Trans. Evol.Comput.*, vol. 12, no. 6, pp. 702–713.

Vlachogiannis, J. G., and K. Y. Lee, (2009), Economic dispatch-A comparative study on heuristic optimization techniques with an improved coordinated aggregation-based PSO, *IEEE Trans. Power Systems*, vol. 24, no. 2, pp. 991-1001.

Xu, B., C. Y. Evrenosoglu, A. Abur and E. Akleman, Interactive evaluation of ATC using a
      graphical user interface,

Zhang, X. P., C. Rehtanz and B. Pal, Flexible ac transmission systems: Modeling and control,
      Springer-Verlag Berlin Heidelberg 2006.

10

Design of a Virtual Lab to Evaluate and Mitigate Power Quality Problems Introduced by Microgeneration

Sonia Pinto, J. Fernando Silva, Filipe Silva and Pedro Frade
DEEC; Instituto Superior Técnico, TULisbon,
Cie3 – Centre for Innovation in Electrical and Energy Engineering
Portugal

1. Introduction

The technological advances of the last decades favored a widespread of power electronics converters in the majority of household appliances, industrial equipment connected to the Low Voltage (LV) grid and, more recently, in distributed power generation, near the consumer – microgeneration (μG).

Most of this electronic equipment is a strong producer of current harmonics, polluting the LV network and generating sensitivity to dips, unbalances and harmonics, being also more sensitive to Power Quality issues. In the future, the massive use of renewable and decentralized sources of energy will probably worsen the problem, increasing Total Harmonic Distortion (THD), RMS voltage values, increasing unbalances and decreasing Power Factor in Low Voltage Networks.

In these and in other Power Quality related issues, power electronics became, to a certain extent, the cause of the problem. However, due to the continuous development of power semiconductors characteristics, less demanding drive circuits, integration in dedicated modules, microelectronic control circuits improvement, allowing their operation at higher frequencies and with higher performance modulation and control methods, power electronics converters also have the potential to become the solution for the problem. Still, even the non polluting grid connected converters are not usually exploited to their full capability as, in general, they are not used to mitigate Power Quality problems.

The smart exploitation of μG systems may become very attractive, using power electronics converters and adequate control strategies to allow the local mitigation of some power quality problems, minimizing the LV grid harmonics pollution (near unitary power factor) and guaranteeing their operation as active power filters (APF).

Based on these new challenges, the main aim of this work is to create a virtual LV grid laboratory to evaluate some power quality indicators, including power electronics based models to guarantee a more realistic representation of the most significant loads connected to the LV grid. The simulated microgenerators are represented as Voltage Source Inverters (VSI) and may be controlled to guarantee: a) near unity power factor (conventional μG); b) local compensation of reactive power and harmonics (active μG).

From the obtained results, active µG have the capability to guarantee an overall Power Quality improvement (voltage THD decrease and Power Factor increase) allowing a voltage THD decrease when compared to voltage THD values obtained with conventional µG.

2. Model of Low Voltage grid

The power electronics based low voltage network model is obtained using the SimPowerSystems Toolbox of Matlab/Simulink. The models include the Medium/Low voltage (MV/LV) transformer, the distribution lines, the most significant electrical loads and the microgenerators connected to the grid.

2.1 Distribution transformer

It is assumed that the distribution MV/LV transformer is ΔYN, with the secondary neutral directly connected to ground. The transformer used in the simulations is fed by a 30kV voltage on MV (medium voltage) and, in LV (Low Voltage) the line/phase voltage is 400V / 230V. The magnetization and the primary and secondary windings reactance and resistance are calculated from the transformer manufacturer no-load, short-circuit and nominal load tests [Elgerd, 1985].

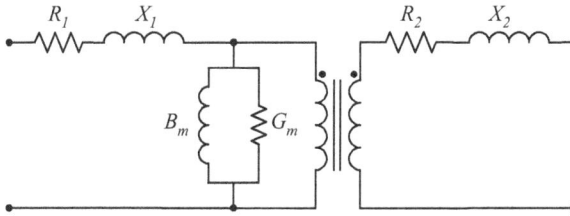

Fig. 1. Equivalent single phase model of a distribution transformer

From the no-load test, applying the nominal voltage U_n to the secondary side of the transformer, and leaving the primary side open, it is possible to obtain the transformer magnetizing current I_m. As the series impedance is much lower than the magnetizing impedance, it is assumed that the iron losses are nearly equal to the no-load losses P_0. Then, from the nominal voltage U_n, the magnetizing current I_m and the no load losses P_0, it is possible to determine the transformer magnetizing reactance and resistance. The magnetizing conductance is given by (1).

$$G_m = \frac{P_0}{U_n^2} \tag{1}$$

The magnetizing resistance R_m (2) is obtained from the magnetizing conductance G_m (1).

$$R_m = \frac{1}{G_m} \tag{2}$$

From the magnetizing current I_m and the magnetizing conductance R_m it is possible to determine the magnetizing susceptance B_m (3):

$$B_m = -\sqrt{\left(\frac{I_m}{U_n}\right)^2 - G_m^2} \qquad (3)$$

The magnetizing reactance X_m is given by (4):

$$X_m = \frac{1}{B_m} \qquad (4)$$

The magnetizing impedance is much higher than the series branch impedances (Fig. 1). Then, from the short-circuit test, it is possible to obtain the short-circuit impedance Z_{cc} (5) and the total resistance R_t (6) from the transformer primary and secondary windings, knowing the short-circuit voltage U_{cc}, necessary to guarantee the current nominal value I_n and the short-circuit losses P_{cc}.

$$Z_{cc} = \frac{U_{cc}}{I_n} \qquad (5)$$

$$R_t = \frac{P_{cc}}{I_n^2} \qquad (6)$$

Then, from (5) and (6) it is possible to determine the leakage reactance X_t (7):

$$X_t = \sqrt{Z_{cc}^2 - R_t^2} \qquad (7)$$

The resistance and leakage reactance from the primary and secondary windings may be assumed to be equal. Then:

$$R_1 = R_2 = \frac{R_t}{2} \qquad (8)$$

$$X_1 = X_2 = \frac{X_t}{2} \qquad (9)$$

In this work a 400kVA 30kV/400V distribution transformer (base values S_b=400kVA, U_b=30kV, $I_b = S_b / (\sqrt{3}\, U_b)$) is used. From the no-load test a magnetizing current I_m=2.9% and no-load losses of P_0=1450W are considered. From the short-circuit test it is assumed U_{cc}=4.5%, with nominal current I_n (1 pu) and short-circuit losses P_{cc}=8.8 kW.

2.2 Distribution cables

The distribution cables models are based on the π model (Fig. 2) and their section is chosen according to the current nominal values. The series resistance and inductance and the shunt admittance may be obtained from the manufacturers values depending on the cables section and length.

In LV distribution networks four-wire cables are used (three phase conductors and a neutral conductor insulated separately), all enclosed by an outer polyethylene insulation mantle. Usually the conductors are sector shaped. The shunt and series impedance are determined by the physical construction of the cable.

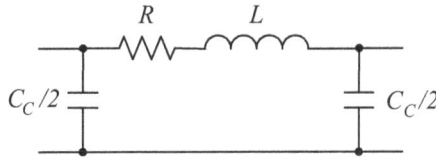

Fig. 2. π model of the distribution electrical network

Based on the single phase model of Fig.2, the model of a three phase distribution cable is obtained, Fig. 3 [Ciric et all 2003], [Ciric et all 2005].

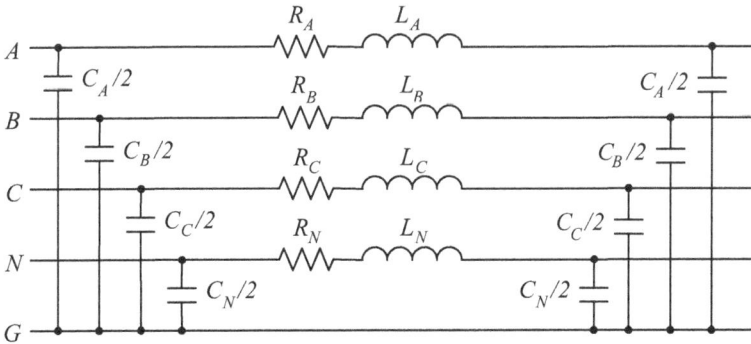

Fig. 3. Modified π model of the distribution electrical network

The series resistance R (Ω/km) [Jensen et all, 2001] depends on the cable internal resistance, on the ground resistance (there is no screen and the current diverted to ground must be included in the model) and on the proximity effect resistance. The skin effect and the proximity effect result in the increase of the conductors resistance.

The cable apparent inductance L_s depends on the self inductance, on the mutual inductance and on the inductance due to non-ideal ground.

The cable shunt admittance depends on the capacitances between conductors and on the conductors to ground capacitances [Jensen et all, 2001].

In overhead lines only the series impedance is considered. The capacitance is usually negligible.

Both for underground cables and overhead lines, the length should be adequate to guarantee their protection, according to the manufacturer values, and to assure that despite the voltage drops, the compliance with RMS voltage standard values [EN 50160] is always guaranteed.

2.3 Linear loads

Linear loads are represented as simple resistances (R) and inductances (RL). Resistive loads may be used to simulate incandescent lamps or conventional heaters, whether inductive loads may be used to simulate refrigerators, according to the measurements performed with a FLUKE 435 and shown in figure 4.

a)   b)

c)

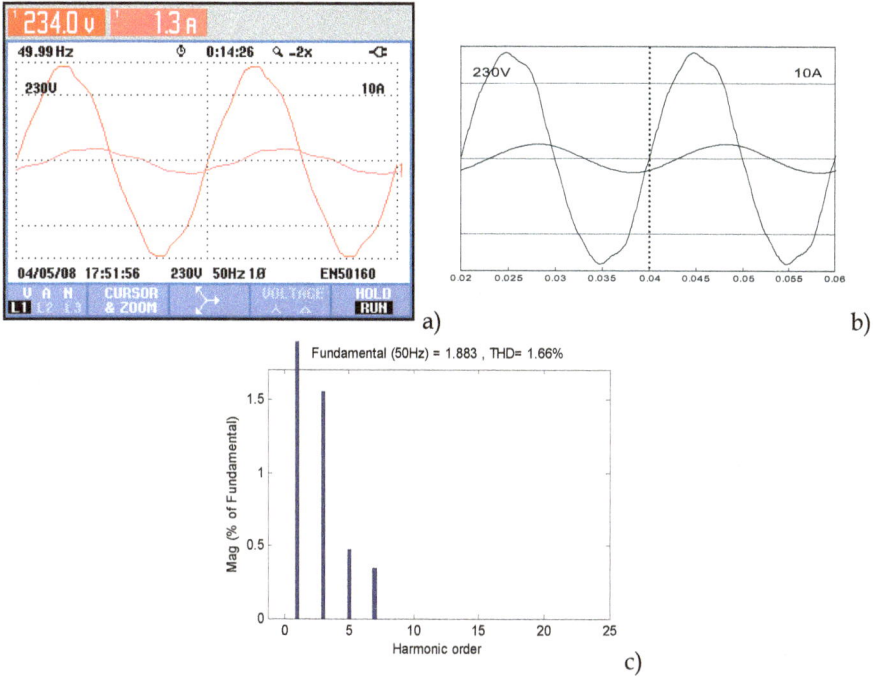

Fig. 4. Grid voltage and current waveform obtained for a refrigerator: a) Measured with a Fluke 435, THD_i=10.8% and PF=0.57; b) Obtained with the simulated model, considering THD_v=5%; c) Simulated current harmonics, THD_i=1.66% and PF=0.57

2.4 Nonlinear loads

Nonlinear loads are assumed to be mainly represented as diode rectifiers and are divided in three groups depending on their rated power.

The first group includes low power electronic equipment as TV sets, DVD players or computers. Usually, these electronic apparatus have isolated DC supplies connected to the grid through single phase rectifiers and they can be modelled as their first stage converter: a single phase rectifier feeding a DC $R_o//C_o$ load (Fig. 5) [Mohan et all, 1995].

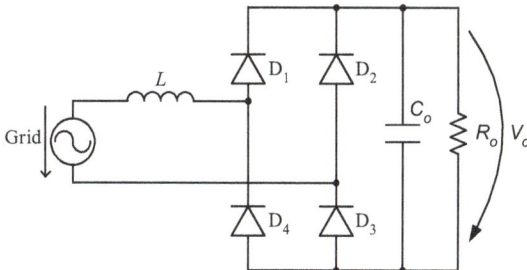

Fig. 5. Single phase rectifier as a model for the majority of electronic apparatus

Fig. 6 shows the voltage and current measurements obtained for a TV set and the equivalent simulated waveforms.

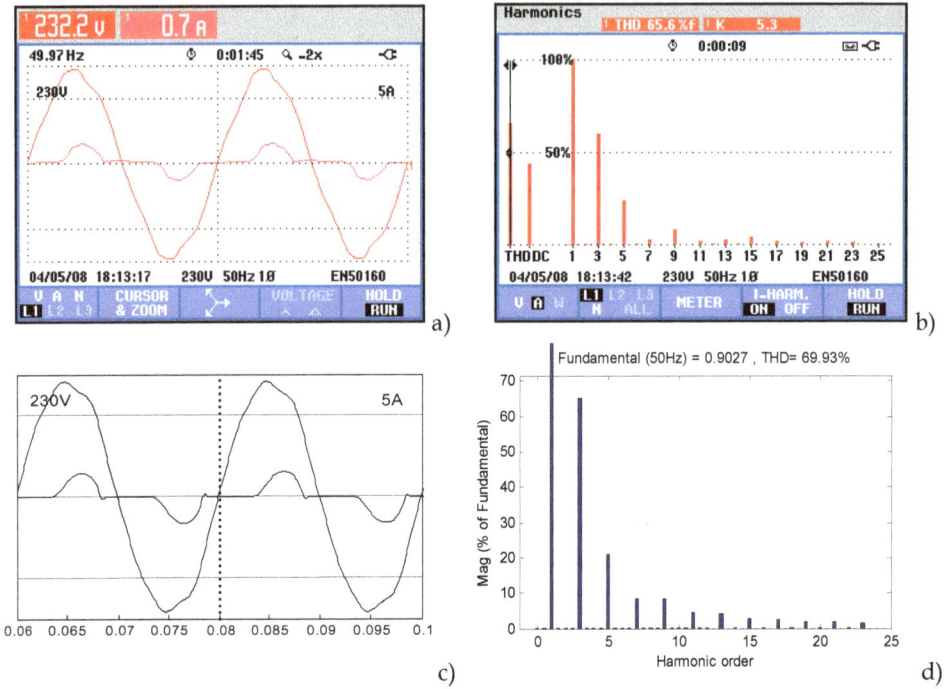

a)

b)

c)

d)

Fig. 6. Grid voltage and current waveform obtained for a TV set; a) b) Measured with a Fluke 435, THD$_i$=65.6% and PF=0.75; c) Obtained with the simulated model, considering voltage THD$_v$=5%; d) Simulated current harmonics, THD$_i$=69.9% and PF=0.76

Fig. 7 shows the voltage and current measurements obtained for a washing machine and the equivalent simulated waveforms.

The virtual lab models of these non linear loads are sized based on their rated power. Then, assuming an adequate DC voltage $V_{o_{av}}$, the value of the equivalent output resistance R_o is obtained from (10). For the TV set an output voltage average value $V_{o_{av}}$=300V is assumed.

$$R_o \approx \frac{V_{o_{av}}^2}{P} \tag{10}$$

The capacitor C_o is designed to limit the output voltage ripple ΔV_o. Also, it depends on the output voltage average value $V_{o_{av}}$, on the equivalent output resistance R_o, and on the time interval when all the diodes are OFF (approximately equal to one half of the grid period Δt=10ms). In the simulations, the ripple is assumed to be lower than ΔV_o=50V.

$$C_o \approx \frac{V_{o_{av}}}{R_o} \frac{\Delta t}{\Delta V_o} \tag{11}$$

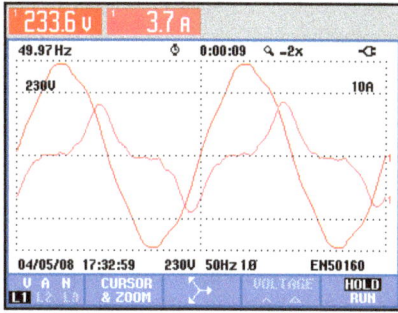
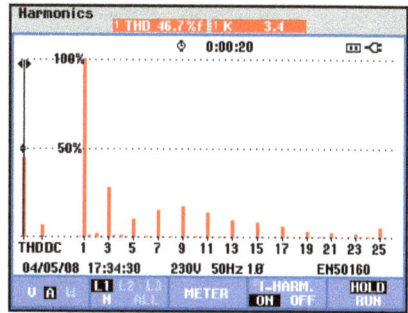
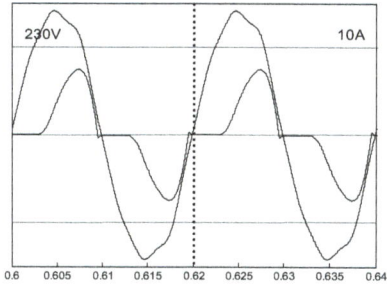
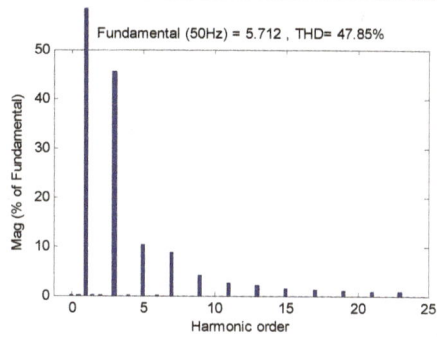
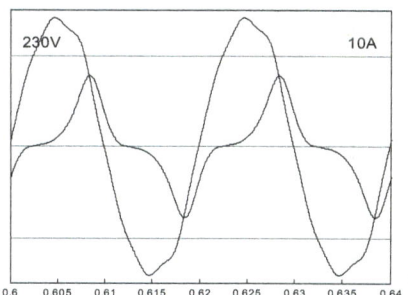
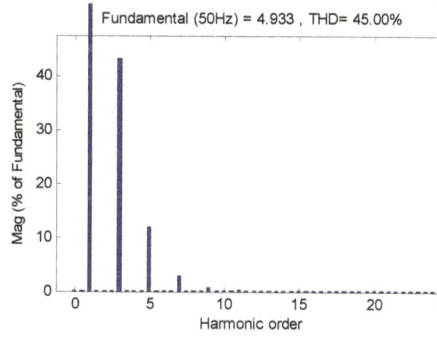

Fig. 7. Grid voltage and current waveform obtained for a washing machine; a) b) Measured with a Fluke 435, THD_i=46.7%; c) Obtained with the simulated model, considering voltage THD_v=5%; d) Simulated current harmonics, THD_i=47.85% and PF=0.76; e) Obtained with the simulated model, considering voltage THD_v=5% and a saturated inductance; f) Simulated current harmonics, THD_i=45% and PF=0.5

To smooth the current absorbed from the LV network, the rectifier is connected to the grid through a filtering inductance, which is calculated as a percentage of the output load impedance (3), where f represents the grid frequency and k is a constant, usually k=0.03 for lower power equipment as TV sets.

$$L_R = \frac{k\,R_o}{2\pi f} \tag{12}$$

As an example, with the designed model it is possible to obtain current waveforms similar to those measured on a TV set (Fig. 6), using the previously calculated values of R_o, C_o and L_R and assuming $P=150W$.

For other higher power household appliances as modern washing or dishwashing machines, a similar model may be used but the average rated power P should be higher, as well as the input filtering inductance. The voltage and current measurements obtained for a washing machine are shown in Fig. 7 a) b) and the equivalent simulated waveforms are shown in figures 7 c) d) where $P=1kW$, and the filtering inductance is obtained from (12) assuming $k=0.1$. Comparing figures 7 b) and 7 d) the measured and simulated currents THD as well as the harmonic contents are similar. Still, the current waveforms of Fig. 7 a) c) present some differences. To obtain similar current waveforms, the saturation effect of the input inductance should be considered, as shown in Fig. 7 c) d).

Even though the majority of LV grid connected loads are single phase, there may be a few three phase loads, as welding machines or three phase drives in small industries. Again, this equipment may be represented as their first stage converter, usually a three phase diode rectifier feeding an equivalent $R_{o3}//C_{o3}$ load (Fig. 8).

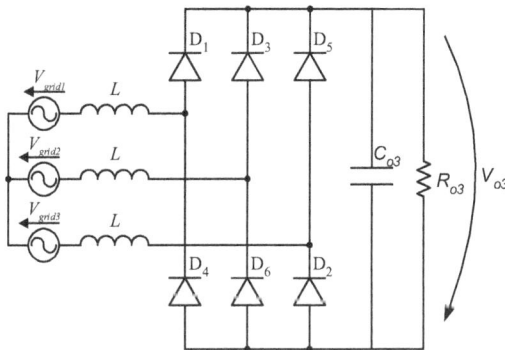

Fig. 8. Three phase rectifier as a model for an electronic equipment of a small industry

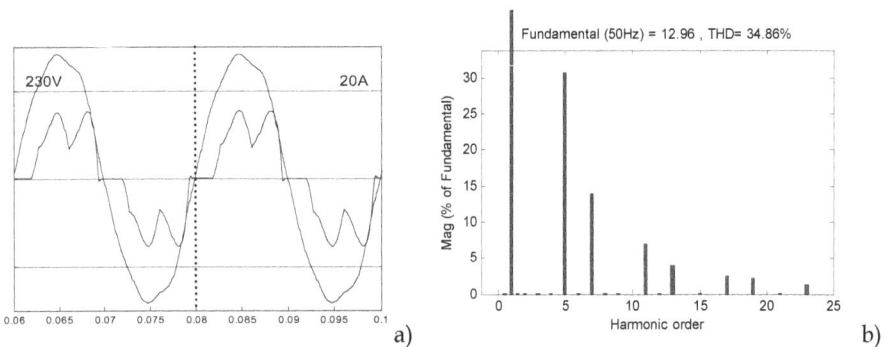

Fig. 9. a) Grid voltage and current waveform obtained for a three phase rectifier obtained with the simulated model, considering voltage $THD_v=5\%$; b) Current harmonics and $THD_i=34.86\%$, PF=0.91

In this model the equivalent output load may be calculated from (1) assuming P=6kW and $V_{o_{av}}$=520V. The output filter capacitor is obtained from (2) considering Δt=3.3ms (in a 6 pulse rectifier Δt=$T/6$). The input filtering inductance is obtained from (3) considering k=0.03. Fig. 9 shows the voltage and current waveforms obtained with the designed model.

2.5 Conventional single phase microgenerators

Microgenerators are connected to the LV grid through single phase VSI (voltage source inverters) (Fig. 10) [Pogaku et all, 2007] and they are designed to guarantee the compliance with international standards (as EN 50438) and to have characteristics similar to the authorized equipment (maximum rated power, current THD and input power factor).

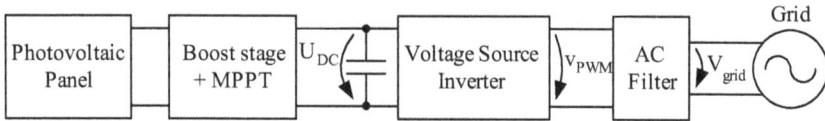

Fig. 10. Block diagram of a conventional µG

For simplicity reasons and minimization of simulation times, the microgenerators are simulated considering only the grid connection stage, as current controlled inverters fed by a DC voltage source U_{DC} (Fig. 11).

It is assumed that the VSI is connected to the grid through a filtering inductance designed to guarantee a current ripple lower than ΔI_{grid}. To minimize filtering, a three level PWM is used. Then, the inductance L_L (Fig. 11) is calculated according to (13), where U_{DC} is the DC link voltage, f_s is the switching frequency and ΔI_{grid} is the current ripple.

$$L_L = \frac{U_{DC}}{4 f_s \, \Delta I_{grid}} \tag{13}$$

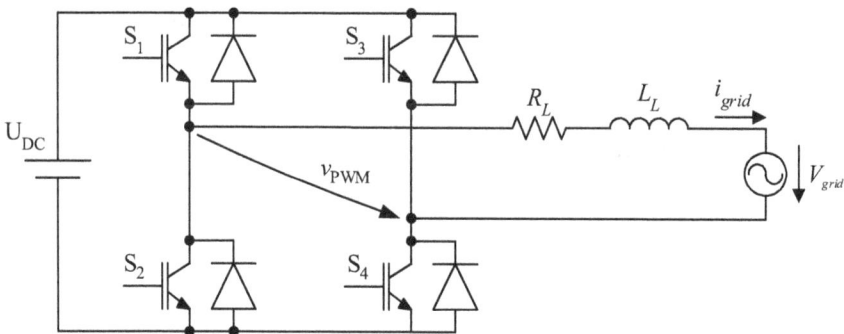

Fig. 11. Model of the single phase microgenerator

The VSI is controlled using a linear control approach, assuming that the maximum power is supplied to the grid and guaranteeing that the current injected in LV grid has a nearly unitary power factor.

Generally, the association of the modulator and the power converter may be represented as a first order model (14), with a gain K_D and a dominant pole dependent on the average delay time T_d (usually one half of the switching period $T_d = T_s/2$) [Rashid, 2007].

$$G_C(s) = \frac{v_{PWM_{av}}(s)}{u_c(s)} \approx \frac{K_D}{sT_d + 1} \tag{14}$$

The incremental gain K_D (15) depends on U_{DC} voltage and on the maximum value $u_{c_{max}}$ of the triangular modulator voltage.

$$K_D = \frac{U_{DC}}{u_{c_{max}}} \tag{15}$$

To control the current injected in the LV grid it is usual to choose a PI compensator (to guarantee fast response times and zero steady-state error to the step response). The block diagram of the current controller is then represented in Fig. 12, where α_i represents the gain of the current sensor.

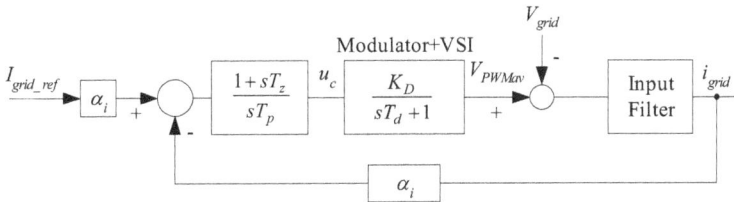

Fig. 12. Block diagram of the current controlled VSI

To design the current controller it is then necessary to obtain the closed loop transfer function of the whole system. To guarantee some insensitivity to the disturbance introduced by the grid voltage V_{grid}, it is assumed that the disturbance is known (is the grid voltage). For simplicity in the controller design, it is considered that the µG sees an equivalent resistance $R_0 = V_{grid}/i_{grid}$ connected to its terminals. From the controller point of view, this results in $R = R_L + R_0$. Then, making the compensator zero T_z coincident with the pole introduced by the input filter $T_z = L_L/R$, the second order transfer function of the current controlled VSI is obtained from (16).

$$G_{cl}(s) = \frac{i_{grid}(s)}{i_{grid_{ref}}(s)} = \frac{\dfrac{K_D \alpha_i}{T_p T_d R}}{s^2 + \dfrac{1}{T_d}s + \dfrac{K_D \alpha_i}{T_p T_d R}} \tag{16}$$

The transfer function (16) is then compared to the second order transfer function (17) written in the canonical form.

$$G_2(s) = \frac{\omega_n^2}{s^2 + 2\xi\omega_n s + \omega_n^2} \tag{17}$$

From (16) and (17), assuming a damping factor $\xi = \sqrt{2}/2$, the value of T_p is obtained from (18).

$$T_p = \frac{2K_D \alpha_i T_d}{R} \tag{18}$$

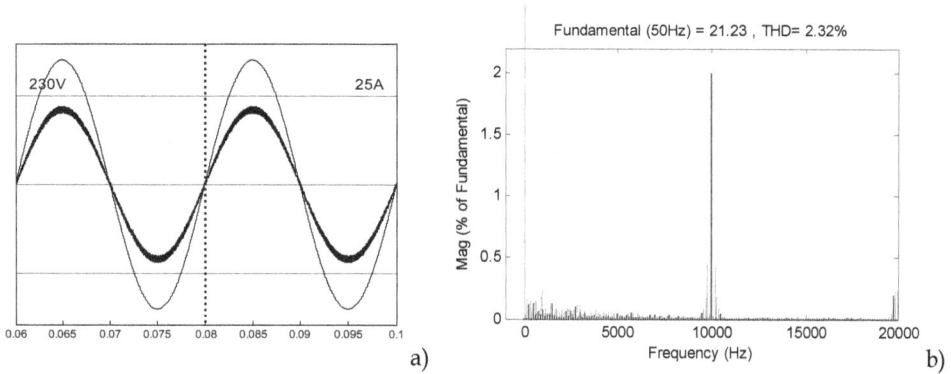

Fig. 13. a) Current and voltage waveforms of a single phase VSI obtained with the simulated model; b) Current harmonics and $THD_i=2.33\%$, PF=-0.999

Figure 13 shows the results obtained for the proposed μG model, assuming that the μG apparent power is $S=3450VA$, the DC voltage is $U_{DC}=400V$, the switching frequency is near 10kHz and $\Delta I_{grid}<0.1\, I_{grid}$.

The μG power factor is negative, even though nearly unitary as the displacement factor between the voltage and the current is 180°. The current THD is lower than 3%. However, considering only the first 50 harmonics, as in most power quality meters, the current THD decreases to $THD_i=0.35\%$ These results are according to the manufacturers values, guaranteeing the compliance with international standards.

Even though these microgenerators are designed to present high power quality parameters (high power factor and low current THD), still they are not usually exploited to their full extent as in general, they are sized and the controllers are designed only to minimize the impact on the LV grid. The mitigation of Power Quality issues is not considered.

As an example, consider a small LV grid, as the one represented in figure 14, with a μG and a non-linear load.

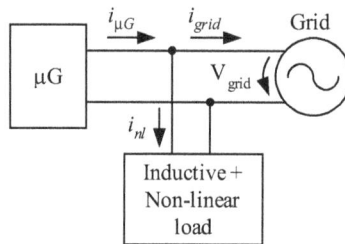

Fig. 14. Example of a small LV grid with a μG and a non-linear load

Using the previously designed μG the current $i_{μG}$ (Fig. 14) will be equal to the one obtained in Fig. 13. The non-linear load current i_{nl} is represented in Fig. 15 and is characterized by THD_i=47.55%.

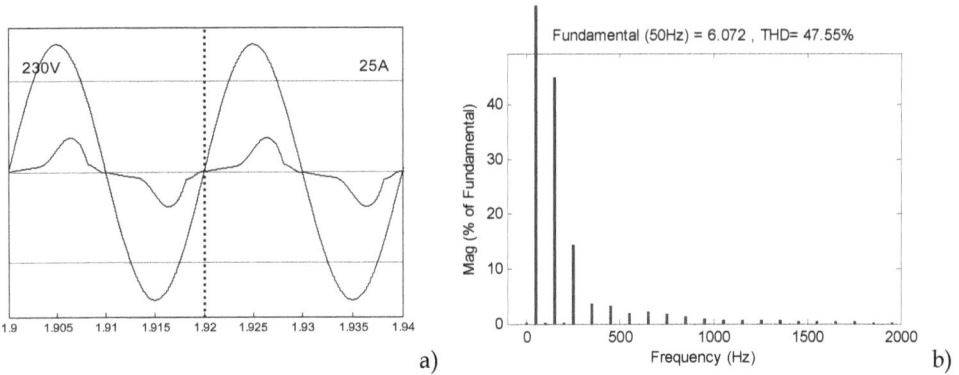

Fig. 15. a) Grid voltage V_{grid} and current waveform i_{nl} obtained for the non-linear load; b) Current harmonics and THD_i=47.55%, PF=0.15

The grid current i_{grid} is represented in Fig. 16 and, as a result of the non-linear load THD_i=18.79%.

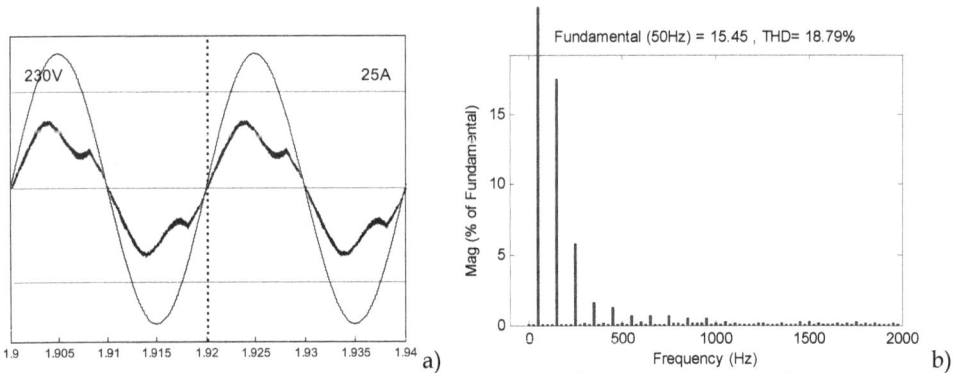

Fig. 16. a) Waveforms of grid voltage V_{grid} and current i_{grid}; b) Current harmonics and THD_i=18.79%

From this example it is possible to conclude that even though the μG injects nearly sinusoidal currents in the grid (Fig. 13), still it is not capable of guaranteeing sinusoidal currents when other nonlinear loads are connected to the grid.

2.6 Active microgenerators

To minimize some power quality problems as current and voltage THD, an active μG is included in this Lab (Fig. 17). Even though using the same power electronics converters as the conventional μG, with adequate control strategies and adequate filtering, it is possible to guarantee its operation as active power filter (APF), allowing the local mitigation of some power quality issues, as current THD, reducing the LV grid harmonic pollution (and near unitary power factor).

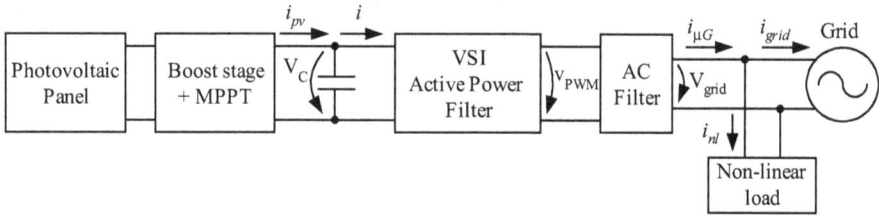

Fig. 17. Block diagram of an active μG

Based on the conventional μG model (Fig. 11), the proposed active μG is simulated according to Fig. 18, considering the DC link filtering stage and the disturbance introduced by the current i_{pv} of the photovoltaic panel + boost stage.

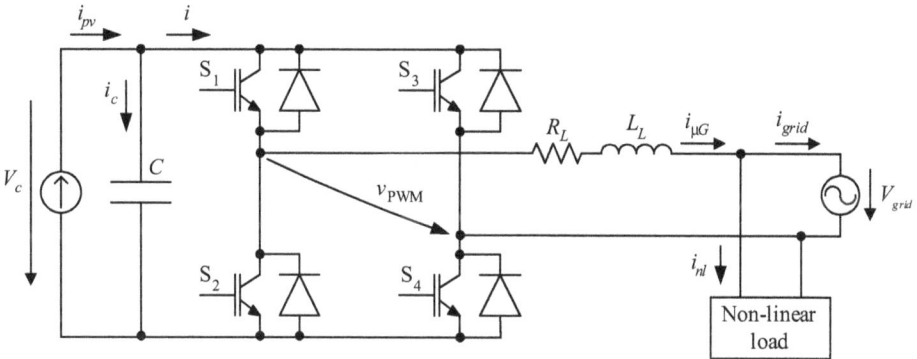

Fig. 18. Model of the single phase active microgenerator

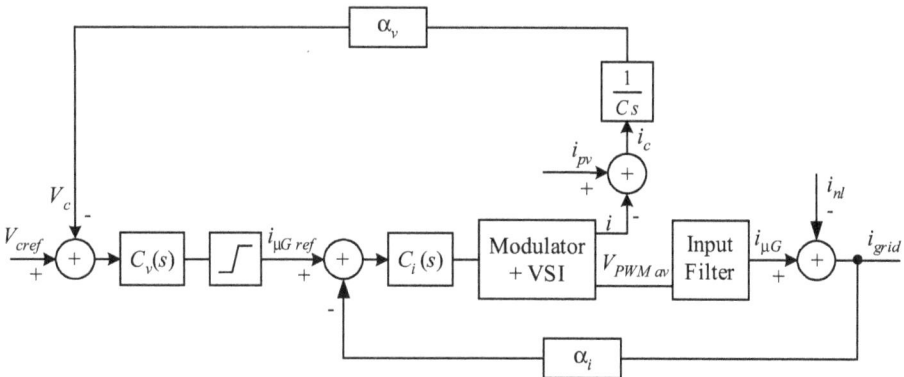

Fig. 19. Diagram block of the DC voltage controller and of the grid current controller to guarantee active filtering of the current harmonics introduced by the non-linear load

Assuming that the V_c voltage dynamics in the DC link is considerably slower than the dynamics of the microgenerator AC current $i_{\mu G}$, then the active µG current $i_{\mu G}$ and the voltage V_c may be controlled according to the diagram block of Fig. 19.

The active µG current controller design is equal to the design of the controller used for the conventional µG. Then, neglecting the high frequency poles, the current controlled system may be represented according to (19), where the controller gain G_i (20) is obtained from the input/output power constraint, where V_{max} represents the amplitude of the grid voltage.

$$\frac{i(s)}{i_{\mu Gref}(s)} \approx \frac{\dfrac{G_i}{\alpha_i}}{T_{dv}\, s + 1} \tag{19}$$

$$G_i(s) \approx \frac{V_{max}}{2 V_c} \tag{20}$$

Then, the current controlled system may be represented as a current source (19), as shown in Fig. 20.

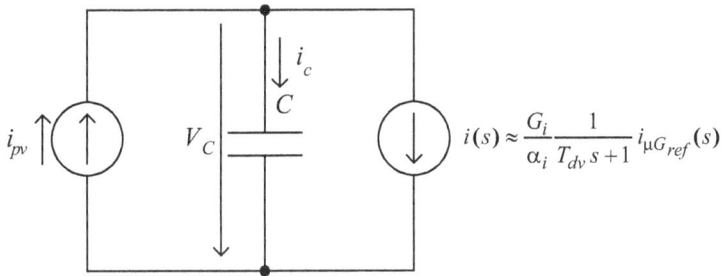

Fig. 20. Simplified block diagram used to design the voltage controller

From Fig. 20, the block diagram of the DC voltage controller is obtained and represented in figure 21.

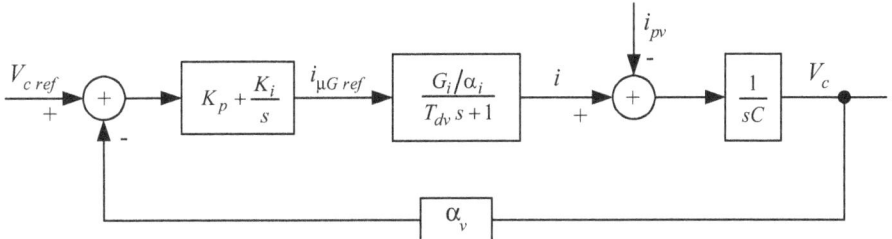

Fig. 21. Block diagram of DC stage voltage controller

From Fig. 21, the voltage response to the disturbance introduced by the photovoltaic panel is given by (21):

$$\left.\frac{v_c(s)}{i_{pv}(s)}\right|_{v_{cref}=0} = \frac{\dfrac{1}{sC}}{1+\alpha_v\left(K_p+\dfrac{K_i}{s}\right)\dfrac{G_i}{\alpha_i}\dfrac{1}{T_{dv}s+1}\dfrac{1}{sC}} \tag{21}$$

Simplifying (21) it is possible to obtain the transfer function in the canonical form (22).

$$\left.\frac{v_c(s)}{i_{pv}(s)}\right|_{v_{cref}=0} = \frac{s\dfrac{\alpha_i}{T_{dv}C\alpha_i}(T_{dv}s+1)}{s^3+\dfrac{1}{T_{dv}}s^2+\dfrac{\alpha_v G_i K_p}{T_{dv}C\alpha_i}s+\dfrac{\alpha_v G_i K_i}{T_{dv}C\alpha_i}} \tag{22}$$

From the final value theorem (23), the response to the disturbance introduced by i_{pv} current is zero, meaning that in steady-state, the PI controller guarantees the minimization of the disturbances.

$$\lim_{s\to 0}\left.\frac{v_c(s)}{i_{pv}(s)}\right|_{v_{cref}=0} = 0 \tag{23}$$

To determine the PI controller parameters, the denominator of (22) is compared to the third order polynomial (24).

$$P_3(s) = s^3 + 1.75\,\omega_0\,s^2 + 2.15\omega_0^2\,s + \omega_0^3 \tag{24}$$

Then:

$$\begin{cases} 1.75\omega_0 = \dfrac{1}{T_{dv}} \\[2mm] 2.15\omega_0^2 = \dfrac{\alpha_v G_i K_p}{T_{dv}C\alpha_i} \\[2mm] \omega_0^3 = \dfrac{\alpha_v G_i K_i}{T_{dv}C\alpha_i} \end{cases} \tag{25}$$

Solving (25), the proportional gain K_p and the integral gain K_i are obtained:

$$\begin{cases} K_p = \dfrac{2.15\,C\alpha_i}{\alpha_v G_i T_{dv}(1.75)^2} \\[3mm] K_i = \dfrac{C\alpha_i}{\alpha_v G_i (1.75)^3(T_{dv})^2} \end{cases} \tag{26}$$

Assuming that the dynamics of V_c voltage is considerably slower than the dynamics of the microgenerator AC current $i_{\mu G}$, then the pole T_{dv} is assumed to be $T_{dv}\approx 2T$, where T is the grid period.

Figures 22 to 24 show the results obtained for the proposed active μG model, assuming that the μG apparent power is S=3450VA, the DC voltage is controlled to be V_c=400V, the semiconductors switching frequency is near 10kHz and $\Delta I_{grid}<0.1\,I_{grid}$. The DC link capacitor is C=2.7mF, guaranteeing a voltage ripple lower than 5%. The results obtained for the non-linear load are those presented in figure 15.

From Fig. 22 it is possible to conclude that the proposed active µG acts as an active power filter, guaranteeing nearly sinusoidal grid currents. Comparing the results of Fig. 22 and Fig. 16, there is a clear reduction of the grid current THD. This reduction will become more obvious for more complex grids and highly non-linear loads.

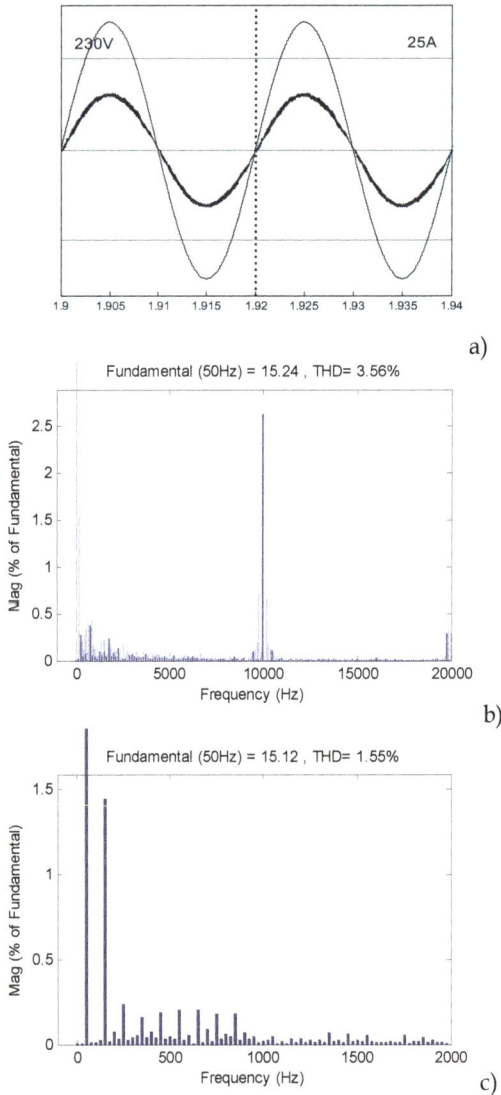

Fig. 22. a) Waveforms of grid voltage and current; b) Grid current harmonics and THD_i=3.56%, PF=0.9999.; c) Grid current harmonics and THD_{i50}=1.55% (considering only till the 50th order harmonic)

To guarantee nearly sinusoidal grid currents, the μG current will be the one presented in Fig. 23.

The average value of the capacitor voltage is V_c=400V, as shown in Fig. 24.

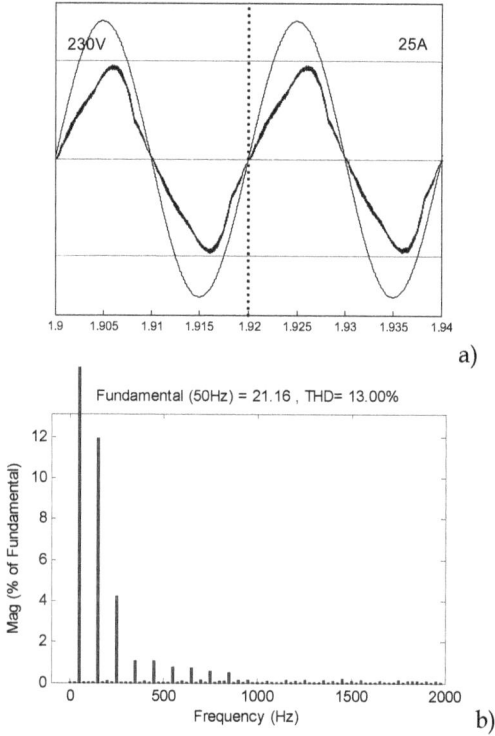

a)

b)

Fig. 23. a) Waveforms of grid voltage and μG current; b) μG current harmonics and THD$_i$=13%

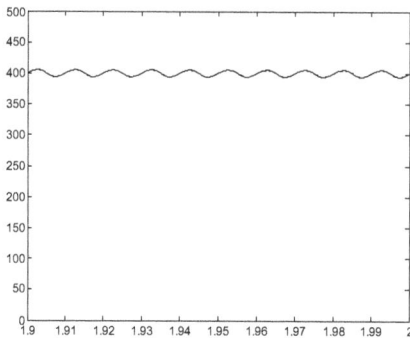

Fig. 24. Waveform of the DC link capacitor voltage

The proposed models will be further tested in a low voltage grid.

3. Modelling the Low Voltage grid

The performance of microgenerators can be compared in this virtual lab using the designed low voltage grid model with six clusters of loads (Fig. 25). It is assumed that 85% of these loads are non-linear and 15% are linear. Also, on the transformer Medium Voltage side the 5th and 7th harmonics are considered. At the Low Voltage side it is assumed that the voltage RMS value is 2.5% above the nominal value.

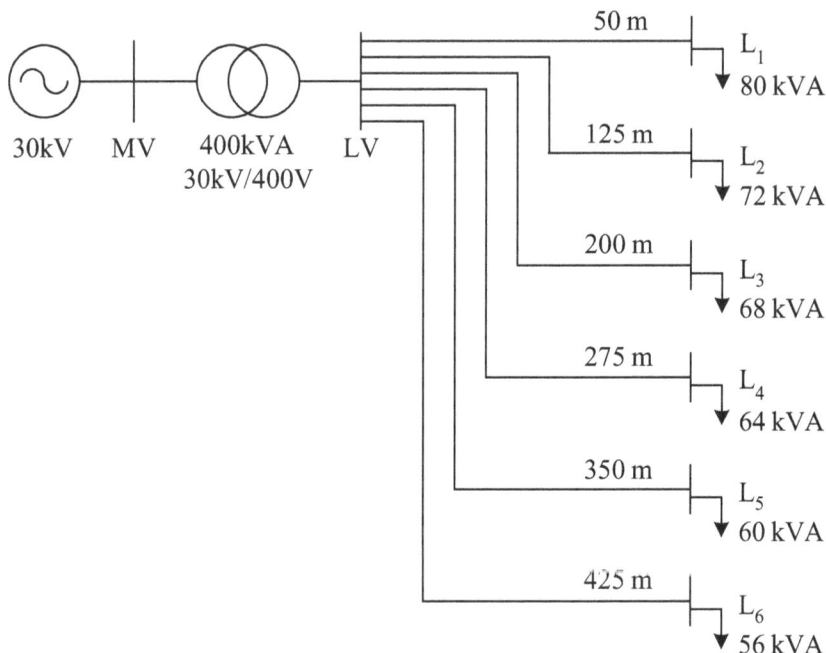

Fig. 25. Topology of the simulated LV grid

The simulations are carried out assuming two different load scenarios:
a. Distribution transformer at 15% of its nominal power (S_N) (nearly no load scenario, assuming 15% of values represented in Fig. 25);
b. Distribution transformer at 85% of its nominal power (full load scenario, assuming 85% of values represented in Fig. 25).
Each one of these scenarios is tested:
a. without µG;
b. with conventional µG;
c. with active µG.
It is assumed that the microgeneration total power never exceeds 25% of the transformer rated power S_N.
Figure 26 presents the results obtained without µG, assuming that the transformer may be at 15 % or at 85 % of its rated power S_N. The measurements of phase voltages and currents are carried out on the transformer LV side for each one of the groups of loads L_1 to L_6.

No load scenario (15% S_N) – phase A
No load scenario (15% S_N) – phase B
No load scenario (15% S_N) – phase C
Full load scenario (85% S_N) - phase A
Full load scenario (85% S_N) - phase B
Full load scenario (85% S_N) - phase C

Voltage THD

Current THD

Power Factor

Voltage RMS value

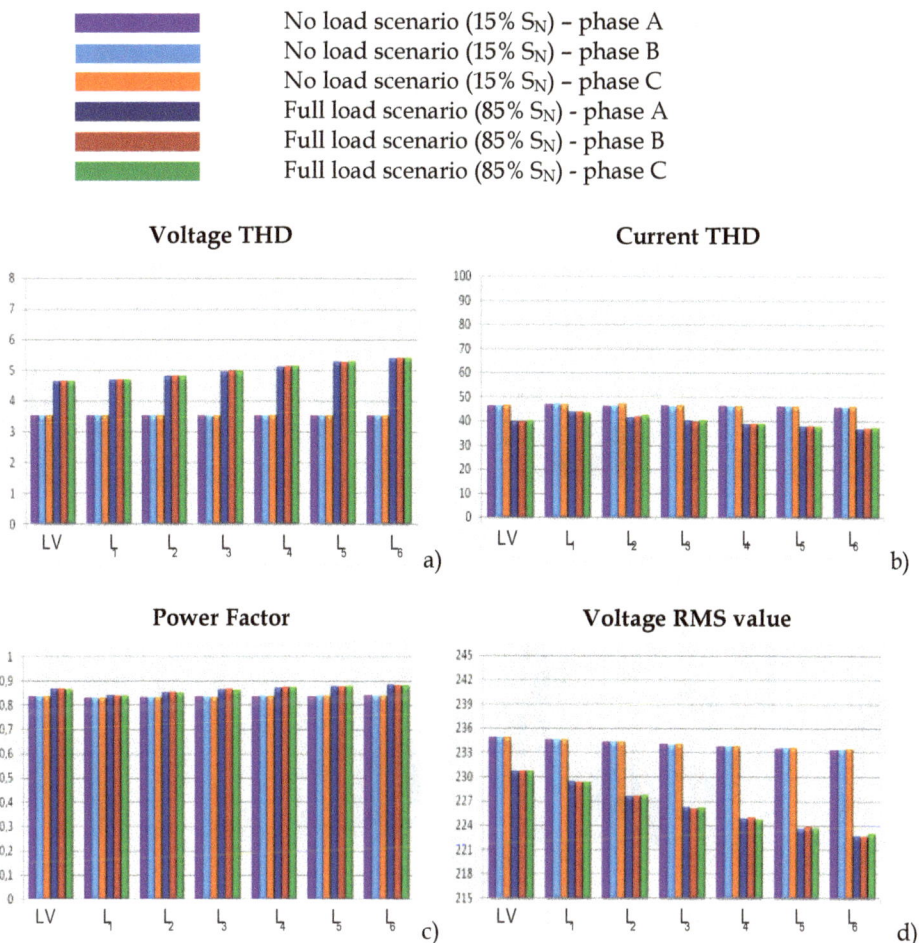

Fig. 26. Results obtained for 15 % and 85 % of the transformer rated power, without μG. Measurements carried out on the transformer LV side for each one of the groups of loads L1 to L6: a) Voltage THD; b) Current THD; c) Power Factor; d) Value of RMS voltage

Fig. 26 shows that the voltage THD increases more than 50% (as in load 6) from the no-load (15% S_N) to the full load (85% S_N) scenario. As the percentage of linear and non-linear loads is nearly equal for both scenarios, the current THD does not present significant changes (it even decreases slightly in the full load scenario). Also, the Power Factor results are similar for both scenarios, even though slightly lower for the no-load scenario. As for the load voltages RMS values, higher loads result in higher voltage drops. Also, as the transformer to load distance increases, the voltage drop increases as well.

Figure 27 presents the results obtained with μG assuming that the transformer is at 15 % of its rated power S_N (no load scenario). The measurements of phase voltages and currents are carried out on the transformer LV side for each one of the groups of loads L_1 to L_6.

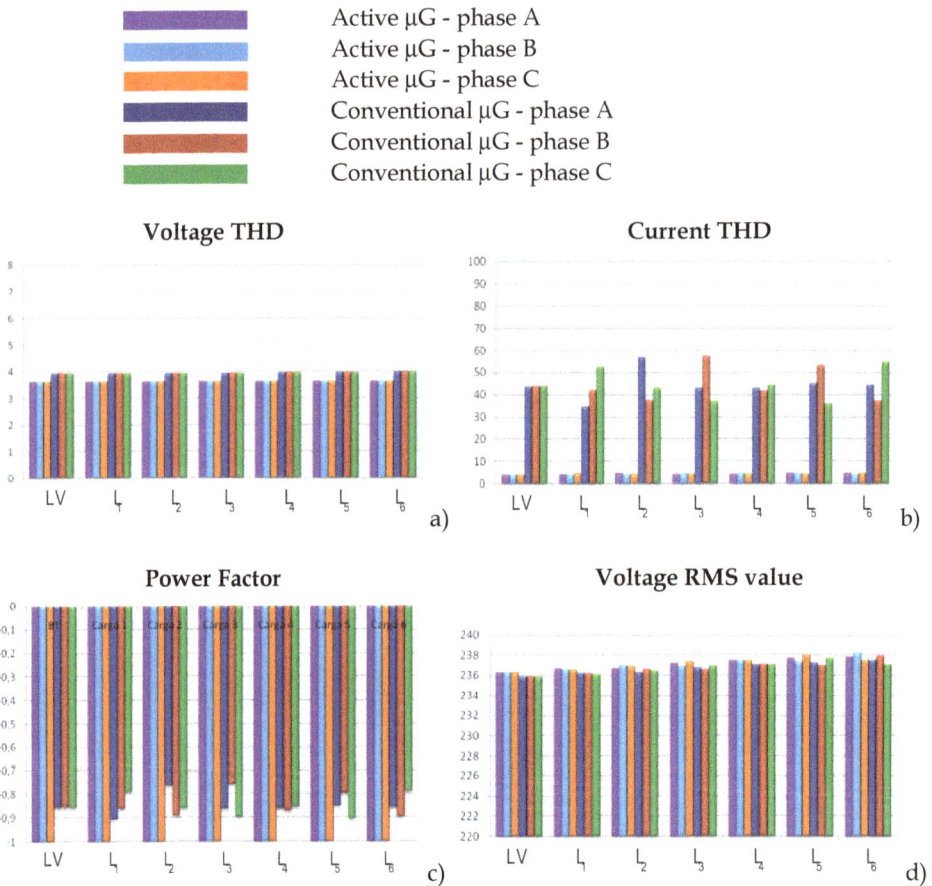

Active μG - phase A
Active μG - phase B
Active μG - phase C
Conventional μG - phase A
Conventional μG - phase B
Conventional μG - phase C

Voltage THD

Current THD

a)

b)

Power Factor

Voltage RMS value

c)

d)

Fig. 27. Results obtained for 15% S_N of conventional μG or active μG: a) Voltage THD; b) Current THD; c) Power Factor; d) Value of RMS voltage

From the results obtained for the first scenario (15% S_N) (Fig. 27), the use of active μG guarantees a clear improvement of voltage and current THD, when compared to the conventional μG. Also, the use of active μG guarantees near unity power factor, even though it is negative. This results from the fact that the power flows from the microgenerators to the transformer, instead of flowing from the transformer to the loads.

Figure 28 presents the results obtained with μG assuming that the transformer is at 85 % of its rated power S_N (full load scenario). The measurements of phase voltages and currents are carried out on the transformer LV side for each one of the groups of loads L_1 to L_6.

The results obtained for the full load scenario (85% S_N) (Fig. 28) show the improvement introduced by active μG in voltage THD (Fig 28a), as well as current THD (Fig 28b) and power factor (Fig 28c). From Fig. 28 active microgeneration allows a voltage THD reduction up to 30%, when compared to conventional microgeneration. Also, comparing with the values obtained without microgeneration (Fig. 26) it is possible to conclude that conventional microgeneration slightly increases voltage THD, while active microgeneration reduces voltage THD.

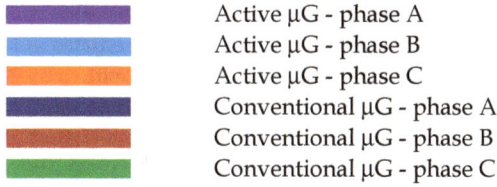

- Active μG - phase A
- Active μG - phase B
- Active μG - phase C
- Conventional μG - phase A
- Conventional μG - phase B
- Conventional μG - phase C

Voltage THD

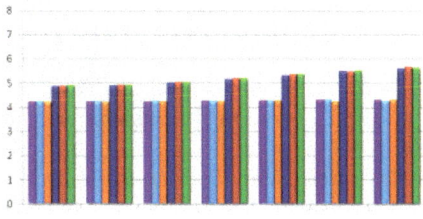

a)

Current THD

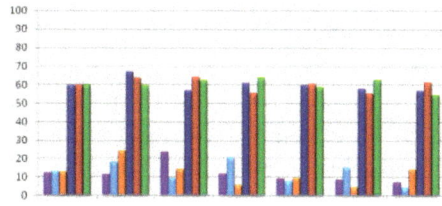

b)

Power Factor

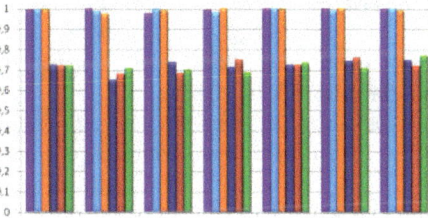

c)

Voltage RMS value

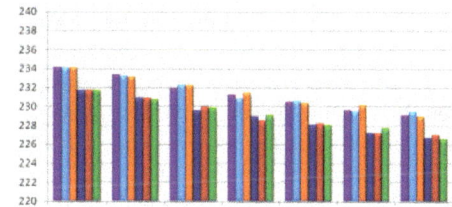

d)

Fig. 28. Results obtained for 85% S_N of conventional μG or active μG; a) Voltage THD; b) Current THD; c) Power Factor

4. Conclusions

In this paper a virtual lab was designed to evaluate and mitigate some power quality problems introduced by μG. The virtual lab includes the Medium/Low voltage (MV/LV) transformer, the distribution lines, linear and non-linear loads, conventional μG and active μG. To validate the designed models, the current waveforms and distortion obtained for each one of the virtual lab loads were compared to those measured with the most used electric and electronic equipment, showing that the obtained results are similar.

The μG model is simulated based on its final stage converter, a single phase inverter, while the active μG also includes high order harmonics compensation, to perform as an active power filter.

Using the proposed models a small low voltage grid model with six clusters of loads is designed to evaluate the impact of conventional µG and active µG on Power Quality for a no-load and a full load scenario.

From the results obtained with the virtual lab LV grid, it was possible to conclude that conventional µG slightly increases voltage THD, while active µG reduces voltage THD (up to 30% when compared to voltage THD values obtained with conventional µG), guaranteeing an overall Power Quality improvement (Power Factor increase).

Even though the µG total power never exceeds 25% of the transformer rated power S_N, with a high percentage of non linear loads, as the one considered in the proposed virtual lab LV grid model (85% of the transformer rated power), the active µG presents promising results and it can be concluded that it may be a solution to mitigate some power quality problems.

5. References

Ciric, R. M.; Ochoa, L. F.; Padilla-Feltrin, A.; Nouri, H.; *Fault Analysis in Four Wire Distribution Networks*; IEE Proceedings on Generation, Transmission and Distribution, Vol. 152, No 6, November 2005.

Ciric, R. M.; Padilla-Feltrin, A.; Ochoa, L. F.; *Power Flow in Four-Wire Distribution Networks – General Approach*; IEEE Transactions on Power Systems, Vol. 18, No 4, November 2003.

Elgerd, O.; *Electric Systems Theory: an Introduction*, 2nd ed., 1985, International Student Edition, Mc Graw Hill, ISBN 0-07-Y66273-8, Singapore.

EN 50160, *Voltage Characteristics of Electricity Supplied by Public Distribution Networks*, European Standard EN 50160, 2001.

EN 50438, *Requirements for the Connection of Micro-Generators in Parallel with Public Low Voltage Distribution Networks*, European Standard EN 50438, 2007.

Jensen, M. H.; Bak-Jensen, B.; *Series Impedance of the Four-Wire Distribution Cable with Sector-Shaped Conductors*, Proc. of PPT 2001, IEEE Porto Power Tech Conference, Porto, Portugal, September 2001.

Jensen, M. H.; Bak-Jensen, B.; *Shunt Admittance of the Four-Wire Distribution Cable with Sector-Shaped Conductors*, Proc. of AUPEC'2001, Australasian University Power Engineering Conference, Perth, Australia, September 2001.

Mohan, N.; Undeland, T.; Robbins, W.; *Power Electronics: Converters, Applications and Design*, 2nd Edition, 1995, John Wiley and Sons, ISBN 0-471-58408-8, USA.

Pogaku, N.; Prodanovik, M.; Green, T.; Modeling, *Analysis and Testing of Autonomous Operation of an Inverter Based Microgrid*, IEEE Transactions on Power Electronics, Vol. 22, No 2, March 2007.

Rashid, M.; *Power Electronics Handbook*, 2nd edition, 2007, Academic Press, Elsevier, ISBN 13: 978-0-12-088479-7, ISBN 10: 0-12-088479-8, USA.

Part 4

Industrial Applications

11

Some Basic Issues and Applications of Switch-Mode Rectifiers on Motor Drives and Electric Vehicle Chargers

C. M. Liaw and Y. C. Chang
National Tsing Hua University, National Chung Cheng University
Taiwan

1. Introduction

Switch-mode rectifier (SMR) or called power factor corrected (PFC) rectifier (Erickson & Maksimovic, 2001; Mohan et al, 2003; Dawande & Dubey, 1996) has been increasingly utilized to replace the conventional rectifiers as the front-end converter for many power equipments. Through proper control, the input line drawn current of a SMR can be controlled to have satisfactory power quality and provide adjustable and well-regulated DC output voltage. Hence, the operation performance of the followed power electronic equipment can be enhanced. Taking the permanent-magnet synchronous motor (PMSM) drive as an example, field-weakening and voltage boosting are two effective approaches to enhance its high-speed driving performance. The latter is more effective and can avoid the risk of magnet demagnetization. This task can naturally be preserved for a PMSM drive being equipped with SMR.

Generally speaking, a SMR can be formed by inserting a suitable DC-DC converter cell between diode rectifier and output capacitive filter. During the past decades, there already have a lot of SMRs, the survey for single-phase SMRs can be referred to the related literatures. Since the AC input current is directly related to the pulse-width modulated (PWM) inductor current, the boost-type SMR possesses the best PFC control capability subject to having high DC output voltage limitation. In a standard multiplier based high-frequency controlled SMR, its PFC control performance is greatly affected by the sensed double-frequency voltage ripple. In (Wolfs & Thomas, 2007), the use of a capacitor reference model that produces a ripple free indication of the DC bus voltage allows the trade off regulatory response time and line current wave shape to be avoided. A simple robust ripple compensation controller is developed in (Chen et al, 2004), such that the effect of double frequency ripple contaminated in the output voltage feedback signal can be cancelled as far as possible. In (Li & Liaw, 2003), the quantitative digital voltage regulation control for a zero-voltage transition (ZVT) soft-switching boost SMR was presented. As to (Li & Liaw, 2004b), the robust varying-band hysteresis current-controlled (HCC) PWM schemes with fixed and varying switching frequencies for SMR have been presented. In (Chai & Liaw, 2007), the robust control of boost SMR considering nonlinear behavior was presented. The adaptation of voltage robust compensation control is made according to the observed nonlinear phenomena. The development and control for a SRM drive with front-end boost SMR were presented in (Chai & Liaw, 2009). In (Chai et al, 2008), the novel random switching approach was developed for effectively reducing the acoustic noise of a low-frequency switching employed in a PMSM drive. In the bridgeless SMRs developed in (Huber et al, 2008), the higher efficiency is achieved by reducing loop diode voltage drops.

In some occasions, the galvanic isolation of power equipment from AC source is required. In (Hsieh, 2010), a single-phase isolated current-fed push pull (CFPP) boost SMR is developed, and the comparative evaluation for the PMSM drive equipped with standard, bridgeless and CFPP isolated boost SMRs is made.

From input-output voltage magnitude relationship, the buck-boost SMR is perfect in performing power factor correction control (Erickson & Maksimovic, 2001; Matsui et al, 2002). And it is free from inrush current problem owing to its indirect energy transfer feature. However, the traditional non-isolated buck-boost SMR possesses some limitations: (i) without isolation; (ii) having reverse output voltage polarity; (iii) discontinuous input and output currents; and (iv) having relatively high voltage and current stresses due to zero direct power transfer. As generally recognized, the use of high-frequency transformer isolated buck-boost SMR can avoid some of these limitations. The performance comparison study among Cuk, single ended primary inductor converter (SEPIC), ZETA and flyback SMRs in (Singh et al, 2006) concludes that the flyback SMR is the best one in the control performance and the required number of constituted component. In (Lamar et al, 2007), in addition to the power rating limits, the limitations of flyback SMR in PFC characteristics and output voltage dynamic response are discussed.

In (Papanikolaou et al, 2005), the design of flyback converter in CCM for low voltage application is presented. In the power circuit developed in (Lu et al, 2003), a dual output flyback converter is employed to reduce the storage capacitor voltage fluctuation against input voltage and load changes of flyback SMR in DCM. Similarly, two flyback converters are also used in the flyback SMRs developed in (Zheng & Moschopoulos, 2006) and (Mishra et al, 2004) to achieve direct power transfer and improved voltage regulation control characteristics. As to the single-stage SMR developed in (Lu et al, 2008), it combines a boost SMR front-end and a two-switch clamped flyback converter. Similarly, an intermediate energy storage circuit is also employed. In (Rikos & Tatakis, 2005), a new flyback SMR with non-dissipative clamping is presented to obtain high power factor and efficiency in DCM. The proposed clamping circuit utilizes the transformer leakage inductance to improve input current waveform. In (Jang et al, 2006), an integrated boost-flyback PFC converter is developed. The soft switching of all its constituted switches is preserved to yield high efficiency. On the other hand, the improved efficiency of the flyback converter presented in (Lee et al, 2008) is obtained via the use of synchronous rectifier.

It is known that digital control for power converter is a trend to promote its miniaturization. In (Newsom et al, 2002), the control scheme realization is made using off-the-shelf digital logic components. And recently, the VLSI design of system on chip application specific integrated circuit (SoC-ASIC) controller for a double stage SMR has also been studied in (Langeslag et al, 2007). It consists of a boost SMR and a flyback DC-DC converter. The latter is controlled using valley-switching approach operating in quasi-resonant DCM, which has fixed on-time and varying off-time according to load.

As far as the switching control strategies are concerned, they can be broadly categorized into voltage-follower control (Erickson & Madigan, 1990) and current-mode control (Backman & Wolpert, 2000). The former belongs to open-loop operation under DCM, and thus the current feedback control is not needed. As to the latter, the multiplier-based current control loop is necessary to achieve PFC control. Basically, the commonly used PWM switching control approaches for a flyback SMR include peak current control (Backman & Wolpert,

2000), average current control, charge control and its modifications (Tang et al, 1993). In the peak current controlled flyback converter presented in (Backman & Wolpert, 2000), the proper choice of magnetizing inductance is suggested to reduce the distortion of input current. In (Tang et al, 1993; Larouci et al, 2002), after turning on the switch at clock, the switch is turned off as the integration of switch current is equal to the control voltage. As to (Buso et al, 2000), a modified nonlinear carrier control approach is developed to avoid the sense of AC input voltage. For easily treating the dynamic control of a single-stage PFC converter, its general dynamic modeling and controller design approaches have been conducted in (Uan-Zo-li et al, 2005). In addition, there were also some special control methods for flyback SMR. See for example, a simplified current control scheme using sensed inductor voltage is developed in (Tanitteerapan & Mori, 2001). In (Y.C. Chang & Liaw, 2009a), a flyback SMR in DCM with a charge-regulated PWM scheme is developed.

For a SMR, the nonlinear behavior and the double-frequency voltage ripple may let the closed-loop controlled SMR encounter undesired nonlinear phenomena (Orabi & Ninomiya, 2003). The key parameters to be observed in nonlinear behavior of a SMR will be the loading condition, the value of output filtering capacitor and the voltage feedback controller parameters. In the flyback SMR developed in (Y.C. Chang & Liaw, 2009a), the simple robust control is proposed to avoid the occurrence of nonlinear phenomena, and also to improve the SMR operating performance.

Random PWM switching is an effective means to let the harmonic spectrum of a power converter be uniformly distributed. Some typical existing studies concerning this topic include the ones for motor drives (Liaw et al, 2000), DC-DC converters (Tse et al, 2000), SMRs (Li & Liaw, 2004b; Chai et al, 2008), etc. In the flyback SMR developed by (Y.C. Chang & Liaw, 2011), to let the harmonic spectrum be dispersdly distributed, a random switching scheme with fixed turn-on period and varying turn-off period is presented.

Although flyback SMR possesses many merits, it suffers from the major limitation of having limited power rating. To enlarge the rating, the parallel of whole isolated converter of flyback SMR was made in (Sangsun & Enjeti, 2002). In the existing interleaved flyback converters, the researches made in (Forest et al, 2007, 2009) are emphasized on the use of intercell transformers. However, the typical interleaving of flyback SMR requires multiple switches and diodes, which increases the cost and complexity of power circuit. For a single-phase flyback SMR, the major DC output voltage ripple is double line frequency component. Hence PWM interleaving control is not beneficial in its ripple reduction. Moreover, the power limitation of flyback transformer is more critical than the other system active components. It follows that sole parallel of transformer (Manh & Guldner, 2006; Inoue et al, 2008) will be the convenient way to enlarge the rating of whole flyback SMR. In (Y.C. Chang & Liaw, 2009b), the rating enlargement is made by parallel connection of transformer.

For the power equipments with higher ratings, the three-phase SMR is a natural choice for higher rated plants. The systematic surveys for the existing three-phase SMRs can be found in (Hengchun et al, 1997; Shah et al, 2005). Similar to transformers, three-phase SMRs can also be formed using multiple single-phase SMR modules via proper connection (Hahn et al, 2002; Li & Liaw, 2004c). For simplicity and less stringent performance, the three-phase single-switch (3P1SW) SMR will be a good choice. In the 3P1SW SMR presented in (Chai et al, 2010), a robust current harmonic cancellation scheme and a robust voltage control scheme are developed. The undesired line current and output voltage ripples are regarded as disturbances and they are reduced via robust controls. In voltage control, a feedback controller is augmented with a simple robust error canceller. The robust cancellation weighting factor is automatically tuned according to load level to yield compromised voltage and power quality control performances.

Similar to single-phase bridgeless SMRs (Zhang et al, 2000; Youssef et al, 2008), there were also some researches being emphasized on the development of three phase bridgeless SMRs (Reis et al, 2008; Oliverira et al, 2009). In (Wang, 2010), a bridgeless DCM three phase SMR is developed and used as a front-end AC-DC converter for the SRM drive.

As generally recognized, soft-switching can be applied for various converters to reduce their switching lossess, voltage stresses and electromagnetic interference. The applications of soft-switching in 3P1SW SMRs have also been conducted in (Gataric et al, 1994; Ueda et al, 2002). For the 3P1SW SMR operating under DCM, only the zero-current switching (ZCS) at turn-off is effective in reducing its switching losses. In (Wang, 2010), the zero-current transition (ZCT) (Gataric et al, 1994) is utilized to the developed 3P1SW to achieve the ZCS of the main switch at turn-off. In realization, an auxiliary resonant branch is added, and the proper switching signals are generated for the main and auxiliary switches. The soft-switching can be achieved without adding extra sensors. And also in (Wang, 2010), the comparative performance evaluation is made for the SRM drive powered using standard 3P1SW SMR, ZCT 3P1SW SMR and bridgeless DCM three phase SMR.

2. Power factor correction approaches

For facilitating the research made concerning power quality, the commonly referred harmonic standard is first introduced. Then the possible power factor correction approaches are described to comprehend their comparative features.

2.1 Harmonic ccurrent emission standard

IEC 61000-3-2 (previously, IEC-555) is the worldwide applied harmonic current emission standard. This standard specifically limits harmonics for equipments with an input current up to 16A, connected to 50Hz or 60Hz, 220V to 240V single phase circuit (two or three wires). The IEC 61000-3-2 standard distinguishes the loads into four classes with different harmonic limits (Erickson & Maksimovic, 2001; Mohan et al, 2003). From the contents one can find that for the equipments below 600W, the harmonic limits of Class A are larger than those of Class D. This advantage will be more significant for lower power level. Taking the third harmonic under 100W as an example, the limit in Class A is 2.3A compared to 0.34A in Class D. Power converter can apply Class D or Class A regulation depending on its input current wave shape. The peaky line drawn current of a diode rectifier with larger filtering capacitor definitely belongs to Class D. However, if the simple low-frequency switching SMR (Chai et al, 2008) is employed, the modified line drawn current may fall into Class A and thus possesses the advantage mentioned above.

2.2 Possible power factor correction methods

Depending on rating, schematic and control complexities, control performance and cost, there are many possible power factor correction approaches. The suited and cost effective one can be chosen according to the desired performance for specific application.

2.2.1 Passive filter

Various series L-C resonant trap filters are connected across the line terminal to attenuate the specific order harmonics. This approach is simple, rugged, reliable and helpful in reducing EMI. However, it is bulky and cannot completely regulate nonlinear loads, and it is needed the redesign adapted to load changes.

2.2.2 Active power filter

Compared with passive filter, active power filter (APF) has the higher control ability to compensate load reactive and harmonic current components. According to the types of connections, active power filters can be categorized into series, shunt and hybrid types (Erickson & Maksimovic, 2001; Mohan et al, 2003). Taking the shunt type active power filter as an example, a controlled current is generated from the APF to compensate the load ripple current as far as possible.

2.2.3 Passive PFC circuits

Fig. 1(a) shows the sketched key waveforms of a full-bridge rectifier with large and small filtering capacitors. One can be aware that if a very small filtering capacitor is employed, the line drawn power quality is improved, and thus the Class A rather than the Class D is applied. However, the effects of DC-link voltage ripple should be considered in making the control of the followed power stage. Recently, to reduce the rectified DC voltage ripple, some plants employ the valley-fill filter as shown in Fig. 1(b) (Farcas et al, 2006).

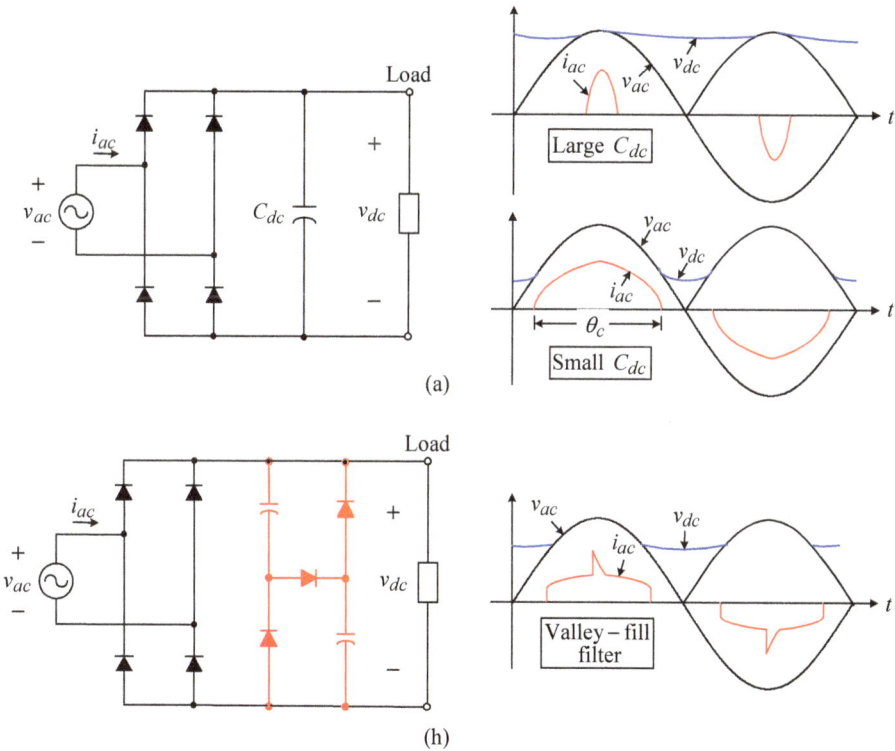

Fig. 1. Some passive PFC circuits: (a) rectifier with small filtering capacitor; (b) rectifier using valley-fill filter

2.2.4 Switch-mode rectifier

The SMRs possess many categories in circuit topology and switching control approaches. A single-phase boost-type SMR is shown in Fig. 2(a), and the typical waveforms of i_{ac} using low-frequency (LF) and high-frequency (HF) switchings are sketched in Figs. 2(b) and 2(c). The features of HF-SMR comparing to LF-SMR are: (i) more complicated in control; (ii) high control performances in line drawn current, power factor and output voltage; (iii) lower efficiency. More detailed survey for SMRs will be presented in the latter paragraphs.

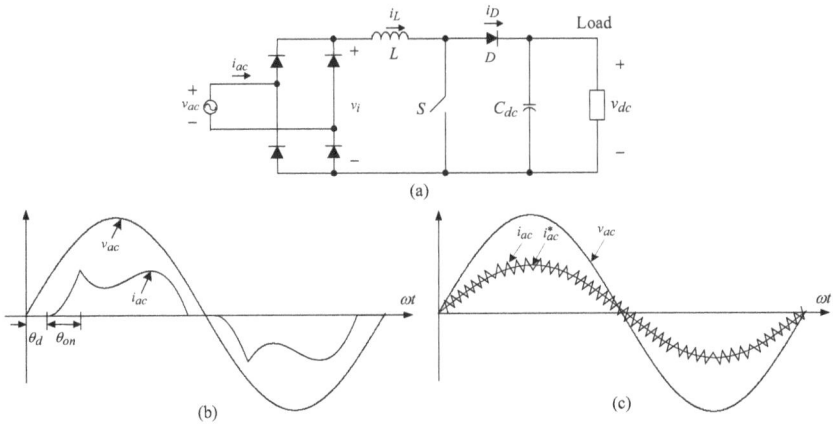

Fig. 2. Boost-type SMR: (a) circuit; (b) sketched key waveforms for low-frequency switching; (c) sketched waveforms for high-frequency switching

3. Classification of SMRs

Basically, a SMR is formed by inserting a suited DC/DC converter between diode rectifier and capacitive output filter, under well regulated DC output voltage, the desired AC input line drawn power quality can be achieved. The existing SMRs can be categorized as:

1. **Schematics**

a. Single-phase or three-phase: each category still possesses a lot of types of SMR schematics. The three-phase SMR will be a natural choice for larger power plants.
b. Non-isolated or isolated: although the former SMR is simpler and more compact, the latter one should be used if the galvanic isolation from mains is required. See for example, the flyback SMR is gradually employed in communication distributed power architecture as a single-stage SMR front-end, or called silver box, to establish -48V DC-bus voltage.
c. Voltage buck, boost or buck/boost: depending on the input-output relative voltage levels, suited type of SMR and its control scheme should be chosen. Basically, the boost-type SMR possesses the best current control ability subject to having high DC output voltage level.
d. Single-stage or multi-stage: generally speaking, the stage number should be kept as small as possible for achieving higher efficiency and system compactness. Hence, single-stage SMR is preferable if possible.

e. One-quadrant or multi-quadrant: multiple quadrant SMR may possess reverse power flow from DC side to AC source, such as the regenerative braking of a SMR-fed AC motor drive can be performed by sending braking energy back to the utility grid.

f. Hard-switching or soft-switching: Similarly, suited soft switching technique can also be applied to reduce the switching loss, switching stress and EMI of a SMR (Li & Liaw, 2003; Wang, 2010).

2. Control methods

a. Low-frequency control: only v-loop is needed and only one switching per half AC cycle is applied. It is simple but has limited power quality characteristics.

b. High-frequency control- voltage-follower control: without current control loop, only some specific SMRs operating in DCM possess this feature, see for example, buck-boost SMR and flyback SMR.

c. High-frequency control- standard control: it belongs to multiplier-based current-mode control approach with both v- and i- control loops.

3.1 Single-phase SMRs

The typical existing single-phase SMR circuits include: (a) boost SMR; (b) buck SMR; (c) buck- boost SMR; (d) Ćuk SMR; (e) SEPIC SMR with coupled inductors; (f) SEPIC SMR; (g) ZETA SMR; (h) buck-boost cascade SMR; (i) boost-buck hybrid SMR; (j) flyback SMR; (k) isolated Ćuk SMR; and (l) isolated ZETA SMR. Some comments are given for these circuits: (i) The SMRs of (a) to (i) belong to non-isolated types, whereas (j) to (l) are isolated ones; (ii) Among the non-isolated SMRs, the boost-type SMR possesses the best PFC control performance, since its AC input current is directly related to the switched inductor current; (iii) The circuits of (d), (i) and (k) possess the common features of having both continuous input and output currents, and hence needing less stringent filter design requirement.

In addition to the SMRs of (j) to (l) mentioned above, some isolated SMRs specifically for PMSM drives (Singh B. & Singh S., 2010) include: (a) push-pull buck; (b) push-pull boost; (c) half-bridge buck; (d) half-bridge boost; (e) full-bridge buck; (f) full-bridge boost. The push-pull boost SMR possesses excellent PFC control ability and high voltage boost ratio.

3.2 Three-phase SMRs

Detailed surveys for the existing three-phase SMR circuits can be referred to (Hengchun et al, 1997; Shah & Moschopoulos, 2005). The complexities of schematic and control mechanism depend on the control ability and the desired performances. Some commonly used boost-type SMRs are briefly introduced as followed.

3.2.1 Three-leg six-switch standard SMR

The standard three-phase six-switch SMR (Hengchun et al, 1997; Shah & Moschopoulos, 2005) possesses four operation quadrants and high flexibility in power conditioning control. For a motor drive equipped with such SMR, it may possess regenerative braking ability. However, the switch utilization ratio of this SMR is low, and its control is complicated.

3.2.2 Four-leg eight-switch SMR

In the four-leg three-phase SMR (Zhang et al, 2000) with eight switches, the additional fourth leg can be arranged to regulate the imbalance caused by source voltage and switching operation, and it can provide fault tolerant operation.

3.2.3 Three-switch Vienna SMR

The Vienna three-phase SMR (Youssef et al, 2008) uses only three switches to achieve good current command tracking control. It can be regarded as a simplified version of three single-phase PFCs connected to the same intermediate bus voltage. The major features of this SMR are: (i) three output voltage levels ($0.5v_o$, v_o, $-0.5v_o$) providing larger switching control flexibility; (ii) lower switch voltage rating, $0.5\,v_o$ rather than v_o; and (iii) lower input current distortion. However, it has only unidirectional power flow capability, and needs complicated power switch and two serially connected capacitors. The specific power switch (VUM 25-05) for implementing this SMR is avaiable from IXYS Corporation, USA.

3.2.4 Single-switch SMR

The three-phase single-switch SMR (3P1SW) possesses the simplest schematic and control scheme. By operating it in DCM, the PFC is naturally preserved without applying current PWM control. However, it possesses the limits: (i) Having higher input peak current and switch stress; (ii) The input line current contains significant lower-frequency harmonics with the orders of $6n \pm 1$, n=1, 2, ..., and the dominant ones are the 5th and 7th harmonics. Thus suitably designed AC-side low-pass filter is required to yield satisfactory power quality; (iii) The line drawn power quality is limited, typically the power factor is slightly higher than 0.95; (iv) Similarly, this 3P1SW SMR possesses only one-quadrant capability.

To improve the input power quality of this three-phase single-switch SMR, many existing researches have been conducted, see for example: (i) Fifth-order harmonic band-stop filtering; (ii) Harmonic-injection approach; (iii) Variable switching frequency controls; (iv) Passive filtering and input current steering; (v) Optimum PWM pattern; and (vi) Injected PWM robust compensation control. In (Chai et al, 2010), the robust current harmonic cancellation scheme is developed to yield improved line drawn power quality. The robust cancellation weighting factor is automatically tuned according to load level.

3.2.5 Two-switch SMR

This SMR (Badin & Barbi, 2008) is constructed by two serially connected DC/DC boost converter cells behind the rectifier. It possesses only unidirectional power flow capability. The boost converters are applied to shape the input currents, and the current injection device is used to inject the third-harmonic currents in front of the diode bridge to improve the line drawn power quality. This converter uses fewer switches but possesses higher input current harmonics.

3.2.6 Modular connection using single-phase SMRs

Similar to three-phase transformers, three-phase SMRs can also be formed by suitable connection of multiple single-phase modules (Hahn et al, 2002; Li & Liaw, 2004c). Fig. 3(a) shows a Y-connected three-phase boost-type SMR. For Δ-connected three-phase SMR, when one module is faulted, the remaining two modules can continuously provide DC power output subject to the reduction of rating.

3.2.7 Bridgeless SMR

As shown in Fig. 3(b) (Reis et al, 2008; Oliverira et al, 2009), the SMR uses three diodes and three switches rather than using diode bridge rectifier. Obviously, one diode drop is eliminated in each line-current path resulting to increase the efficiency compared to single-switch SMR. However, two additional power switches are employed.

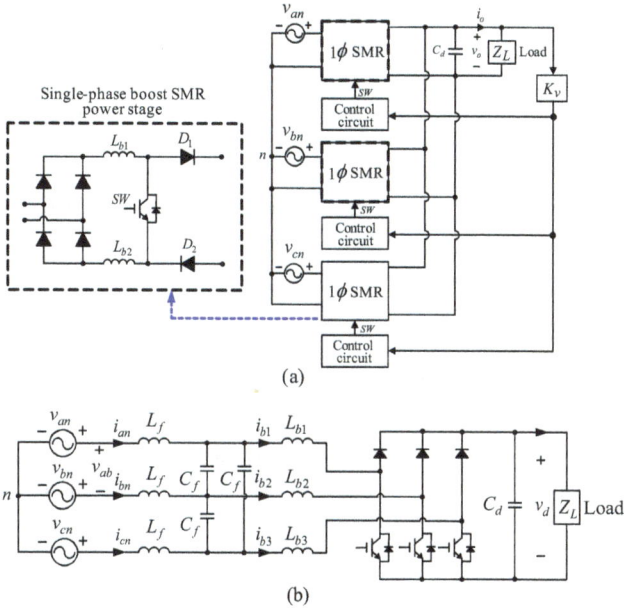

(a)

(b)

Fig. 3. Two types of SMRs: (a) modular connection of three single-phase SMRs; (b) bridgeless DCM three-phase SMR

3.3 Three-phase single-switch ZCT SMRs

The soft-switching SMRs using auxiliary switching circuit can be generally classified into zero-voltage-transition (ZVT) and zero-current-transition (ZCT). The choice depends on the semiconductor devices to be used. The ZVS approaches are generally recommended for MOSFET. On the other hand, ZCS approaches are effective for IGBT. Some existing soft-switching SMRs are introduced as follows:

3.3.1 Classical three-phase single-switch ZCT SMR

The classical 3P1SW ZCT SMR (Wang et al, 1994) is simple in structure and easy to realize. However, the auxiliary switch is not operated on ZCS at turn-off. The efficiency is limited.

3.3.2 Modified three-phase single-switch ZCT SMR

In the modified 3P1SW ZCT SMR presented in (Das & Moschopoulos, 2007). The addition of the transformer in the auxiliary circuit let the circulating energy from the auxiliary circuit be transferred to the output. Hence it possesses higher efficiency than the classical type.

3.3.3 Three-phase three-switch bridgeless ZCT SMR

As to the three-phase bridgeless ZCT SMR (Mahdavi & Farzanehfard, 2009), the auxiliary circuit provides soft-switching condition through ZCT approach for all semiconductor devices without any extra current and voltage stress.

4. Operation principle and some key issues of SMR

4.1 Single-phase SMRs

Fig. 4 shows the conceptual configuration of a single-phase SMR. The AC source input voltage is expressed as $v_{ac} = V_m \sin \omega t = \sqrt{2} V_{ac} \sin \omega t$. If the AC input current i_{ac} can be regulated to be sinusoidal and kept in phase with v_{ac}, then the ideal SMR is similar to an emulated resistor with the effective resistance of R_e viewing from the utility grid. In reality, the double line frequency output voltage ripple always exists for an actual SMR with finite value of output filtering capacitor. This ripple may contaminate to distort the current command, and hence to worsen the power quality control performance. The output power $p(t)$ of the SMR shown in Fig. 4 can be expressed as:

$$p(t) = P_{ac} = \frac{V_m^2}{R_e} \sin^2 \omega t = \frac{V_m^2}{2R_e}(1 - \cos 2\omega t) = \frac{V_{ac}^2}{R_e} - \frac{V_{ac}^2}{R_e} \cos 2\omega t$$

$$= \frac{V_d^2}{R_d} - \frac{V_{ac}^2}{R_e} \cos 2\omega t \triangleq P_d + P_{ac2}$$

(1)

where P_d and P_{ac2} respectively denote the output DC and the double-frequency power components. From the average power invariant property in (1), one can obtain the following equivalent resistance transfer relationship:

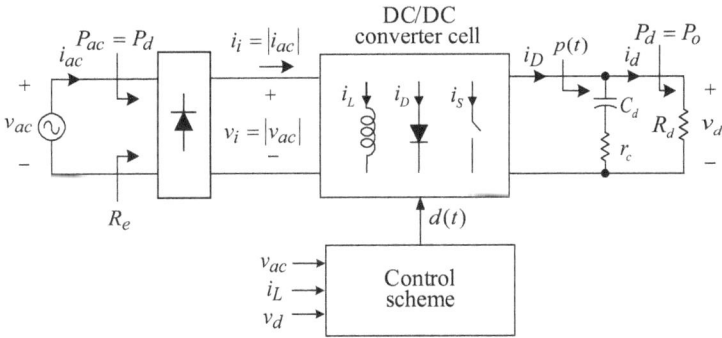

Fig. 4. Conceptual configuration of a single-phase SMR

$$\frac{V_d}{V_{ac}} = \sqrt{\frac{R_d}{R_e}}$$

(2)

By neglecting the capacitor ESR r_c in Fig. 4, the current $i_d(t)$ can be found from (1):

$$i_d(t) \cong \frac{p(t)}{V_d} = \frac{V_{ac}^2}{R_e V_d}(1 - \cos 2\omega t) = I_d + i_{d2}$$

(3)

The AC component i_{d2} is approximately regarded flowing through the capacitor:

$$C_d \frac{d\Delta v_d(t)}{dt} = -\frac{V_{ac}^2}{R_e V_d} \cos 2\omega t$$

(4)

Then the voltage ripple $\Delta v_d(t)$ can be found by integrating the above equation:

$$\Delta v_d(t) = \frac{1}{C_d}\int -\frac{V_{ac}^2}{R_e V_d}\cos 2\omega t\, dt = \frac{1}{2\omega C_d V_d}\frac{V_{ac}^2}{R_e}\sin 2\omega t = \frac{V_d}{2\omega C_d}\frac{1}{R_d}\sin 2\omega t \tag{5}$$

From (5) one can get the peak to peak value of output ripple voltage:

$$\Delta v_d = \frac{V_d}{\omega\, C_d R_d} \tag{6}$$

4.2 Three-phase SMRs
4.2.1 Three-Phase Single-Switch (3P1SW) SMR
For a well-regulated three-phase single-switch (3P1SW) DCM SMR shown in Fig. 5, it can be regarded as a loss-free emulated resistor R_e viewing from the phase AC source with line drawn current having dominant 5th and 7th harmonics (Chai et al, 2010). Hence, the three-phase line drawn instantaneous power can be approximately expressed as:

$$
\begin{aligned}
p_{ac} &= v_{an}i_a + v_{bn}i_b + v_{cn}i_c = \frac{V_m^2\sin\omega t}{R_e}[\sin\omega t - \frac{1}{5}\sin 5\omega t - \frac{1}{7}\sin 7\omega t] + \\
&\quad \frac{V_m^2\sin(\omega t - 2\pi/3)}{R_e}[\sin(\omega t - 2\pi/3) - \frac{1}{5}\sin 5(\omega t - 2\pi/3) - \frac{1}{7}\sin 7(\omega t - 2\pi/3)] + \\
&\quad \frac{V_m^2\sin(\omega t + 2\pi/3)}{R_e}[\sin(\omega t + 2\pi/3) - \frac{1}{5}\sin 5(\omega t + 2\pi/3) - \frac{1}{7}\sin 7(\omega t + 2\pi/3)] \\
&= \frac{3V_m^2}{2R_e} - \frac{3V_m^2}{35R_e}\cos 6\omega t \triangleq P_{ac} + \delta p_{ac}
\end{aligned}
\tag{7}
$$

where P_{ac} = average AC power, δp_{ac} = ripple AC power. By neglecting all power losses, one has $P_o = P_d$, i.e.,

$$P_{ac} = \frac{3V_m^2}{2R_e} = \frac{V_d^2}{R_d} \tag{8}$$

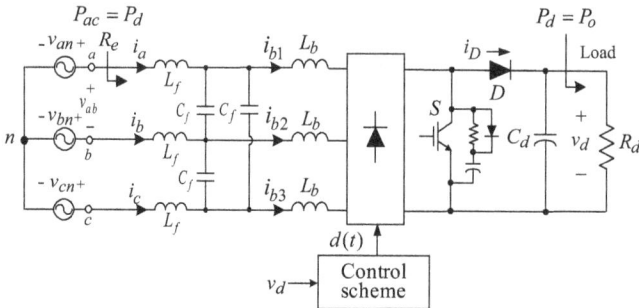

Fig. 5. Conceptual configuration of a three-phase DCM SMR

Then from (7) and (8), the AC charging current flowing the output filtering capacitor is:

$$C_d \frac{dv_d}{dt} = -\frac{2V_d}{35R_d} \cos6\omega t \qquad (9)$$

Thus one can derive the peak-to-peak output voltage ripple:

$$\Delta v_d = \frac{2V_d}{105\omega R_d C_d} \qquad (10)$$

4.2.2 Three-phase three-switch and six-switch SMRs
For the Vienna SMR and three-phase six-switch standard SMR with ideal current mode control, the three-phase line drawn currents will be balanced without harmonics. Hence, from (7) one can find that the DC output voltage ripple will be nearly zero.

4.3 Some key issues of SMR
Taking the DSP-based single-phase standard boost SMR as an example, some key issues are indicated in Fig. 6. In power circuit, the ripples and ratings of the constituted components must be derived, and accordingly the components are properly designed and implemented. Some typical examples can be referred to (Li & Liaw, 2003; Chai & Liaw, 2007; Y.C. Chang & Liaw, 2009a; H.C. Chang & Liaw, 2009).
As to the control scheme, the sensed inductor current and output voltage should be filtered. The feedback controller must first be properly designed considring the desired perfromance and the effects of comtaiminated noises in sensed variables. For satisfying more strict control requirements, in addition to the basic feedback controls, the robust tracking error cancellation controls (Chai & Liaw, 2007; Y.C. Chang & Liaw, 2009a) can further be added. In making DSP-based digital control, the sampling rates are selected according to the achievable loop dynamic response. Other issues may include: (a) random switching to yield spread harmonic spectral distribution (Li & Liaw, 2004b; Chai & Liaw, 2008; Y.C. Chang & Liaw, 2011); (b) the effects of DC-link ripples on the motor drive operating performance (Chai & Liaw, 2007, 2009; Chai et al, 2010); (c) rating enlargement via parallel connection of transformers (Y.C. Chang & Liaw, 2009b) and SMR modules (Li & Liaw, 2004a).

5. Comparative evaluation of three single-phase boost SMRs

Three single-phase boost SMRs are comparatively evalued their prominences experimentally in serving as front-end AC/DC converters of a PMSM drive. For completeness, the traditional diode rectifier is also included as a reference.

5.1 Standard single-phase boost SMR
5.1.1 System configuration
The power circuit and control scheme of the developed SMR are shown in Fig. 6, wherein the two robust controllers are removed. This control system belongs to multi-loop configuration consisting of inner RC-CCPWM scheme and outer voltage loop. The low-pass filtering cut-off frequencies for the sensed current and voltage are respectively set as $f_{ci} = 12\text{Hz}$ and $f_{cv} = 600\text{Hz}$. And the digital control sampling rates of the two loops are chosen as $f_{si} = f_s = 25\text{kHz}$ and $f_{sv} = 2.5\text{kHz}$.

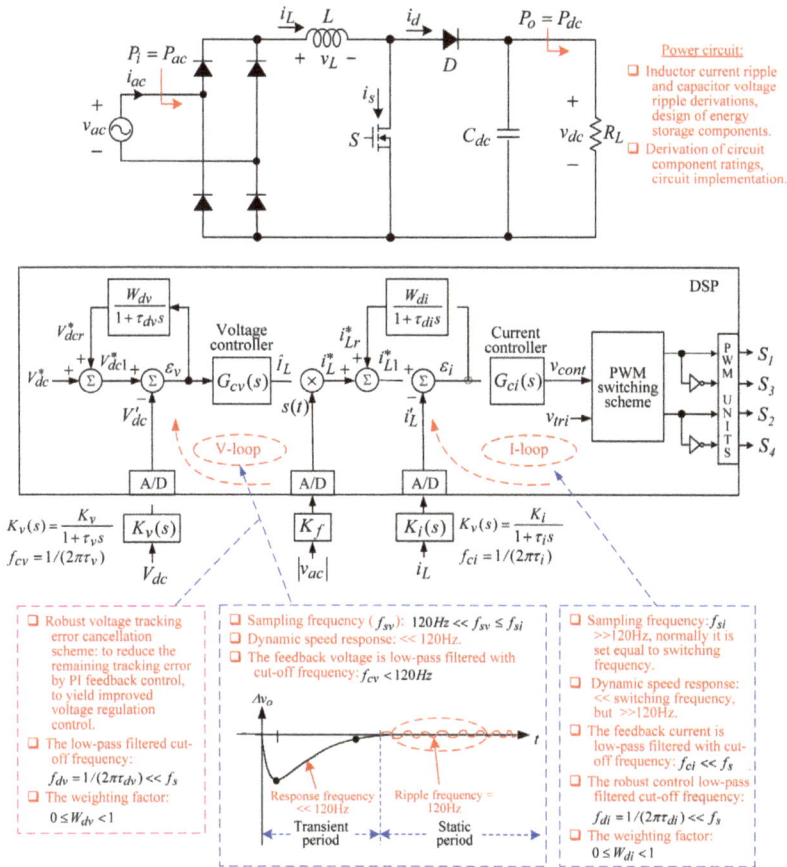

Fig. 6. Key issues of a DSP-based single-phase standard boost SMR

The system variables and specifications of the established SMR are given as follows:

AC input voltage: $V_{ac} = 110V \pm 10\% / 60Hz$.

DC output: $V_{dc} = 300V \sim 350V$ ($\geq 110V \times 1.1 \times \sqrt{2} = 171V$), $P_{dc} = 1500W$.

Switching frequency: $f_s = 25kHz$.

Efficiency: $\eta \geq 90\%$. Power factor: $PF \geq 0.95$ (Lagging).

5.1.2 Design of circuit components

The design of energy storage inductor, output filtering capacitor and power devices for this type of SMR are made according to the given specifications.

a.   Boosting inductor

Some assumptions are made in performing the inductor design: (i) continuous conduction mode (CCM); (ii) all constituted components are ideal; (iii) $v_{dc} = V_{dc} = 350V$; (iv) $v_{ac} \triangleq \sqrt{2}V_{ac}\sin\omega t$, $V_{ac,\min} = 110V \times 0.9 = 99V$, $\hat{V}_{ac,\min} = \sqrt{2}V_{ac,\min} = 140V$; (v) the inductor current ripple is treated at $\omega t = 0.5\pi$, since at which the current ripple is maximum.

The maximum inductor current occurred at $\omega t = 0.5\pi$ can be calculated as

$$\left(\hat{i}_L\right)_{max} = \frac{P_{dc}}{\hat{V}_{ac,min}\eta} \times 2 = \frac{1500}{110 \times \sqrt{2} \times 0.9 \times 0.9} \times 2 = 23.81\text{A} \tag{11}$$

Let the inductor current ripple be:

$$\Delta i_L = \frac{\hat{V}_{ac,min}DT_s}{L} \leq 0.1\left(\hat{i}_L\right)_{max} = 2.38\text{A} \tag{12}$$

The instantaneous duty ratio at $\omega t = 0.5\pi$ can be found as:

$$D = \frac{V_{dc} - \hat{V}_{ac,min}}{V_{dc}} = \frac{350 - 140}{350} = 0.6 \tag{13}$$

Hence from (12) and (13), the condition of boosting inductance L is obtained as:

$$L \geq \frac{\hat{V}_{ac,min}D}{f_s \Delta i_L} = 1.41\text{mH} \tag{14}$$

The inductor L is formed by serially connected two available inductors L_1 and L_2. The measured inductances using HIOKI 3532-50 LCR meter are L_1 = (2.03mH, ESR = 210m Ω at 60Hz, and 1.978mH, ESR= 5.68 Ω at 25kHz) and L_2 = (2.11mH, ESR= 196m Ω at 60Hz, and 1.92mH, ESR= 62 Ω at 25kHz). Hence $L = L_1 + L_2$ =4.14mH, which is suited here.

b. Output capacitor

By choosing the output filtering capacitor $C_d = 2200\mu F/450V$, the peak-to-peak output voltage ripple can be found as:

$$\Delta V_{dc} = \frac{V_{dc}}{\omega_1 R_L C_d} = \frac{P_{dc}}{\omega_1 C_d V_{dc}} = \frac{1500}{2\pi \times 60 \times 2200 \times 10^{-6} \times 350} = 5.17\text{V} \tag{15}$$

c. Power semiconductor devices

The maximum current of the main switch S and the diode D is $\left(\hat{i}_L\right)_{max} + 0.5\Delta i_L = 25\text{A}$, which is calculated from (11) and (12), and their maximum voltage is 350V. Accordingly, the MOSFET IXFK44N80P (IXYS) (800V, ID= 44A (continuous), IDM = 100A (pulsed)) and the fast diode DSEP60-06A (IXYS) (600V, average current IFAVM = 60A) are chosen for implementing the main switch S and all diodes respectively.

5.1.3 Control schemes
Current controller:
he current feedback controller $G_{ci}(s)$ in Fig. 6 is chosen to be PI-type:

$$G_{ci}(s) = K_{Pi} + \frac{K_{Ii}}{s} \tag{16}$$

The upper limit of the P-gain is first determined based on large-signal stability at switching frequency:

$$\frac{dv_{cont}}{dt} = K_{Pi} K_i \frac{|V_{dc} - |v_{ac}||}{L} < \frac{dv_{tri}}{dt} \tag{17}$$

The parameters of the developed SMR shown in Fig. 6 are set as: $V_{dc} = 300\text{V}$, $K_i = 0.04\text{V/A}$, $f_s = 25\text{kHz}$, $L = 4.14\text{mH}$ and $dv_{tri}/dt = 25\text{kV/sec}$. Using the given data, the upper value of the K_{Pi} can be found from (17) to be $K_{Pi} < \bar{K}_{Pi} = 8.625$ ($v_{ac} = 0$ is set here). Accordingly $K_{Pi} = 4.0$ is set.

In making the determination of integral gain, the magnitude frequency response of the loop gain $LG(s = j\omega) \triangleq i_L(s) / \varepsilon_i(s)\big|_{s=j\omega}$ is measured using the HP 3563A control systems analyzer as shown in Fig. 7, wherein V_{inj} denotes an injected swept sine signal. Fig. 8 shows the measured magnitude frequency response of the loop gain. The measurement conditions are set as: (i) $v_{inj,peak} = 10\text{mV}$; (ii) swept sine frequency range is from 400Hz to 11kHz; (iii) the voltage loop is opened, and the current command is set as $I_L^* = \hat{I}_L \times |v_{ac}|$ with $\hat{I}_L = 8\text{A}$; (iv) $R_L = 200\Omega$; (v) $v_{ac} = 110\text{V}_{rms}$; (vi) the current feedback controllers are set as $K_{Pi} = 4$ and $K_{Ii} = 45000$. The measured result in Fig. 8 indicates that the crossover frequency is $f_c = 1.47\text{kHz} < f_s / 2$, which is reasonable for a ramp-comparison current-controlled PWM scheme. Hence finally,

$$G_{ci}(s) = K_{Pi} + \frac{K_{Ii}}{s} = 4 + \frac{45000}{s} \tag{18}$$

If the measurement of loop-gain frequency response is not convenient, one can also use the derived small-signal dynamic model (Chai & Liaw, 2007), or using trail-and-error approach to determine the integral gain.

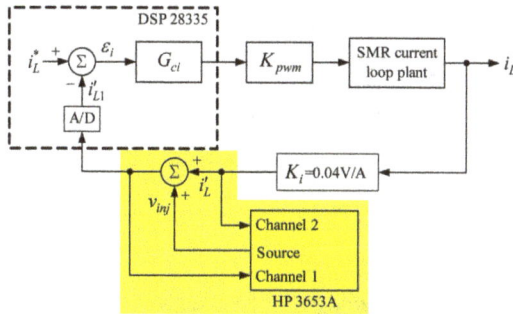

Fig. 7. System configuration in current loop gain measurement

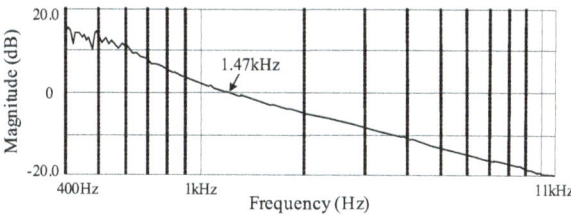

Fig. 8. Measured magnitude frequency response of current loop gain

Voltage controller:
Although the quantitative controller design can be achieved (Y.C. Chang & Liaw, 2009a), the
PI voltage feedback controller is chosen trial-and-error here to be:

$$G_{cv}(s) = K_{Pv} + \frac{K_{Iv}}{s} = 8 + \frac{200}{s} \tag{19}$$

5.1.4 Experimental results

Let V_{ac} =110 V /60Hz and V_{dc} = 300V , the measured efficiencies η , THD_i of i_{ac} and PF at
(R_L = 400Ω , P_{dc} =227.7W) and (R_L = 200Ω , P_{dc} =473.6W) are summarized in Table 1. And
the measured (i_L^* , i_L'') and (v_{ac} , i_{ac}) under (R_L = 200Ω , P_{dc} =473.6W) are shown in Figs. 9(a)
and 9(b). The results indicate that the input current i_{ac} is nearly sinusoidal and kept almost
in phase with the utility voltage v_{ac} . Good line drawn power quality can also be observed
from Table 1.

Load cases Variables	Resistive load (R_L = 400Ω)	Resistive load (R_L = 200Ω)
V_{ac}	110V/60Hz	110V/60Hz
P_{ac}	241.6W	502.2W
V_{dc}	300.8V	300.2V
P_{dc}	227.7W	473.6W
η	94.25%	94.31%
THD_i	6.61%	6.11%
PF (Lagging)	0.992	0.994

Table 1. Measured steady-state characteristics of the standard boost SMR under two loads

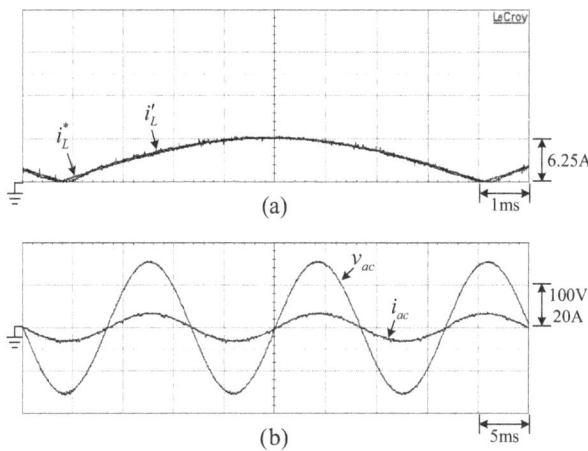

Fig. 9. Measured results of the standard boost SMR at R_L = 200Ω : (a) (i_L^* , i_L''); (b) (v_{ac} , i_{ac})

5.2 Bridgeless boost SMR
5.2.1 System configuration

Fig. 10 shows the bridgeless boost SMR, its control scheme is identical to those shown in Fig. 6 with the two switches being respectively operated in positive and negative half cycles. Although the efficiency of bridgeless SMR can be slightly increased, it possesses the common mode EMI problem due to the large parasitic capacitance between the output and ground, which provides a relatively low impedance path. To reduce this problem, the boosting inductor is divided into two equal inductors, and they are placed at AC source side.

5.2.2 Circuit design

The specifications are identical to those listed above. The two inductors L_1 and L_2 in Sec. 5.1.2 are used here as the two bridgeless boost SMR inductors, i.e., $L_1 = L_2 = 0.5L$.

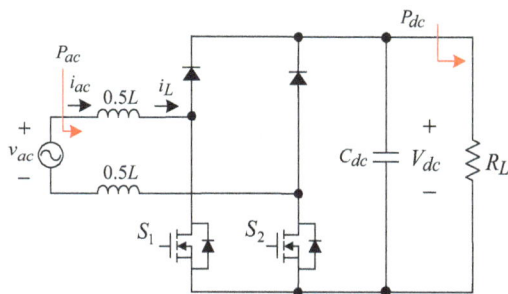

Fig. 10. Schematic and control scheme of the developed bridgeless boost SMR

5.2.3 Control schemes

Current controller: Following the similar process introduced in Sec. 5.1.3 one can get $K_{Pi} < \overline{K}_{Pi} = 8.625$. Hence, $K_{Pi} = 3.0$ is set and the integral gain is chosen via trial-and-error. Finally:

$$G_{ci}(s) = K_{Pi} + \frac{K_{Ii}}{s} = 3 + \frac{2000}{s} \qquad (20)$$

Voltage controller: The PI voltage feedback controller is chosen to be:

$$G_{cv}(s) = K_{Pv} + \frac{K_{Iv}}{s} = 8 + \frac{200}{s} \qquad (21)$$

5.2.4 Experimental results

The measured key waveforms are almost identical to Fig. 9 and are not repeated here. Table 2 lists the measured efficiencies η , THD_i of i_{ac} and PF at two loads. From Tables 1 and 2 one can find the slight higher efficiencies being yielded by the bridgeless SMR.

5.3 Current-Fed Push-Pull (CFPP) isolated boost SMR
5.3.1 System configuration and operation

The power circuit and control scheme of the CFPP isolated boost SMR are shown in Figs. 11(a) and 11(b). In making the analysis, some assumptions are made: (i) all circuit components are ideal; (ii) the active voltage clamp circuits including S_3, S_4 and C_a are neglected; (iii) $v_{in} = |v_{ac}| = V_m \sin \omega t = \sqrt{2} V_{ac} \sin \omega t$; (iv) the circuit is operated under CCM. In the established current-fed push-pull SMR, the duty ratio $D \triangleq t_{on} / T_s$ ($0.5 < D < 1$) is set. The gate signal of S_2 is generated from S_1 by shifting $180°$. Detailed analysis process can be referred to (Hsieh, 2010), only a brief description and some key formulas are given here. During analysis, the voltage transfer ratio from v_{in} to V_{dc} can be derived as:

$$\frac{V_{dc}}{v_{in}} = \frac{N_s}{N_p} \frac{1}{2(1-D)} \tag{22}$$

It should be noted that the duty ratio D is a time varying function for the constant V_{dc} and time varying input DC voltage $v_{in} = |v_{ac}|$. Moreover, the variations of V_{dc} and V_{ac} should be considered in making the derivation of component ratings.

Load Cases Variables	Resistive load ($R_L = 400\Omega$)	Resistive load ($R_L = 200\Omega$)
V_{ac}	110V/60Hz	110V/60Hz
P_{ac}	233.1W	497.5W
V_{dc}	300.6V	300.2V
P_{dc}	223.5W	475.9W
η	95.88%	95.66%
THD_i	6.02%	6.11%
PF (Lagging)	0.996	0.996

Table 2. Measured characteristics of the developed bridgeless boost SMR under two loads

5.3.2 Circuit design

a. Specifications

The system variables and specifications of the established SMR are given as follows:
AC input voltage: $V_{ac} = 110\text{V} \pm 10\% / 60\text{Hz}$.
DC output: $V_{dc} = 300\text{V} \sim 350\text{V}$ ($\geq 110\text{V} \times 1.1 \times \sqrt{2} = 171\text{V}$), $P_{dc} = 1200\text{W}$.
Switching frequency: $f_s = 25\text{kHz}$, Efficiency: $\eta \geq 75\%$, $PF \geq 0.95$ (Lagging).

b. Boosting inductor

To provide magnetization path of the inductor, duty cycle must be greater than 0.5 at any time, and from (22):

$$n \triangleq \frac{N_s}{N_p} \leq \frac{300}{\sqrt{2} \times 110} \frac{2(1-0.5)}{1} \tag{23}$$

Thus the turn ratio can be found to be $n \leq 1.935$. By choosing $n = 1$, the instantaneous duty ratio at $\omega t = 0.5\pi$ can be found from (22) as:

$$D_{max} = 1 - \frac{\hat{V}_{ac,min}}{2V_{dc,max}} = 1 - \frac{140}{2 \times 350} = 0.8 \tag{24}$$

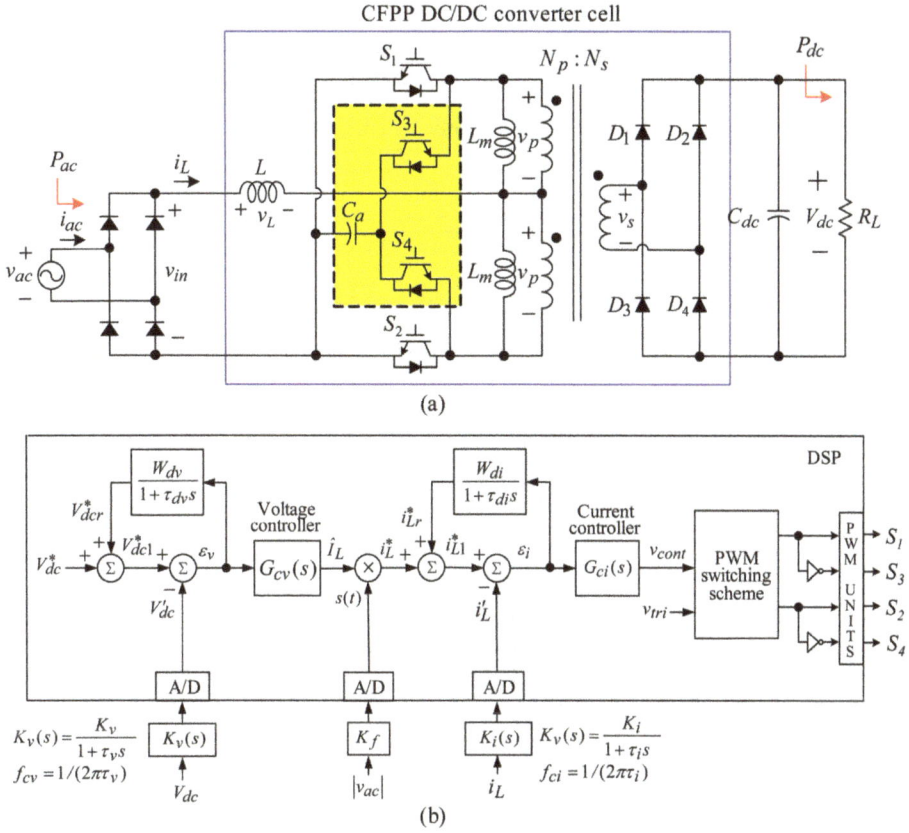

Fig. 11. The current-fed push-pull isolated boost SMR: (a) power circuit; (b) control scheme

The maximum inductor current occurred at $\omega t = 0.5\pi$ can be calculated as:

$$(\hat{i}_L)_{max} = \frac{P_{dc}}{\hat{V}_{ac,min}\eta} \times 2 = \frac{1200}{110 \times \sqrt{2} \times 0.9 \times 0.75} \times 2 = 22.856A \qquad (25)$$

Let the inductor current ripple be:

$$\Delta i_L = \frac{\hat{V}_{ac,min}(D-0.5)T_s}{L} \leq 0.1(\hat{i}_L)_{max} = 2.2856A \qquad (26)$$

The condition of boosting inductance L is obtained as:

$$L \geq \frac{\hat{V}_{ac,min}(D-0.5)}{f_s \Delta i_L} = 0.735mH \qquad (27)$$

The inductances of an available inductor measured using HIOKI 3532-50 LCR meter are $L = (2.03mH, ESR = 210m\Omega$ at 120Hz) and (1.978mH, ESR = 5.68 Ω at 25kHz). Hence this inductor is suited and employed here. Using the inductance of $L = 1.978\text{mH}$ at $f = 25\text{kHz}$, the inductor current ripple given in (26) becomes $\Delta i_L = 0.85\text{A}$.

c.   Output capacitor

The output filtering capacitor $C_d = 2200\mu\text{F}/450\text{V}$ is chosen to yield the following peak-to-peak voltage ripple:

$$\Delta V_{dc} = \frac{V_{dc}}{\omega_1 R_L C_d} = \frac{P_{dc}}{\omega_1 C_d V_{dc}} = \frac{1200}{2\pi \times 60 \times 2200 \times 10^{-6} \times 350} = 4.13\text{V} \qquad (28)$$

d.   Power semiconductor devices

From (25) and (26), the maximum current flowing through the switches and all the diodes can be calculated as $i_{S,\max} = (\hat{i}_L)_{\max} + 0.5\Delta i_L = 22.856 + 0.5 \times 0.85 = 23.28\text{A}$. The maximum voltage for the switches is 700V which is found from Fig. 11, and voltage rating for the load side rectifier diodes is 350V. Accordingly, the IGBT K40T120 (Infineon) (1200V, ID= 40A ($100°\text{C}$, continuous), IDM = 105A (pulsed)) and the fast diode DSEP60-06A (IXYS) (600V, average current IFAVM = 60A) are chosen for implementing the switches and all the diodes, respectively.

e.   Transformer design

The AMCU series UU core AMCU-80 manufactured by AMOSENSE Cooperation is used to wind the push-pull transformer here. The designed results (Hsieh, 2010) are summarized as followed. To lower the core loss, B = 0.25T is set, and thus the maximum flux density variation will be $\Delta B = 2 \times 0.25\text{T} = 0.5\text{T}$. From Faraday's law, the turns of the primary side can be expressed as follows:

$$N_p = \frac{n \times V_{dc.\max}(1 - D_{\min})T_s}{A_e \times \Delta B} \qquad (29)$$

The known parameters in (29) are: $n = 1$, $V_{dc,\max} = 350\text{V}$, $D_{\min} = 0.5$, $A_e = 5.21\text{cm}^2$, $T_s = 40\mu\text{s}$. Hence $N_p = 26.87$ is found, and $N_p = N_s = 32$ are chosen here. The measured parameters of the designed transformer at $f = 25\text{kHz}$ using HIOKI 3532-50 LCR meter are: $L_m = 1.086\text{mH}$, $L_{ls1} = 10.795\mu\text{H}$, $L_{ls2} = 8.838\mu\text{H}$, ESR = 20.4Ω, where L_{ls1} and L_{ls2} denote the leakage inductances of the two transformer primary windings.

f.   Active voltage clamp

As generally known that a current-fed push-pull boost converter may possess serious problems due to the voltage spikes caused by transformer leakage inductances, the problems lie in having lower efficiency and increased voltage stress of power switches. The active voltage clamp circuit (Kwon, 2008; Sangwon & Sewan, 2010) is used to solve this problem. As shown in Fig. 11(a), the active voltage clamp circuit consists of two auxiliary switches (S_3, S_4) and one capacitor C_a. These two auxiliary switches are switched in complement fashion to the two main switches (S_1, S_2) but with a small dead-time. The used components for active voltage clamp circuit are: $C_a = 0.4\mu\text{F}/1000\text{V}$, S_3 and S_4 are IGBT K40T120 (Infineon) (1200V, ID= 40A ($100°\text{C}$, continuous), IDM = 105A (pulsed)), the dead-time $t_d = 1\mu\text{s}$ is set here.

5.3.3 Controller design of CFPP Isolated boost SMR

a.   Current controller

Similarly, the upper value of K_{Pi} can be found from (17) to be $K_{Pi} < \overline{K}_{Pi} = 1.53$ (V_{dc} is replaced by v_p and $v_{ac} = 0$ is set). Accordingly $K_{Pi} = 0.5$ is set. Then the integral gain is chosen via trial-and-error, and finally it is found that:

$$G_{ci}(s) = K_{Pi} + \frac{K_{Ii}}{s} = 0.5 + \frac{2500}{s} \tag{30}$$

The robust current tracking error cancellation controller shown in Fig.11(b) is not applied here.

b.   Voltage controller

The voltage loop dynamic model and the proposed feedback control scheme are shown in Fig. 12, the SMR is reasonably represented by a first-order process in main dynamic frequency range. The voltage feedback sensing factor is set as $K_v = 0.002\,\mathrm{V/V}$. The desired voltage response due to a step load power change is also sketched in Fig. 12, which possesses the key features: (i) no overshoot and steady-state error; (ii) the typical key response points indicated in Fig. 12 are: ($t_1 = t_f$, $\Delta V_{dc1} = 0.5\Delta v_{om}$), ($t_2 = t_m$, $\Delta V_{dc2} = \Delta v_{om}$), ($t_3 = t_{re}$, $\Delta V_{dc3} = 0.1\Delta v_{om}$), with t_f = fall time, t_m = the time at which maximum dip being occurred, t_{re} = restore time, Δv_{om} = maximum voltage dip.
For the ease of implementation, the PI voltage feedback controller is chosen:

$$G_{cv}(s) = \frac{K_{Pv}s + K_{Iv}}{s} \tag{31}$$

The quantitative design technique presented in (Y.C. Chang & Liaw, 2009a) is applied to here to find the parameters of $G_{cv}(s)$ to have the desired regulation response shown in Fig. 12. The details are neglected and only a brief description is given here.

a.   Dynamic model estimation

   i.   Let the $G_{cv}(s) = K_{Pv} + K_{Iv}/s = 6 + 10.5/s$ be arbitrary set, and the SMR is normally operated at the chosen operating point ($V_{dc}^* = 300\mathrm{V}$, $P_{dc} = 302.8\mathrm{W}$).

   ii.   A step load resistor change of $R_L = 300\Omega \to 200\Omega$ ($\Delta P_{dc} = 149.3\mathrm{W}$, $P_{dc} = 302.8\mathrm{W} \to 452.1\mathrm{W}$) is applied and the response of V_{dc} is recorded. By choosing three typical response points as indicated in Fig. 12 to be ($-4.4\mathrm{V}$,27.5ms), ($-7.6\mathrm{V}$,55ms) and ($-1\mathrm{V}$,1500ms), through careful derivation (Y.C. Chang & Liaw, 2009a), one can obtain the estimated dynamic model parameters are obtained:

$$a = 7.95\,,\ b = 2975.71\,,\ K_{pl} = 0.00084542 \tag{32}$$

b.   Controller design

At the given operating point ($V_{dc} = 300\mathrm{V}$, $R_L = 300\Omega$), the voltage regulation control specifications are defined as: $\Delta V_{dc,\max} = 5.0\mathrm{V}$, $t_{re} = 800\,\mathrm{ms}$ for a step load power change of $\Delta P_{dc} = 149.3\mathrm{W}$. Following the quantitative design process presented in (Y.C. Chang & Liaw, 2009a) one can solve to obtain:

$$G_{cv}(s) = K_{Pv} + \frac{K_{Iv}}{s} = 10.0036 + \frac{33.8878}{s} \qquad (33)$$

The simulated and measured output voltage responses (not shown here) are confirmed their closeness and satisfying the specified control specifications.

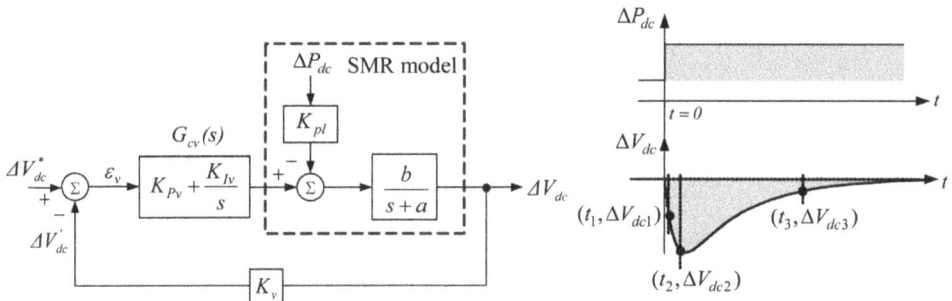

Fig. 12. The established current-fed push-pull boost SMR control scheme and the desired regulation response

c.   Robust voltage error cancellation controller

A simple robust voltage error cancellation controller (RVECC) presented in (Chai & Liaw, 2007) is applied here to enhance the SMR voltage regulation control robustness. In the control system shown in Fig. 11(b), a robust compensation control command V^*_{dcr} is generated from the voltage error ε_v through a weighting function $W_d(s) = W_d / (1 + \tau_d s)$ with W_d being a weighting factor. The low pass filter process with cut-off frequency $f_{cd} = 1 / (2\pi\tau_d) = 120\text{Hz}$ ($\tau_d = 0.0013263$) is used to reduce the effects of high-frequency noises on dynamic control behavior.

From Fig. 11(b) one can derive that the original voltage tracking error $\varepsilon_v = V^*_{dc} - V'_{dc}$ will be reduced to

$$V^*_{dc} - V'_{dc} = (1 - \frac{W_d}{1 + \tau_d s})\varepsilon_v \approx (1 - W_d)\varepsilon_v , \; 0 \le W_d < 1 \qquad (34)$$

where the approximation is made for the main dynamic signals. Hence the original voltage error can be reduced by a factor of $(1 - W_d)$ within main dynamic frequency range. The selection of W_d must be made considering the compromise between control performance and effects of system noises.

Figs. 13(a) and 13(b) show the simulated and measured output voltage responses by PI control without ($W_d = 0$) and with ($W_d = 0.5$) robust control due to a step load power change of $\Delta P_{dc} = 149.3\text{W}$ ($P_{dc} = 302.8\text{W} \rightarrow 452.1\text{W}$, $V^*_{dc} = 300\text{V}$). The results show that they are very close and the effectiveness of robust control in the improvement of voltage regulation response.

Let $V_{ac} = 110\text{V}/60\text{Hz}$ and $V^*_{dc} = 300\text{V}$, and the PI feedback and robust controls are all operated, the measured steady-state characteristics at ($R_L = 400\Omega$, $P_{dc} = 234.6\text{W}$), ($R_L = 200\Omega$, $P_{dc} = 465.8\text{W}$), ($R_L = 133\Omega$, $P_{dc} = 623.8\text{W}$) and ($R_L = 100\Omega$, $P_{dc} = 908.5\text{W}$) are summarized in Table 3. And the measured (i_L^*, i_L) and (v_{ac}, i_{ac}) under ($R_L = 100\Omega$, $P_{dc} = 908.5W$) are shown in Figs.14(a) and 14(b), respectively. From the results, one can find that the developed SMR possesses good power quality control performances under wide load range. The lower efficiencies compared with the previous two types of SMRs are observed, this is mainly due to the addition of isolated HF transformer.

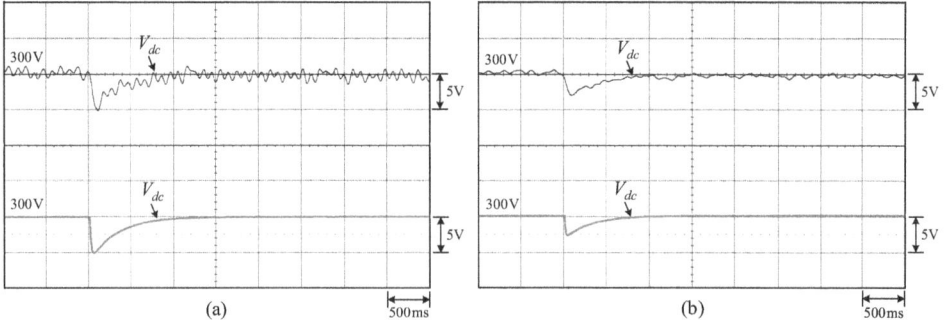

Fig. 13. Measured (upper) and simulated (lower) V_{dc} by PI control without ($W_d = 0$) and with ($W_d \neq 0$) robust control due to a step load power change of $\Delta P_{dc} = 149.3W$ ($P_{dc} = 302.8W \rightarrow 452.1W$, $V_{dc}^* = 300V$): (a) $W_d = 0$; (b) $W_d = 0.5$

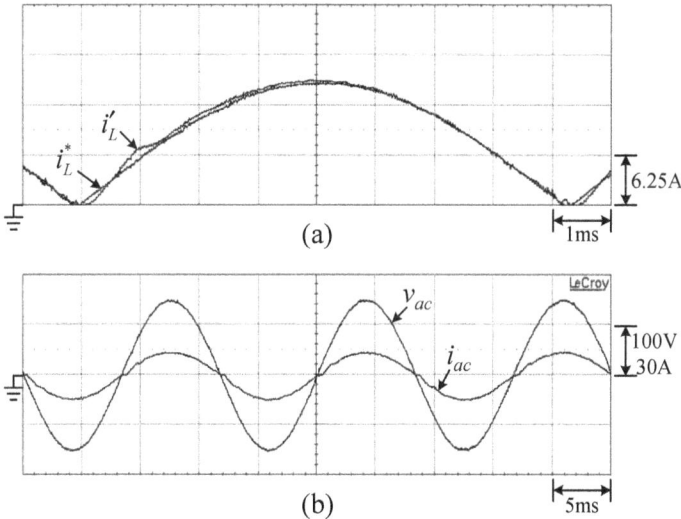

Fig. 14. Measured steady-state results of the current-fed push-pull boost SMR at $V_{ac} = 110V/60Hz$ and $R_L = 100\Omega$: (a) (i_L^*, i_L'); (b) (v_{ac}, i_{ac})

5.4 Evaluation for the PMSM drive with different front-end AC/DC converters
Figs. 15(a) to 15(d) show the standard PMSM drives equipped with different front-end AC/DC converters. In making experimental works, the following inputs are set: (i) Diode

rectifier: $V_{ac} = 220\text{V}/60\text{Hz}$, V_{dc} will vary with loading conditions; (ii) SMRs: $V_{ac} = 110\text{V}/60\text{Hz}$, $V_{dc} = 300\text{V}$ with satisfactory regulation control, the switching frequency $f_s = 25\text{kHz}$ is set. The measured results are summarized as follows:

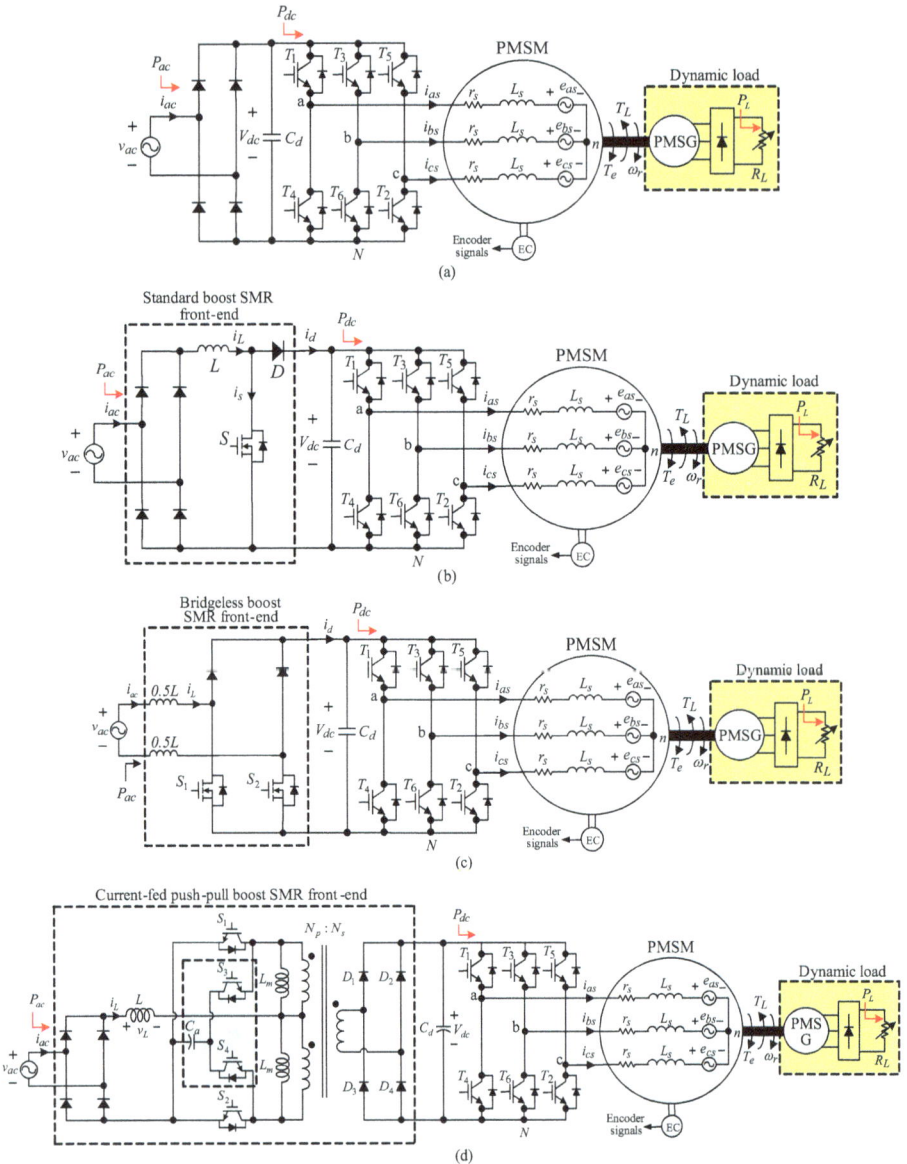

Fig. 15. The circuit configuration of established standard PMSM drive with SMR front-end: (a) diode rectifier front-end; (b) standard boost SMR front-end; (c) bridgeless boost SMR front-end; (d) current-fed push-pull boost SMR front-end

Load Cases Variables	Resistive load ($R_L = 400\Omega$)	Resistive load ($R_L = 200\Omega$)	Resistive load ($R_L = 133\Omega$)	Resistive load ($R_L = 100\Omega$)
V_{ac}	110V/60Hz	110V/60Hz	110V/60Hz	110V/60Hz
P_{ac}	298.6W	557.4W	697.4W	1082.2W
V_{dc}	301.2V	300.8V	300.5V	300.3V
P_{dc}	234.6W	465.8W	623.8W	908.5W
η	78.57%	83.57%	89.45%	83.95%
THD_i	8.83%	6.62%	6.53%	3.82%
PF (Lagging)	0.997	0.998	0.998	0.998

Table 3. Measured characteristics of the current-fed push-pull boost SMR under four loads

5.4.1 Diode rectifier front-end

The measured (ω_r^*, ω_r), (H_A, i_{as}^*, i_{as}') and (v_{ac}, i_{ac}) at ($\omega_r^* = 2000\text{rpm}$, $R_L = 44.7\Omega$) and ($\omega_r^* = 3000\text{rpm}$, $R_L = 44.7\Omega$) are shown in Fig. 16(a) and Fig. 16(b), and the corresponding steady-state characteristics are listed in Table 4. One can notice the normal operation of the PMSM drive under $\omega_r^* = 2000\text{rpm}$. However, the results in Fig. 16(b) indicate that large tracking errors exist in speed and phase current under $\omega_r^* = 3000\text{rpm}$. This is mainly due to the insufficient DC-link voltage ($V_{dc} = 278.4\text{V}$) established by rectifier for encountering the back-EMF effect. In addition, the peaky i_{ac} leads to poor power factor and high THD_i.

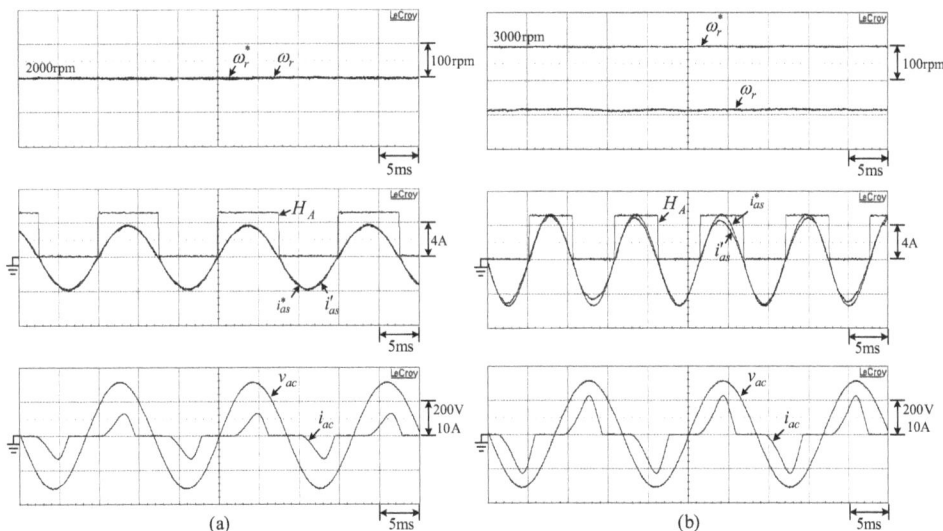

Fig. 16. Measured (ω_a^*, ω_r),(H_A, i_{as}^*, i_{as}') and (v_{ac}, i_{ac}) of the standard PMSM drive with diode rectifier front-end at: (a) ($V_{ac} = 220\text{V}/60\text{Hz}$, $\omega_r^* = 2000\text{rpm}$, $R_L = 44.7\Omega$); (b): ($V_{ac} = 220\text{V}/60\text{Hz}$, $\omega_r^* = 3000\text{rpm}$, $R_L = 44.7\Omega$)

Cases / Variables	$\omega_r^* = 2000\text{rpm}$	$\omega_r^* = 3000\text{rpm}$
V_{ac}	220V/60Hz	220V/60Hz
P_{ac}	527.5W	969.4W
V_{dc}	295.9V	278.4V
P_{dc}	366.7W	706.3W
η	69.52%	72.86%
THD_i	72.95%	65.54%
PF (Lagging)	0.751	0.758

Table 4. Measured characteristics of the standard PMSM drive fed by diode rectifier front-end under two speeds ($V_{ac} = 220\text{V}/60\text{Hz}$, $R_L = 44.7\Omega$)

5.4.2 Three boost SMR front-ends

(i) Standard boost SMR: Fig. 17(a), Fig. 17(b) and Table 5; (ii) Bridgeless boost SMR: Fig. 18(a), Fig. 18(b) and Table 6; (iii) CFPP boost SMR: Fig. 19(a), Fig. 19(b) and Table 7. The results indicate that for all cases, the close winding current tracking performances are obtained, and thus good line drawn power quality characteristics are achieved.

Further observations find that: (i) the efficiencies of bridgeless SMR are slightly higher than those of standard boost SMR; (ii) the efficiencies of the CFPP SMR are lower than the other two SMRs. This is mainly due to the increased losses in the high-frequency transformer.

Fig. 20(a) and Fig. 20(b) show the measured (ω_r^* , ω_r), (i_{qs}^* , i_{qs}') and V_{dc} of the whole PMSM drive with CFPP boost SMR front-end at ($V_{dc} = 300\text{V}$, $\omega_r = 2400\text{rpm}$, $R_L = 44.7\Omega$) due to a step speed command change of 100rpm and due to a step load resistance change from $R_L = 75\Omega$ to $R_L = 44.7\Omega$. The results indicate that good speed tracking and regulating responses are obtained by the developed SMR-fed PMSM drive. And the DC-link voltages V_{dc} are well regulated under these two cases.

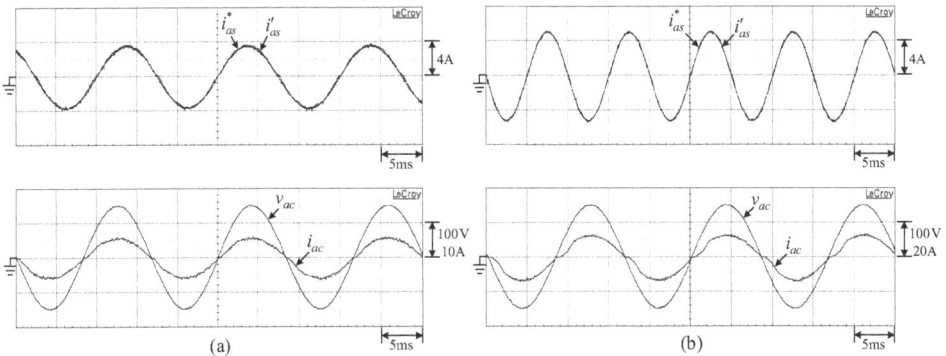

Fig. 17. Measured (ω_r^* , ω_r), (H_A , i_{as}^* , i_{as}') and (v_{ac} , i_{ac}) of the standard PMSM drive with standard boost SMR front-end at: (a) ($V_{ac} = 220\text{V}/60\text{Hz}$, $\omega_r^* = 2000\text{rpm}$, $R_L = 44.7\Omega$); (b): ($V_{ac} = 220\text{V}/60\text{Hz}$, $\omega_r^* = 3000\text{rpm}$, $R_L = 44.7\Omega$)

Cases / Variables	$\omega_r^* = 2000\text{rpm}$	$\omega_r^* = 3000\text{rpm}$
V_{ac}	110V/60Hz	110V/60Hz
P_{ac}	546.5W	1145.3W
V_{dc}	300.6V	300.3V
P_{dc}	365.9W	791.2W
η	66.95%	69.08%
THD_i	8.493%	7.085%
PF (Lagging)	0.998	0.997

Table 5. Measured characteristics of the standard PMSM drive fed by standard boost SMR front-end under two speeds ($V_{dc} = 300\text{V}$, $R_L = 44.7\Omega$)

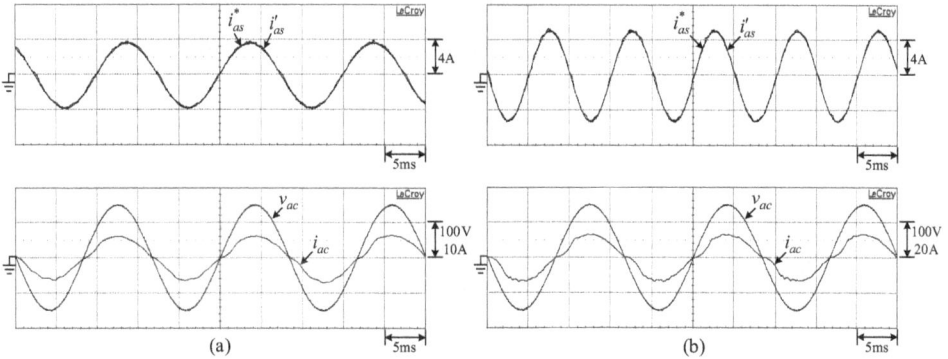

Fig. 18. Measured (ω_r^* , ω_r), (H_A , i_{as}^* , i_{as}') and (v_{ac} , i_{ac}) of the standard PMSM drive with bridgeless boost SMR front-end at: (a) ($V_{ac} = 220\text{V}/60\text{Hz}$, $\omega_r^* = 2000\text{rpm}$, $R_L = 44.7\Omega$); (b): ($V_{ac} = 220\text{V}/60\text{Hz}$, $\omega_r^* = 3000\text{rpm}$, $R_L = 44.7\Omega$)

Cases / Variables	$\omega_r^* = 2000\text{rpm}$	$\omega_r^* = 3000\text{rpm}$
V_{ac}	110V/60Hz	110V/60Hz
P_{ac}	543.9W	1142.4W
V_{dc}	300.4V	300.1V
P_{dc}	364.8W	790.3W
η	67.07%	69.18%
THD_i	8.238%	7.022%
PF (Lagging)	0.998	0.996

Table 6. Measured characteristics of the standard PMSM drive fed by bridgeless boost SMR front-end under two speeds ($V_{dc} = 300\text{V}$, $R_L = 44.7\Omega$)

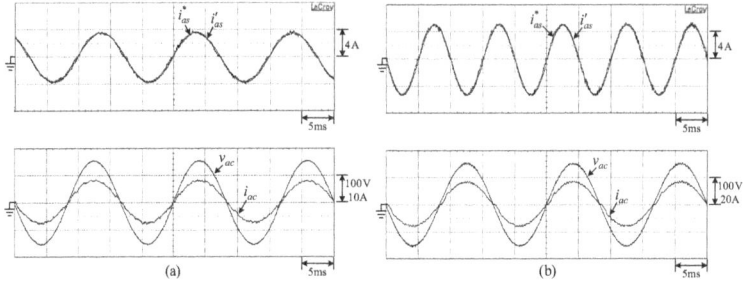

(a)   (b)

Fig. 19. Measured (ω_r^*, ω_r), (H_A, i_{as}^*, i_{as}') and (v_{ac}, i_{ac}) of the standard PMSM drive with current-fed push-pull SMR front-end at: (a) ($V_{ac} = 220\text{V}/60\text{Hz}$, $\omega_r^* = 2000\text{rpm}$, $R_L = 44.7\Omega$); (b): ($V_{ac} = 220\text{V}/60\text{Hz}$, $\omega_r^* = 3000\text{rpm}$, $R_L = 44.7\Omega$)

Cases \\ Variables	$\omega_r^* = 2000\text{rpm}$	$\omega_r^* = 3000\text{rpm}$
V_{ac}	110V/60Hz	110V/60Hz
P_{ac}	597.8W	1231.8W
V_{dc}	301.2V	300.7V
P_{dc}	365.3W	792.1W
η	61.10%	64.30%
THD_i	3.90%	4.02%
PF (Lagging)	0.998	0.998

Table 7. Measured characteristics of the standard PMSM drive fed by current-fed push-pull boost SMR front-end under two speeds ($V_{dc} = 300\text{V}$, $R_L = 44.7\Omega$)

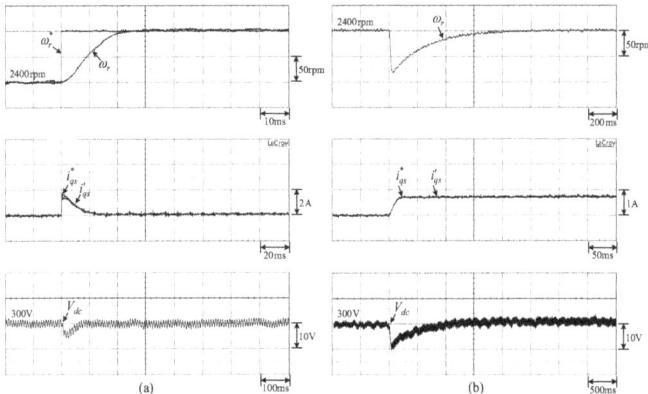

(a)   (b)

Fig. 20. Measured (ω_r^*, ω_r), (i_{qs}^*, i_{qs}') and V_{dc} of the whole PMSM drive with current-fed push-pull boost SMR front-end at ($V_{dc} = 300\text{V}$, $\omega_r^* = 2400\text{rpm}$, $R_L = 44.7\Omega$): (a) due to a step speed command change of 100rpm: (b) due to a step resistive load change from $R_L = 75\Omega$ to $R_L = 44.7\Omega$

At three cases of ($\omega_r = 1500rpm$, $R_L = 13.2\Omega$, $P_d = 2135W$), ($\omega_r = 1000rpm$, $R_L = 22\Omega$, $P_d = 1396W$) and ($\omega_r = 100rpm$, $R_L = 3.4\Omega$, $P_d = 807W$), and the SMR robust voltage control scheme and the SRM drive control schemes are normally operated, the measured power quality characteristics of the established SMR without ($W_h = 0$) and with ($W_h = 1.0$) current harmonic compensation are listed in Table 8.The results show that the fundamental and all other harmonic currents are all reduced and the efficiency of the SMR is increased accordingly by the harmonic compensation approach. Moreover, the line drawn power quality improvements at all cases are also obtained.

$P_d = 2.135kW$, $\omega_r = 1500rpm$, $R_L = 13.2\Omega$								
	PF	P_{ac} (kW)	THDi (%)	I_{a1} (Arms)	I_{a5} (Arms)	I_{a7} (Arms)	I_{a11} (Arms)	Efficiency (%)
$W_h = 0$	0.953	2.414	18.82	5.18	0.81	0.55	0.08	88.44
$W_h = 1.0$	0.968	2.362	10.33	5.01	0.42	0.31	0.04	90.39
$P_d = 1.396kW$, $\omega_r = 1000rpm$, $R_L = 22\Omega$								
$W_h = 0$	0.941	1.598	19.13	3.58	0.62	0.29	0.05	87.36
$W_h = 1.0$	0.951	1.579	11.01	3.03	0.28	0.18	0.04	88.41
$P_d = 0.807kW$, $\omega_r = 100rpm$, $R_L = 3.4\Omega$								
$W_h = 0$	0.935	0.959	23.47	2.16	0.44	0.25	0.08	84.15
$W_h = 1.0$	0.943	0.932	19.54	1.91	0.33	0.17	0.06	86.59

Table 8. The measured power quality characteristics under SRM drive active load at various power levels without and with current harmonic compensation

6.2 Switch-Mode Rectifier based EV battery charger
6.2.1 System configuration
A battery powered SRM drive for electric vehicle propulsion is shown in Fig. 23(a) (H.C. Chang & Liaw, 2009). In driving mode, the switches are set as: $S_m \to$ M and $S_d \to$ closed. The SRM (DENSEI company Japan) is rated as 4-phase, 8-6, 48V, 6000rpm, 2.3kW. The components S_b, D_b, L_b and C_d in Fig. 23(a) form a DC/DC boost converter. The nominal battery voltage is $V_b = 12 \times 4 = 48V$, it is boosted and establishes the DC-link voltage with $48V \leq V_{da} \leq 72V$. During demagnetization of each communication stroke, the winding energies can be directly sent back to the battery bank via the diodes D_1, D_3, D_5 and D_7.

In charging mode, the switches in Fig. 23(a) are set as: $S_m \to$ C and S_d permanently off. With the insertion of off-board part, a buck-boost SMR based charger is formed and drawn in Fig. 23(b) with the employed embedded motor drive components being highlighted. The diode D_e is added to avoid the short circuit of battery when Q_6 is turned on. The inductances of the first two motor windings are used as the input filter components during each half AC cycle. And the third motor winding inductance is employed as the energy storage component of the SMR.

The SMR control scheme shown in Fig. 23(b) consists of outer charging control scheme and inner current controlled PWM scheme. Initially, the battery is charged in constant current mode to let the batteries be charged under maximum current (0.25C/9.5A) until the condition of $V_b \geq 52V \, (=13V \times 4)$ reaches. Then the charging enters constant voltage floating mode.

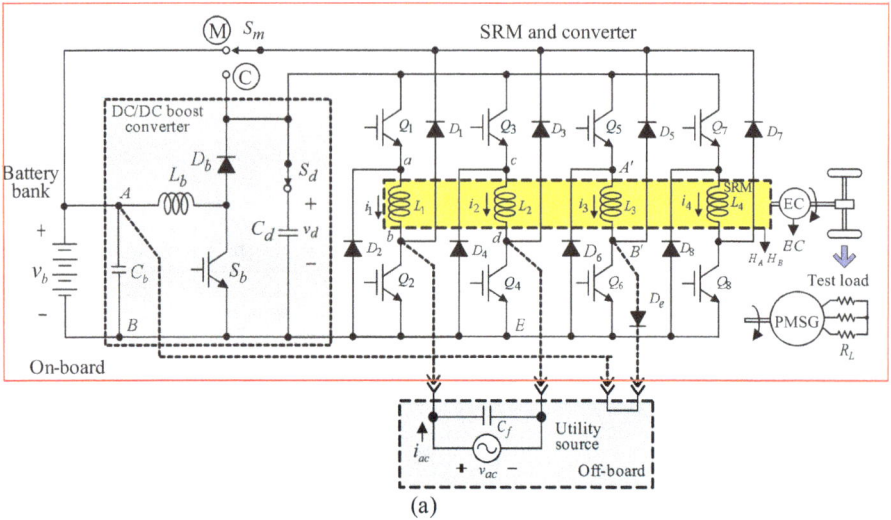

(a)

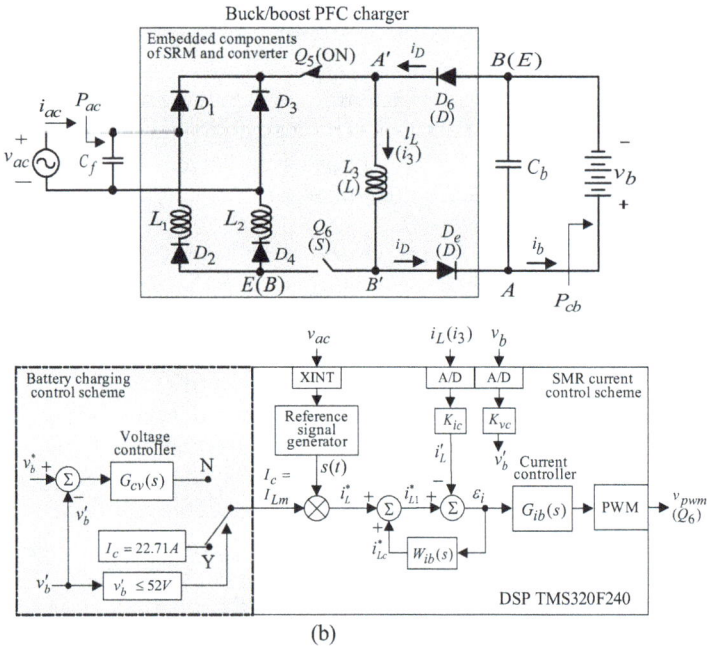

(b)

Fig. 23. A battery powered SRM drive with voltage boosting for electric vehicle propulsion: (a) system configuration; (b) schematic and control scheme of the formed on-board buck-boost SMR based battery charger in idle status

6.2.2 Performance evaluation

The derivation of circuit component ratings and the implementation affairs of the SRM drive shown in Fig. 23(a) and Fig. 23(b) can be referred to (H.C. Chang & Liaw, 2009). Some results concerning charging mode are observed here. Under the constant current charging mode with $I_{Lm} = I_c = 21.3A$ and switching frequency $f_s = 12.5kHz$, the measured i_L and its command i_L^* by PI control and robust control ($W_{ib} = 0.85$) are shown in Fig. 24(a) ($v_b = 47.65V$, $I_b = 9.45A$). The close inductor current tracking control is observed from the results. Fig. 24(b) shows the corresponding AC source voltage v_{ac} and current i_{ac} (PF = 0.989, $THD_i = 4.23$ %). Good line drawn waveform and power quality by the buck-boost SMR charger can be seen from the results.

To observe the effects of switching frequency on the SMR control performance, Table 9 lists measured performance parameters corresponding to the switching frequencies of $f_s = 12.5kHz$, $15kHz$, $7.5kHz$ and $2.5kHz$ ($W_{ib} = 0.85$). Some facts are observed from the results: (i) The current loop is normally operated at each switching frequency; and (ii) As the switching frequency becomes smaller, the inductor current will gradually become partial and then total discontinuous current mode (DCM) within the AC cycle. Accordingly, the power quality becomes worse.

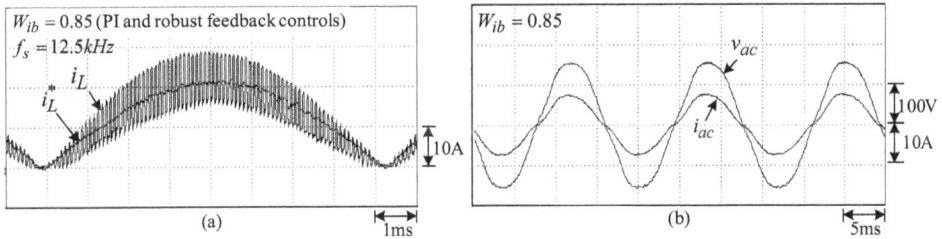

Fig. 24. Measured results of the buck-boost SMR based charger in constant-current charging mode at ($v_b = 47.65V$, $I_b = 9.45A$): (a) i_L and i_L^* by PI and robust controls ($W_{ib} = 0.85$, $f_s = 12.5kHz$); (b) v_{ac} and i_{ac} (PF=0.989, $THD_i = 4.23\%$)

f_s / Variables	2.5kHz	7.5kHz	12.5kHz	15kHz
P_{ac} (W)	579.7	563.4	547.8	560.6
I_{ac} (A)	5.56	5.15	5.11	5.14
P_b (W)	388.4	427.3	453.3	459.4
V_b (V)	46.32	47.12	47.65	47.73
I_b (A)	8.22	9.02	9.45	9.45
$\eta(\%)$	67.00	75.84	82.75	81.95
PF	0.972	0.981	0.989	0.991
THD_{iac} (%)	10.02	7.84	4.23	4.04

Table 9. Measured power quality parameters of the buck-boost SMR based charger under different switching frequencies

6.3 Flyback Switch-Mode Rectifier based auxiliary plug-in charger

6.3.1 System configuration

Fig. 25 shows a switched-reluctance generator (SRG) based DC microgrid distributed power system (Y.C. Chang & Liaw, 2011). The SRG establishes a 48V DC-link voltage, and a common 400V DC-grid is formed through a current-fed push-pull (CFPP) DC-DC converter. To preserve the microgrid power quality, a lead-acid battery energy storage system is equipped. The battery bank (48V) is interfaced to the common 400V DC-grid via a bidirectional buck-boost converter. The battery bank can also be charged from the utility grid by a flyback SMR based auxiliary plug-in charger. For the flyback converter employed in the SMR, three paralleled transformers are used to enlarge its power rating. The The specifications of the developed flyback SMR are given as: (i) AC input: 110V/60Hz; (ii) DC output: $V_o = V_b = 48\text{V}/300\text{W}$; and (iii) power factor: PF > 0.97. The flyback SMR is operated under discontinuous current mode using the charge-regulated PWM scheme developed in (Y.C. Chang & Liaw, 2009a). The turn-on time in is set as $t_{on} = dT_s = 9.2\mu s$, which is constant for the employed PWM switching scheme.

6.3.2 Performance evaluation for the flyback SMR based auxiliary plug-in charger

For the battery bank shown in Fig. 25, the constant charging current $I_b = 6\text{A}$ (i.e., $\hat{I}_{Sc} = 6 / K_{ic} = 6 / 1.107 = 5.42\text{A}$) is set. The measured ($v_{ac}$, i_{ac}) and (v_b, i_b) at steady state of $I_b = 6\text{A}$ during charging process are plotted in Figs. 26(a) and 26(b). The measured static characteristics under two charging currents ($I_b = 6\text{A}$ and 5A) are listed in Table 10. The results show that good charging characteristics with satisfactory line drawn power quality are obtained by the developed flyback SMR based auxiliary plug-in charger.

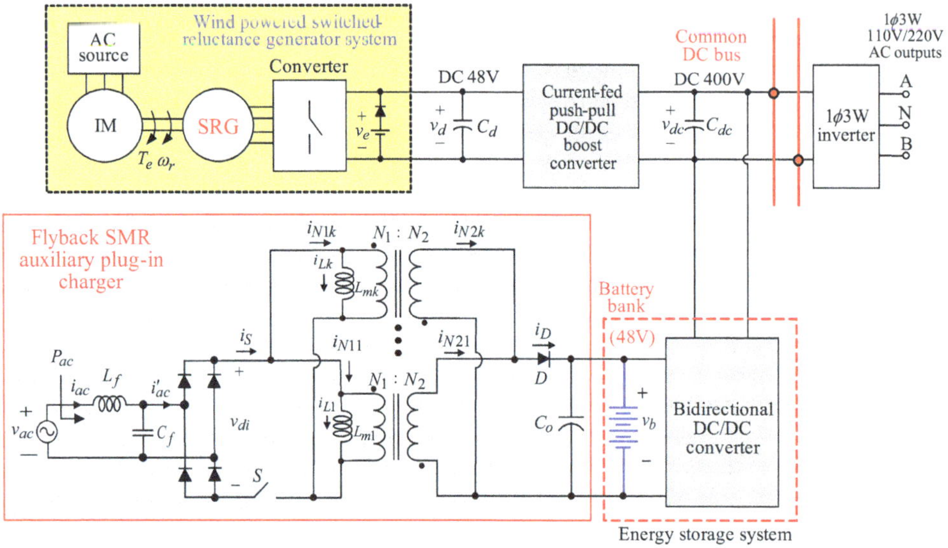

Fig. 25. System configuration of a switched-reluctance generator based DC micro-grid system with flyback SMR auxiliary plug-in charger

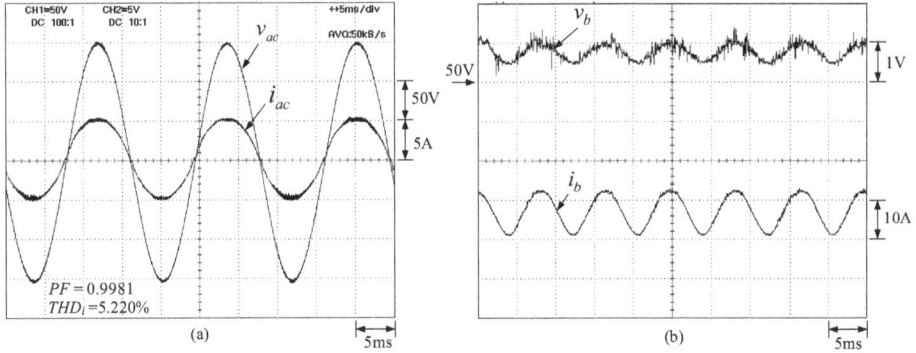

Fig. 26. Measured results of the developed flyback SMR based auxiliary plug-in charger under steady-state charging current of $I_b = 6$A : (a) (v_{ac} , i_{ac}); (b) (v_b , i_b)

I_b	5A	6A
P_b	265.28W	306.56W
P_{ac}	364.17W	415.98W
η	72.85%	73.70%
PF	0.9986	0.9981
THD_i	3.742%	5.220%

Table 10. Measured results of the developed flyback SMR based auxiliary plug-in charger at two charging current levels

7. Conclusions

This article has presented the basic issues of switch-mode rectifiers for achieving batter performance. The schematic type and control scheme should be properly chosen according to the specific application and the desired operation characteristics. The considering issues include input-output relative voltage levels, operation quadrant, galvanic isoltation, phase number, DCM or CCM operation, voltage mode or current mode control, dynamic control requirement, etc. In power circuit establishment, the ratings of circuit components and the ripples of energy storage components should be analytically derived, and accordingly, the constituted components are designed and implemented.

As to the control affairs, the sensed inductor current and output voltage should be filtered with suited low-pass cut-off frequencies. Then the basic feedack controllers are designed considring the desired perfromance and the effects of comtaiminated system noises. If more stringent control requirements are desired. The simple advanced control, such as the robust tracking error cancellation controls (Chai & Liaw, 2007; Y.C. Chang & Liaw, 2009a), can further be applied. Other possible affairs lie in the digital control with properly chosen sampling intervals, random switching, the considrations of DC-link ripple effects on the followed power stage, parallel opeartion to enlarge SMR ratings, etc.

In this article, the applications of various SMRs to PMSM drive, SRM drive, electric vehicle plug-in battery charger and microgrid plug-in battery charger were presented. The treatments of basic issues in power circuit and control scheme have been decribed. And their comparative performances have also been assessed to demonstrate the effectiveness of the introduced issues. It is believed that the cost-effective SMR fed power palnt with satisfactory performance can be achieved if the suggested basic issues can be considered.

8. References

Backman, N. & Wolpert, T. (2000). Simplified Single Stage PFC Including Peak Current Mode Control in a Flyback Converter, *Proceedings of IEEE Telecommunications Energy Conference*, pp. 317-324, ISBN 0-7803-6407-4

Badin, A. A. & Barbi, I. (2008). Unity Power Factor Isolated Three-Phase Rectifier with Split DC-Bus Based on the Scott Transformer. *IEEE Transactions on Power Electronics*, Vol.23, No.3, (May 2008), pp. 1278-1287, ISSN 0885-8993

Buso, S.; Spiazzi, G. & Tagliavia, D. (2000). Simplified Control Technique for High-Power-Factor Flyback Cuk and Sepic Rectifiers Operating in CCM. *IEEE Transactions on Industry Applications*, Vol.36, No.5, (September 2000), pp. 1413-1418, ISSN 0093-9994

Chai, J. Y. & Liaw, C. M. (2007). Robust Control of Switch-Mode Rectifier Considering Nonlinear Behavior. *IET Electric Power Applications*, Vol.1, No.3, (May 2007), pp. 316-328, ISSN 1751-8660

Chai, J. Y.; Ho, Y. H.; Chang, Y. C. & Liaw, C. M. (2008). On Acoustic Noise Reduction Control Using Random Switching Technique for Switch-Mode Rectifiers in PMSM Drive. *IEEE Transactions on Industrial Electronics*, Vol.55, No.3, (March 2008), pp. 1295-1309, ISSN 0278-0046

Chai, J. Y. & Liaw, C. M. (2009). Development Of a Switched-Reluctance Motor Drive with PFC Front-End. *IEEE Transactions on Energy Conversion*, Vol.24, No.1, (March 2009), pp. 30-42, ISSN 0885-8969

Chai, J. Y.; Chang, Y. C. & Liaw, C. M. (2010). On the Switched-Reluctance Motor Drive with Three-Phase Single-Switch-Mode Rectifier Front-End. *IEEE Transactions on Power Electronics*, Vol.25, No.5, (May 2010), pp. 1135-1148, ISSN 0885-8993

Chang, H. C. & Liaw, C. M. (2009). Development of a Compact Switched-Reluctance Motor Drive for EV Propulsion with Voltage Voosting and PFC Charging Capabilities. *IEEE Transactions on Vehicular Technology*, Vol.58, No.7, (March 2009), pp. 3198-3215, ISSN 0018-9545

Chang, Y. C. & Liaw, C. M. (2009). Design and Control for a Charge-Regulated Flyback Switch-Mode Rectifier. *IEEE Transactions on Power Electronics*, Vol.24, No.1, (January 2009), pp. 59-74, ISSN 0885-8993

Chang, Y. C. & Liaw, C. M. (2009). Development and Control of a Flyback Switch-Mode Rectifier with Enlarged Rating via Paralleled Transformers, *Proceedings of Taiwan Power Electronics Conference and Exhibition*, pp. 1287-1293

Chang, Y. C. & Liaw, C. M. (2011). A Flyback Rectifier with Spread Harmonic Spectrum. *IEEE Transactions on Industrial Electronics*, to appear in No.99, ISSN 0278-0046

Chen, H. C.; Li, S. H. & Liaw, C. M. (2004). Switch-Mode Rectifier with Digital Robust Ripple Compensation and Current Waveform Controls. *IEEE Transactions on Power Electronics*, Vol.19, No.2, (March 2004), pp. 560-566, ISSN 0885-8993

Das, P. & Moschopoulos, G. (2007). A Comparative Study of Zero-Current-Transition PWM Converters. *IEEE Transactions on Industrial Electronics*, Vol.54, No.3, (June 2007), pp. 1319-1328, ISSN 0278-0046

Dawande, M. S. & Dubey, G. K. (1996). Single-Phase Switch Mode Rectifiers, *Proceedings of IEEE Power Electronics, Drives and Energy Systems for Industrial Growth*, pp. 637-642, ISBN 0-7803-2795-0

Erickson, R. & Madigan, M. (1990). Design of a Simple High-Power-Factor Rectifier Based on the Flyback Converter, *Proceedings of IEEE Applied Power Electronics Conference and Exposition*, pp. 792-801,

Erickson, R. W. & Maksimovic, D. (2001). *Fundamentals of Power Electronics (2nd)*, Kluwer Academic Publishers, ISBN 0-7923-7270-0, Norwell Massachusetts

Farcas, C.; Petreus, D.; Simion, E.; Palaghita, N. & Juhos, Z. (2006). A Novel Topology Based on Forward Converter with Passive Power Factor Correction, *Proceedings of International Spring Seminar on Electronics Technology*, pp. 268-272, ISBN 1-4244-0551-3

Forest, F.; Laboure, E.; Meynard, T. A. & Huselstein, J. J. (2007). Multicell Interleaved Flyback Using Intercell Transformers. *IEEE Transactions on Power Electronics*, Vol.22, No.5, (September 2007), pp. 1662-1671, ISSN 0885-8993

Forest, F.; Laboure, E.; Gelis, B.; Smet, V.; Meynard, T. A. & Huselstein, J. J. (2009). Design of Intercell Transformers for High-Power Multicell Interleaved Flyback Converter. *IEEE Transactions on Power Electronics*, Vol.24, No.3, (March 2009), pp. 580-591, ISSN 0885-8993

Gataric, S.; Boroyevich, D. & Lee, F. C. (1994). Soft-Switched Single-Switch Three-Phase Rectifier with Power Factor Correction, *Proceedings of IEEE Applied Power Electronics Conference and Exposition*, pp. 738-744, ISBN 0-7803-1456-5

Hahn, J.; Enjeti, P. N. & Pitel, I. J. (2002). A New Three-Phase Power Factor Correction (PFC) Scheme Using Two Single-Phase PFC Modules. *IEEE Transactions on Industry Applications*, Vol.38, No.1, (October 2002), pp. 123-130, ISSN 0093-999

Hengchun, M.; Lee, F. C.; Boroyevich, D. & Hiri, S. (1997). Review of High Performance Three-Phase Power-Factor Correction Circuits. *IEEE Transactions on Industrial Electronics*, Vol.44, No.4, (August 1997), pp. 437-446, ISSN 0278-0046

Hsieh, Y. H. (2010) *Position Sensorless Permanent-Magnet Synchronous Motor Drive with Switch-Mode Rectifier Front-End*, Master Thesis, Department of Electrical Engineering, National Tsing Hua University, Taiwan, ROC

Huber, L.; Jang, Y. & Jovanovic, M. M. (2008). Performance Evaluation of Bridgeless PFC Boost Rectifiers. *IEEE Transactions on Power Electronics*, Vol.23, No.3, (February 2008), pp. 1381-1390, ISSN 1048-2334

Inoue, T.; Hamamura, S.; Yamamoto, M.; Ochi, A. & Sasaki, Y. (2008). AC-DC Converter Based on Parallel Drive of Two Piezoelectric Transformers. *Japanese Journal of Applied Physics*, Vol.47, No.5, (May 2008), pp. 4011-4014, ISSN 0021-4922

Jang, Y.; Dillman, D. L. & Javanovic, M. M. (2006). A New Soft-Switched PFC Boost Rectifier with Integrated Flyback Converter for Stand-By Power. *IEEE Transactions on Power Electronics*, Vol.21, No.1, (January 2006), pp. 66-72, ISSN 0885-8993

Kwon, J. M.; Kim, E. H.; Kwon, B. H. & Nam, K. H. (2008). High Efficiency Fuel Cell Power Conditioning System with Input Current Ripple Reduction. *IEEE Transactions on Industrial Electronics*, Vol.56, No.3, (March 2009), pp. 826-834, ISSN 0278-0046

Lamar, D. G.; Fernandez, A.; Arias, M.; Rodriguez, M.; Sebastian, J. & Hernando, M. M. (2007). Limitations of the Flyback Power Factor Corrector as a One-Stage Power

Supply, *Proceedings of IEEE Power Electronics Specialists Conference*, pp. 1343-1348, ISBN 0275-9306

Langeslag, W.; Pagano, R.; Schetters, K.; Strijker, A. & Zoest, A. (2007). VLSI Design and Application of A High-Voltage-Compatible SoC–ASIC in Bipolar CMOS/DMOS Technology for AC–DC Rectifiers. *IEEE Transactions on Industrial Electronics*, Vol.54, No.5, (October 2007), pp. 2626-2641, ISSN 0278-0046

Larouci, C.; Ferrieux, J. P.; Gerbaud, L.; Roudet, J. & Barbaroux, J. (2002). Control of a Flyback Converter in Power Factor Correction Mode: Compromise Between the Current Constraints and the Transformer Volume, *Proceedings of IEEE Applied Power Electronics Conference and Exposition*, Vol.2, pp. 722-727, ISBN 0-7803-7404-5

Lee, J. J.; Kwon, J. M.; Kim, E. H.; Choi, W. Y. & Kwon, B. H. (2008). Single-Stage Single-Switch PFC Flyback Converter Using a Synchronous Rectifier. *IEEE Transactions on Industrial Electronics*, Vol.55, No.3, (March 2008), pp. 1352-1365, ISSN 0278-0046

Li, S. H. & Liaw, C. M. (2003). Modelling and Quantitative Direct Digital Control for a DSP-Based Soft-Switching-Mode Rectifier. *IEE Proceedings of Electric Power Applications*, Vol.150, No.1, (January 2003), pp. 21-30, ISSN 1350-2352

Li, S. H. & Liaw, C. M. (2004). Paralleled DSP-based Soft Switching-Mode Rectifiers with Robust Voltage Regulation Control. *IEEE Transactions on Power Electronics*, Vol.19, No.4, (July, 2004), pp. 937-946, 0885-8993

Li, S. H. & Liaw, C. M. (2004). On the DSP-Based Switch-Mode Rectifier with Robust Varying-Band Hysteresis PWM Scheme. *IEEE Transactions on Power Electronics*, Vol.19, No.6, (November, 2004), pp. 1417-1425, ISSN 0885-8993

Li, S. H.; & Liaw, C. M. (2004). Development of Three-Phase Switch-Mode Rectifier Using Single-Phase Modules. *IEEE Transactions on Aerospace and Electronic Systems*, Vol.40, No.1, (January 2004), pp. 70-79, ISSN 0018-9251

Liaw, C. M.; Lin, Y. M.; Wu, C. H. & Hwy, K. I. (2000). Analysis, Design and Implementation of a Random Rrequency PWM Inverter. *IEEE Transactions on Power Electronics*, Vol.15, No.5, (September 2000), pp. 843-854, ISSN 0885-8993

Lu, D. D.; Cheng, D. K. & Lee, Y. S. (2003). Single-Switch Flyback Power-Factor-Corrected Ac/dc Converter with Loosely Regulated Intermediate Storage Capacitor Voltage, *Proceedings of 2003 International Symposium on Circuits and Systems*, pp. 264-267, ISBN 0-7803-7761-3

Lu, D. D.; Iu, H. H.; & Pjevalica, V. (2008). A Single-Stage AC/DC Converter with High Power Factor, Regulated Bus Voltage, and Output Voltage. *IEEE Transactions on Power Electronics*, Vol.23, No.1, (January 2008), pp. 218-228, ISSN 0885-8993

Mahdavi, M. & Farzanehfard, H. (2009). Zero-Current-Transition Bridgeless PFC without Extra Voltage and Current Stress. *IEEE Transactions on Industrial Electronics*, Vol.56, No.7, (July 2009), pp. 2540-2547, ISSN 0278-0046

Manh, D. C. & Guldner, H. (2006). High Output Voltage DC/DC Converter Based on Parallel Connection of Piezoelectric Transformers, *Proceedings of International Symposium on Power Electronics, Electrical Drives, Automation and Motion*, pp. 625-628, ISBN 1-4244-0193-3

Matsui, K.; Yamamoto, I; Kishi, T.; Hasegawa, M.; Mori, H. & Ueda, F. (2002). A Comparison Of Various Buck-Boost Converters and Their Application to PFC, *Proceedings of IEEE 2002 28th Annual Conference of the Industrial Electronics Society*, Vol.1, pp. 30-36, ISBN 0-7803-7474-6

Mishra, S. K.; Fernandes, B. G. & Chatterjee, k. (2004). Single Stage Single Switch AC/DC Converters with High Input Power Factor and Tight Output Voltage Regulation, *Proceedings of IEEE 2004 30th Annual Conference of the Industrial Electronics Society*, pp. 2690-2695, ISBN 0-7803-8730-9

Mohan, N.; Undeland, T. M. & Robbins, W. P. (2003). *Power Electronics Converters, Applications and Design (3rd)*, John Wiley & Sons, ISBN 0-4712-2693-9, New York

Newsom, R. L.; Dillard W. C. & Nelms, R. M. (2002). Ditigal Power-Factor Correction for a Capacitor-Charging Power Supply. *IEEE Transactions on Industrial Electronics*, Vol.49, No.5, (October 2002), pp. 1146-1153, ISSN 0278-0046

Olivera, D. S.; Barreto, L.; Antunes, F.; Silva, M.; Queiroz, D. L. & Rangel, A. R. (2009). A DCM Three-Phase High Frequency Semi-Controlled Rectifier Feasible for Power WECS Based on a Permanent Magnet Generator, *Proceedings of IEEE Power Electronics Conference*, pp. 1193-1199, ISBN 2175-8603

Orabi, M. & Ninomiya, T. (2003). Nonlinear Dynamics of Power-Factor-Correction Converter. *IEEE Transactions on Industrial Electronics*, Vol.50, No.6, (December 2003), pp. 1116-1125, ISSN 0278-0046

Papanikolaou, P. N.; Tatakis, C. E. & Kyritsis, A. C. (2005). Design of a PFC AC/DC Flyback Converter for Low Voltage Applications, *Proceedings of 2005 European Conference on Power Electronics and Applications*, pp. 1-10, ISBN 90-75815-09-3

Reis, M. M.; Soares, B.; Barreto, L.; Freitas, E.; Silva, C. E. A.; Bascope, R. T. & Olivera, D. S. (2008). A Variable Speed Wind Energy Conversion System Connected to the Grid for Small Wind Generator, *Proceedings of IEEE Applied Power Electronics Conference and Exposition*, Vol.1, pp. 751-755, ISBN 1048-2334

Rikos, E. J. & Tatakis, E. C. (2005). Single-Stage Single-Switch Isolated PFC Converter with Non-Dissipative Clamping. *IEE Proceedings Electric Power Applications*, Vol.152, No.2, (March 2005), pp. 166-174, ISSN 1350-2352

Sangsun, K. & Enjeti, P. N. (2002). A Parallel-Connected Single Phase Power Factor Correction Approach with Improved Efficiency, *Proceedings of IEEE Applied Power Electronics Conference and Exposition*, Vol.1, pp. 263-269, ISBN 0-7803-7404-5

Sangwon, L. & Sewan, C. (2010). AThree-Phase Current-Fed Push-Pull DC-DC Converter with Active Clamp for Fuel Cell Applications, *Proceedings of IEEE Applied Power Electronics Conference and Exposition*, pp. 1934-1941, ISBN 978-1-4244-4782-4

Shah, J. & Moschopoulos, G. (2005). Three-Phase Rectifiers with Power Factor Correction, *Proceedings of 2005 Canadian Conference on Electrical and Computer Engineering*, pp. 1270-1273, ISBN 0-7803-8885-2

Singh, B.; Singh, B. P.; & Dwivedi, S. (2006). Performance Comparison of High Frequency Isolated AC-DC Converters for Power Quality Improvement at Input AC Mains, *Proceedings of IEEE International Conference on Power Electronics, Drives and Energy Systems*, pp. 1-6, ISBN 0-7803-9772-X

Singh, B. & Singh, S. (2010). Single-Phase Power Factor Controller Topologies for Permanent Magnet Brushless DC Motor Drives. *IET Power Electronics*, Vol.3, No.2, (March 2010), pp. 147-175, ISSN 1755-4535

Tang, W.; Jiang, Y. H.; Verghese, G. C. &Lee, F. C. (1993). Power Factor Correction with Flyback Converter Employing Charge Control, *Proceedings of IEEE Applied Power Electronics Conference and Exposition*, pp. 293-298, ISBN 0-7803-0983-9

Tanitteerapan, T. & Mori, S. (2001). An Input Current Shaping Technique for PFC Flyback Rectifier by Using Inductor Voltage Detection Control Method, *Proceedings of IEEE Region 10 International Conference on Electrical and Electronic Technology*, Vol.2, pp. 799-803, ISBN 0-7803-7101-1

Ting, Q. H. & Lehman,B. (2008). Coupled Input-Series and Output-Parallel Dual Interleaved Flyback Converter for High Input Voltage Application. *IEEE Transactions on Power Electronics*, Vol.23, No.1, (January, 2008), pp. 88-95, ISSN 0885-8993

Tse, K. K.; Chung, H. S. –H.; Hui, S. Y. R. & So, H. C. (2000). A Comparative Investigation on the Use of Random Modulation Schemes for DC/DC Converters. *IEEE Transactions on Industrial Electronics*, Vol.47, No.2, (April 2000), pp. 253-263, ISSN 0278-0046

Uan-Zo-li, A.; Lee, F. C. & Burgos, R. (2005). Modeling, Analysis and Control Design of Single-Stage Voltage Source PFC Converter, *Proceedings of IEEE Industry Applications Conference*, Vol.3, pp. 1684-1691, ISBN 0-7803-9208-6

Ueda, A.; Ito, Y.; Kurimoto, Y. & Torii, A. (2002). Boost Type Three-Phase Diode Rectifier Using Current Resonant Switch, *Proceedings of IEEE Power Conversion Conference*, Vol.1, pp. 13-18, ISBN 0-7803-7156-9

Wang, C. M. (2010) *Development of Switched-Reluctance Motor Drive with Three-Phase Switch-Mode Rectifier Front-End*, Master Thesis, Department of Electrical Engineering, National Tsing Hua University, Taiwan, ROC

Wang, K.; Lee, F. C. & Boroyevich (1994). Soft-Switched Single-Switch Three-Phase Rectifier with Power Factor Correction, *Proceedings of IEEE Applied Power Electronics Conference and Exposition*, pp. 738-744, ISBN 0-7803-1456-5

Wolfs, P. & Thomas, P. (2007). Boost Rectifier Power Factor Correction Circuits with Improved Harmonic and Load Voltage Regulation Responses, *Proceedings of IEEE Power Electronics Specialists Conference*, pp. 1314-1318, ISBN 0275-9306

Youssef, N. B. H.; Al-Haddad, K. & Kanaan, H. Y. (2008). Implementation of a New Linear Control Technique Based on Experimentally Validated Small-Signal Model of Three-Phase Three-Level Boost-Type Vienna Rectifier. *IEEE Transactions on Industrial Electronics*, Vol.55, No.4, (April 2008), pp. 1666-1676, ISSN 0278-0046

Zhang, R.; Lee, F. C. & Boroyevich, D. (2000). Four-Legged Three-Phase PFC Rectifier with Fault Tolerant Capability, *Proceedings of IEEE Power Electronics Specialists Conference*, Vol.1, pp. 359-364, ISBN 0275-9306

Zheng, Y. & Moschopoulos, G. (2006). Design Considerations for a New AC-DC Single Stage Flyback Converter, *Proceedings of IEEE Applied Power Electronics Conference and Exposition*, pp. 400-406, ISBN 0-7803-9547-6

12

Battery Charger with Power Quality Improvement

Dylan Dah-Chuan Lu
School of Electrical and Information Engineering
The University of Sydney
Australia

1. Introduction

Battery storage has long been used in many applications such as portable multimedia player, mobile phone, portable tool, laptop computer, emergency exit sign, uninterruptible power supply and transportation auxiliary supply. Owing to the advancement of material science and packaging technologies, newer batteries with higher energy density and reliability have been produced. Batteries are now being used in higher power applications such as electric vehicles (EV), renewable energy systems and microgrid. Examples of high power batteries are Lithium-ion and Zinc-Bromine which are rated at kilo-watt range and mega-watt range respectively (Roberts, 2009). At such high power level, these batteries will have significant impact on the grid.

Power quality is one of major impacts to the grid when these high power batteries are charging. Since the battery is working at DC level, rectification (i.e., AC to DC conversion) is required. For the traditional design of rectifiers, for example diode-capacitor rectifier and phase-controlled thyristor rectifier, the current drawn by these battery chargers causes high total harmonic distortion (THD) and poor power factor (PF). This results in heating of transformer and cables and tripping of circuit breakers (Bass et al., 2001; Gomez & Morcos, 2003). Switching AC/DC converters with active power factor correction (PFC) is able to reduce THD and improve PF effectively. This technique has been applied to battery charger for electric vehicles (Mi, et al., 2003).

Power electronics enables intelligent control of battery charger such that the power quality of the grid can be improved. One example is the vehicle-to-grid (V2G) reactive power compensation. A mathematical analysis of an electric vehicle charger based on a full-bridge inverter/rectifier and a half-bridge bi-directional dc/dc converter is presented (Kisacikoglu, et al. (2001)). The charger is able to handle different PQ conditions at different operation modes. A relationship between dc link ripple and reactive power flow direction is also derived. The analysis shows that while the charger can achieve reactive power compensation, one has to set a limit on the four-quadrant power transfer of the charger due to the stresses on the components.

Active power filters (APF) have been developed primarily to compensate the harmonic and reactive power components of line current generated by the nonlinear loads and to improve the power quality of the grid (El-Haborouk et al., 2000; Singh et. al., 1999). Current-fed type APF uses an inductor for reactive power compensation while voltage-fed type APF uses a capacitor.

It is possible to integrate an APF function into a battery charger. For example, an uninterruptible power supply (UPS) with integrated APF capability has been proposed (C.-C.& Manjrekar, 2005; Wu & Jou, 1995). In both cases, a voltage-fed type APF is used and the battery is connected in parallel with the capacitor. For UPS, the battery is stationary; it always stays with the power supply system and operates in stand-by mode for emergency situation. For other battery charger such as EV charger, the battery is non-stationary; it only connects to the charger when it needs to be charged. Therefore the configuration where the capacitor is installed in parallel with the battery terminals, as suggested earlier (C.-C.& Manjrekar, 2005; Wu & Jou, 1995), cannot be used. It is because when the battery is removed from the charging terminals, a potential difference between the capacitor and the battery will be created. The worst scenario happens when next time the battery is depleted and putting back to the charging terminals to recharge, it has lower voltage than that of the capacitor. If one simply connects the battery to the charging terminals, a surge discharge current from the capacitor would occur. This will damage the circuitry, connectors and battery due to this high current.

This chapter presents a simple and improved battery charger system with power quality improvement function. It solves the aforementioned parallel capacitor-battery issue by a proposed equal charge concept. And the circuit is simplified by integrating a two-switch dc/dc converter with a full-bridge converter/inverter and using only one inductor. The chapter is organized as follows. The proposed charger and its operation will be described in Section 2. The equal charge concept will be explained in Section 3. Design considerations of the charger will be given in Secion 4. Simulation results will be reported in Section 5 followed by conclusions in Section 6.

2. Proposed battery charger with power quality enhancement

2.1 Circuit description
Fig. 1 shows the proposed battery charger system with power factor correction (PFC) capability. It consists of an integrated full-bridge inverter/converter (S1 to S4), an inductor L_o, a capacitor C_o and a switch (S5). As compared to the two inductors and six switches used in the converter introduced in (Kisacikoglu, et al. (2001)), the proposed converter has fewer component counts. In summary, when charging the battery, it operates as a buck converter with input current shaping for PFC and when discharging the battery, it operates as a boost converter with reactive power compensation.

2.2 Circuit operation and analysis– battery charging
The converter operates as a buck (step-down) converter during charging mode. As the input voltage v_{in} has a general expression of $V_m \sin \omega t$, its value changes from 0V to V_m. Therefore current will flow from the grid to the converter to charge the battery only when the input voltage is higher than the battery voltage V_{batt}. The current flow is controlled by the power switches S1 to S4 operating at high switching frequency and shaped by the inductor L_o. Now suppose at certain instant the input voltage at node A is higher than node B and $v_{in} > V_{batt}$ is satisfied, S1 and S4 turn on to allow input current to flow into the circuit, as shown in Fig. 2(a). The voltage applied across the inductor is $v_{in} - V_{Co}$ and the inductor is charging linearly with a rate equals

$$\frac{di_{Lo}}{dt} = \frac{v_{in} - V_{Co}}{L_o} \tag{1}$$

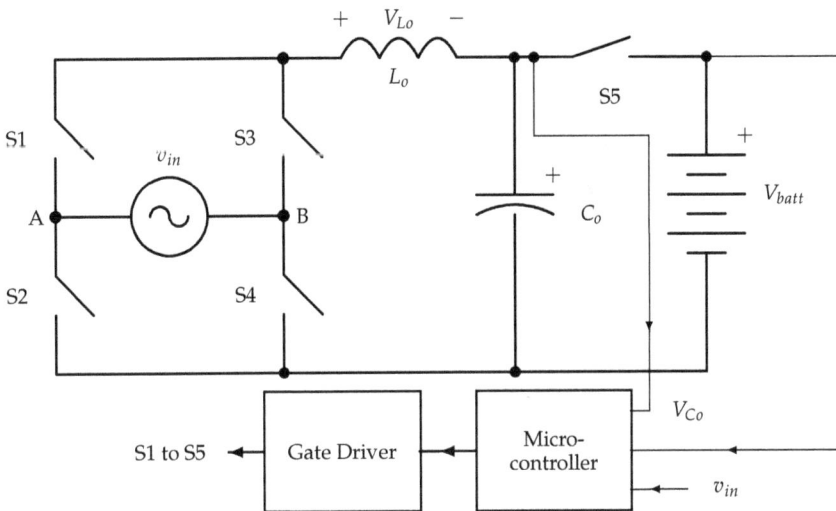

Fig. 1. Proposed battery charger with power quality improvement functions.

The inductor L_o and capacitor C_o ensure the high frequency current ripple to the battery has reduced. After certain interval, we need to reset the inductor to prevent it from saturation. There are a number of ways to discharge the inductor current:

1. Turn on S1 and S2 to provide a free wheeling path with $V_{Lo} = -V_{batt}$

2. Turn on S3 and S4 to provide a free wheeling path with $V_{Lo} = -V_{batt}$

3. Turn on S2 and S3 to provide a discharging path for the inductor with $V_{Lo} = -V_{batt} - v_{in}$

Fig. 2(b) shows the current path for option 2 as described above while Fig. 2(c) shows option 3. Comparing to the first two options, the third option with input voltage putting in series with the battery for discharging of inductor current would achieve a faster response in case a sudden decrease in the output loading condition occurs. But at the same time, comparing to options 1 and 2, option 3 will cause more switching losses because all four switches have to be in action during this mode while for the other two options only three switches are involved. Similiarly for opposite half of the line cycle, i.e., node B has higher potential than node A, and if $v_{in} > V_{batt}$ is satisfied, S2 and S3 turn on to allow input current to flow into the circuit and charge the inductor. For the inductor discharging period, again there are three options to continue the inductor current flow similiar to the previous description.

Apart from charging the battery, the converter in this mode has to provide power factor correction (PFC) according to the international standard such as IEC 61000-3-2 when the converter draws more than 75W of power from the ac line. To achieve PFC, L_o is the main component to shape the input current and it can work in all three modes to achieve the PFC function, i.e. discontinuous conduction mode (DCM), boundary conduction mode (BCM) or continuous conduction mode (CCM). For DCM operation, the input current is shaped

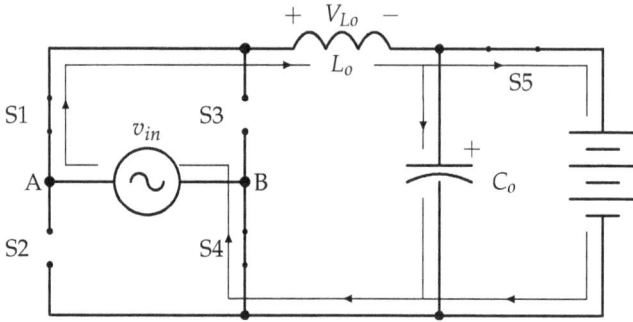

(a) Charging of inductor

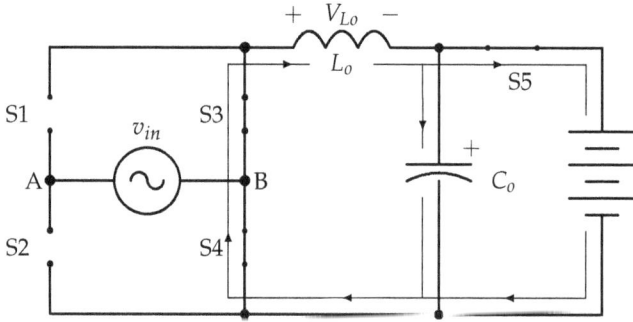

(b) Discharging of inductor through S3 & S4

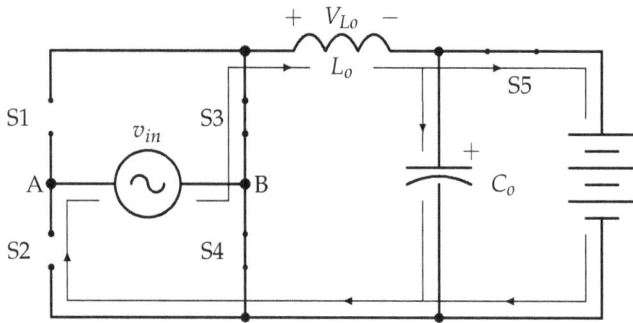

(c) Discharging of inductor through S2 & S3

Fig. 2. Equivalent circuits for charging mode operation

automatically as it is given by

$$i_{in.avg}(t) = \frac{D^2 T_s [v_{in}(t) - V_{batt}]}{2L_o} \tag{2}$$

We can observe from (2) that the average input current, $i_{in,avg}(t)$, of the buck operating mode follows in phase and closely with input voltage v_{in} if duty cycle D is constant but it is negatively offseted so there is a distortion in the current. And the lower the V_{batt}, the better the power factor (PF) this mode can achieve as the conduction angle increases with reducing battery voltage for a given input line voltage. For BCM and CCM operations, the input current has to be sensed and controlled to follow the shape of the input voltage to achieve high PF. A peak current mode controller can be used for both BCM and CCM operations.

2.3 Circuit operation and analysis – battery discharging

The converter operates as a boost (step-up) converter during discharging mode. Unlike the buck mode operation, current from the battery can always flow to the ac line (or grid), v_{in}, via the boost action. Switch S5 remains closed in this mode and the inductor L_o serves as energy storage element as well as shaping the current for reactive power compensation. Suppose at certain instant the potential at node A is higher than node B. To charge L_o, we can turn on either switches pair S1/S2 or switches pair S3/S4. We will discuss what the difference is by switching particular pair soon but suppose at this point we select pair S3/S4. Once the switches pair is turned on, a voltage equals $V_{Lo} = -V_{batt}$ is applied across the inductor. Therefore the inductor current flows from the battery to the switches with a rate equals

$$\frac{di_{Lo}}{dt} = \frac{-V_{batt}}{L_o} \tag{3}$$

Note that the capacitor C_o does help to reduce the current ripple on the battery and serve to provide a fast response as usually the battery is of slow response, in particular to sudden surge of current demand. After a certain interval, the inductor has to be reset. To reset L_o, a voltage which equals $V_{Co} - v_{in}$ needs to apply across the inductor and its rate of discharge equals

$$\frac{di_{Lo}}{dt} = \frac{V_{Co} - v_{in}}{L_o} \tag{4}$$

To achieve this, S3 is turned off and S1 is turned on with S4 remains closed, as shown in Fig. 3(b). From this transition we can observe that two switches are involved. If S1 and S2 were turned on first previously for the inductor charging, then S2 will turn off and S4 will turn on with S1 remains closed for the discharging interval. Hence there are still two switches involved.

Apart from discharging the battery, the converter in this mode is able to improve the power quality of the grid. To achieve high power factor, L_o is the main component to shape the input current and it can work again in all three modes to achieve the PFC function, i.e. DCM, BCM and CCM. The inductor current waveform is shown in Fig. 4. It works in DCM operation. The instantaneous average inductor current is equal to the instantaneous average input current, which is given by

$$i_{ac}(t) = \frac{V_{batt}}{2L_o} d(t)[d(t) + d_1(t)]T_s \tag{5}$$

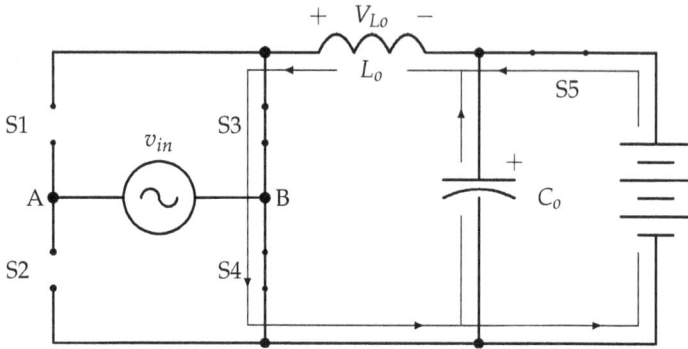

(a) Charging of inductor

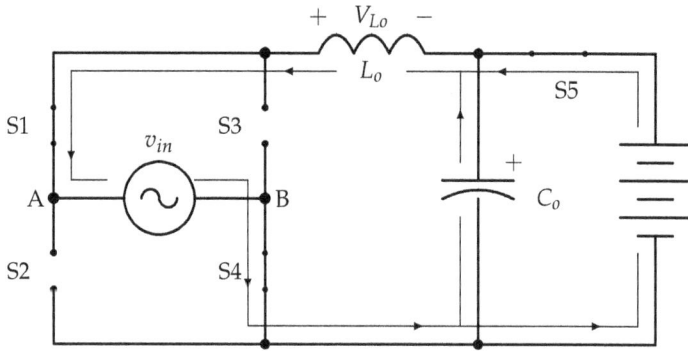

(b) Discharging of inductor

Fig. 3. Equivalent circuits for discharging mode operation

Using voltage-second balance on L_o, the inductor discharging period, $d_1(t)$, is expressed as

$$d_1(t) = \frac{V_{batt}}{v_{in}(t) - V_{batt}} d_1(t) \qquad (6)$$

Therefore the instantaneous average input current has the final form as follows

$$\bar{i_{ac}}(t) = \frac{V_{batt} T_s}{2L_o} d^2(t) \cdot \frac{v_{in}(t)}{v_{in}(t) - V_{batt}} \qquad (7)$$

As it can be seen from (7), the last term of on the right hand side is non-linear due to the time-varying input voltage $v_{in}(t)$. Hence the duty cycle has to vary in response to this varying voltage to maintain high power factor. In order to achieve unity power factor, i.e., $\bar{i_{ac}}(t) =$

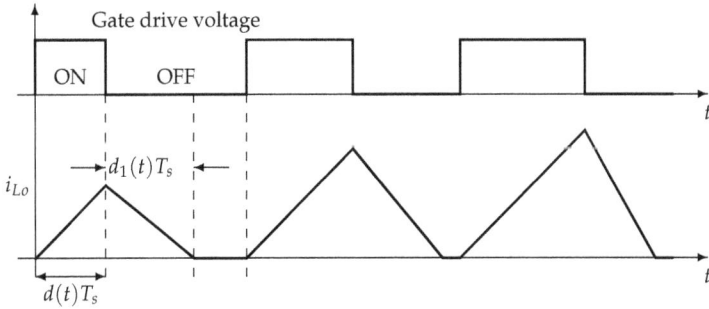

Fig. 4. Gate drive voltage and inductor current waveforms at the discharging mode.

$v_{in}(t)/R_{eq}$ where R_{eq} is the equivalent resistance of the converter, the duty cycle needs to be expressed as

$$d(t) = \sqrt{\frac{1}{v_{in}(t) - V_{batt}}} \qquad (8)$$

Fig. 5 shows the required duty cycle over a half line period, given $V_{batt}=72V$ and $v_{in}(t)=340\sin(\pi t)$. It can be observed from the same figure that at the beginning and near the end of the line cycle, the duty cycle goes to infinity which is impossible in reality. This happens because, as it can be seen from (6) that, when $v_{in}(t) \leq V_{batt}$ is satisfied the boost mode is not operating as the input voltage is not enough to reset the inductor. To prevent this from happening, the simpliest way is to have duty cycle equals zero but this will create current distortion as will be shown from simulation later on.

3. Equal charge concept

3.1 Principle of operation
The equal charge concept is to deal with non-stationary battery where unequal charges are found on the battery and the battery charger. Since the battery will be plugged into the battery charger for charging in parallel, a potential difference between the battery and charger will create a large current (or surge current) flow at the instant of connection. The magnitude of this current depends on the state-of-charge on the battery; it ranges from slightly discharged to fully discharged depending on what type of battery the machine uses and how the battery is used. The larger the potential difference between the two points, the larger the magnitude of the surge current will flow. This could damage the connectors or cabling and other devices in which the surge current passes through. In order to deal with this problem, the equal concept is introduced. The idea is simple: it is to bring the potential difference between the charger and the battery to zero before the electrical connection and as a result there will be no surge current flow. The procedure is explained, with the aid of Fig. 1, as follows: Firstly, switch S5 has to be open. Then the battery is connected to the charger. The voltage of the battery, V_{batt} is sensed and sent to the micro-controller as a reference voltage. The micro-controller compares this reference voltage and the voltage on C_o. If $V_{Co} > V_{batt}$ is satisfied, then the charger operates in discharging mode (i.e., boost mode) to take away some charge on C_o until its voltage is equal to V_{batt}. If $V_{Co} < V_{batt}$ is satisfied, then the charger operates in charging

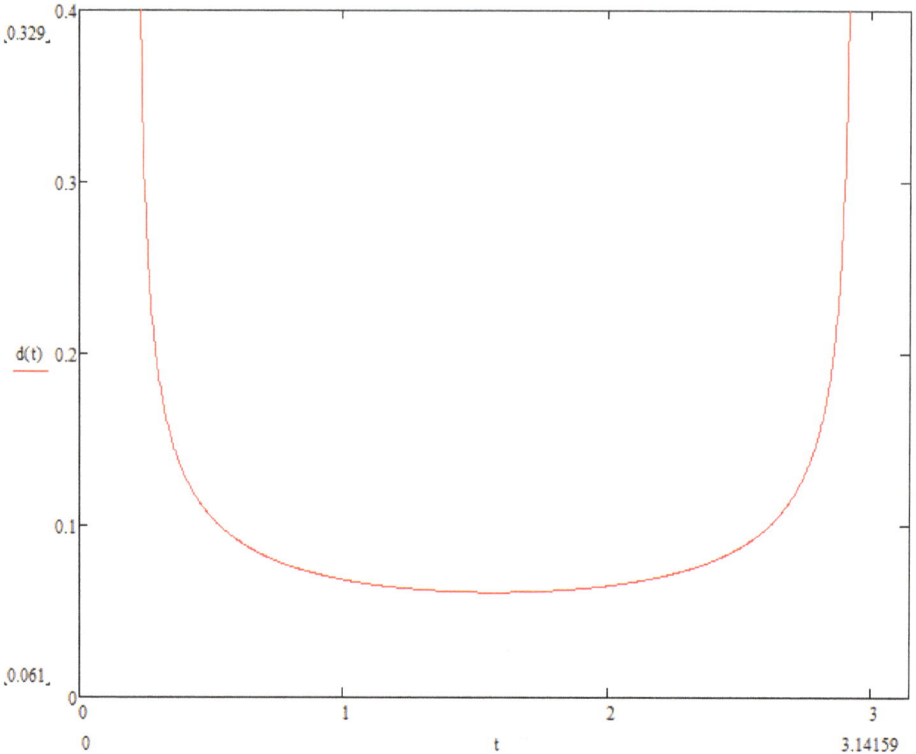

Fig. 5. Duty cycle changes in response to varying input voltage in discharging mode.

mode (i.e., buck mode) to charge up C_o until its voltage is equal to V_{batt}. After the equal charge process has completed, S5 can be closed and the charger operation can be continued.

3.2 Simulation results

To illustrate the effectiveness of the equal charge concept, a series of simulations by a free circuit simulator LTSpiceIV (Linear Technology) have been carried out. The test circuit is based on an open-loop bi-directional dc/dc converter (S1, S3 and L2) with an output capacitor C1, a switch S2 mimiking a relay and a battery model (L1, R1 and V2), as shown in Fig. 6. In the first scenario, the effect of large surge current discharge if the voltages are not matached is shown. The power converter is not in operation but the output voltage at C1 is at 200V initially. The battery before the connection is at 72V. When S2 is closed, a large surge current with peak value at 630A is produced, as shown in Fig. 7. Although it only lasts for a short duration and the current drops to 0A after 5ms, it has enough energy to create a spark at the switch and could melt and wear the joint after a period of operation. In the second scenario, the equal charge concept is realized. The duty cycles of the bi-directional converter are set to produce an output of 74V before connecting the battery, as shown in Fig. 8. After S2 is closed, a maximum current of only 10A is generated and it drops to 5A to charge the battery. Of course one can use the same voltage as that of the battery before battery connection but in this scenario the capability of controlling of the output voltage to control the charging current

PULSE(0 10 100m 100n 100n 10m 1)

V3   **V4**

PULSE(0 10 0 100n 100n 3.6u 10u)

L2   L1

10µH   100nH

V1   SW   C1   R1

200   S3   V5   2200µF   0.1

S1   SW   S2   V2

                                        72

.ic v(n005)=200   PULSE(0 10 0 100n 100n 6.4u 10u)

.tran 0 200m 0 50u

.model sw vswitch(ron=0.1, von=5)

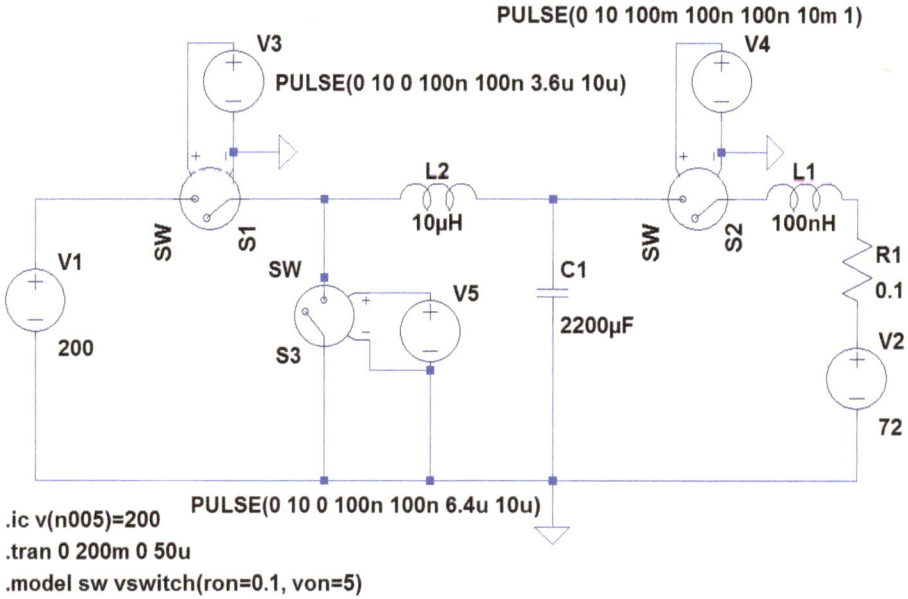

Fig. 6. Test circuit to show equal charge concept.

is demonstrated. Since this converter works in an open-loop condition, the output voltage drops to 73V. But in a closed-loop control condition, the voltage can be regulated. Despite of this, it is still possible to observe that the charging current is limited by the resistance of the relay S2 and the internal resistance of the battery. Therefore the charging current is obtained as $(73-72)V/(0.1+0.1)\Omega = 5A$. In order to control the charging current, another option is to replace the relay by a power bipolar junction transistor (BJT). However, apart from additional control circuit which increases the cost, the necessary saturation voltage across the collector and emittor also imposes a minimum charging current to the battery. Therefore the control is simplified by the proposed voltage control of the converter.

4. Design considerations

4.1 Practical power stage design

The simplified circuit diagram as shown in Fig. 1 and the previous explanation of the battery charging and discharging are based on the fullly bi-directional current blocking capability of the four switches. However both MOSFET and IGBT which are the commonly used power devices cannot be used as such because of their partial current blocking capacbility. For MOSFET, although current can flow in both directions, the body diode cannot be externally controlled. For IGBT, the current can only flow in one direction. In order to achieve the said bi-directional current blocking capability with minimum component count, Fig. 9 shows a practical power stage design. It consists of four IGBTs and two MOSFETs. Two IGBTs (i.e. S1/S2 and S3/S4) form a pair to allow bi-directional current blocking capability. Since an IGBT pair is inserted in each leg and controls the current direction at any time of the input voltage, one can simply use a MOSFET for the other part of each leg instead of another IGBT pair to save part and cost. With this circuit configuration it is convenient to control the current

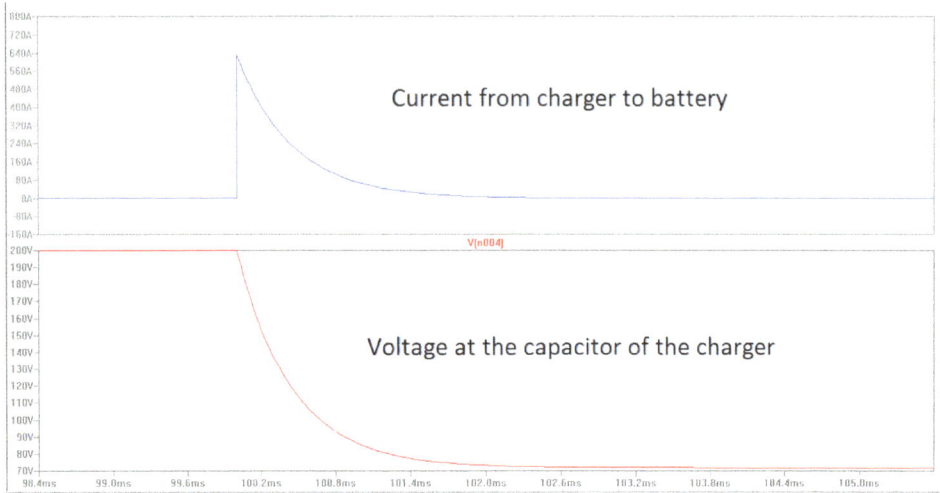

Fig. 7. Output waveforms without equal charge implemented.

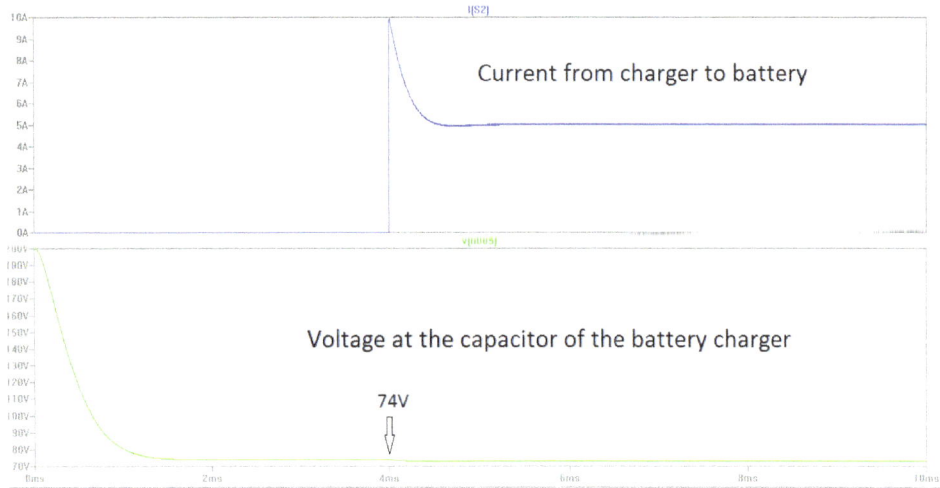

Fig. 8. Output waveforms with equal charge implemented.

direction. During the battery charging phase, switches S2, S4 and S6 (body diode) will come to operation during the positive half line cycle and S1, S3 and S5 (body diode) for the negative half cycle. During the battery discharging phase, switches S1, S3 and S6 will come to operation during the positive half line cycle while S2, S4 and S5 for the negative half cycle.

4.2 Current control

For the control stage as shown in Fig. 1, only voltage sensing at the output of the converter and the battery are shown. However, in practice it is useful to include a current sensing device to measure the battery current. It is because as mentioned earlier different battery has different charging/discharging profile. The current sensor and the micro-controller will work together

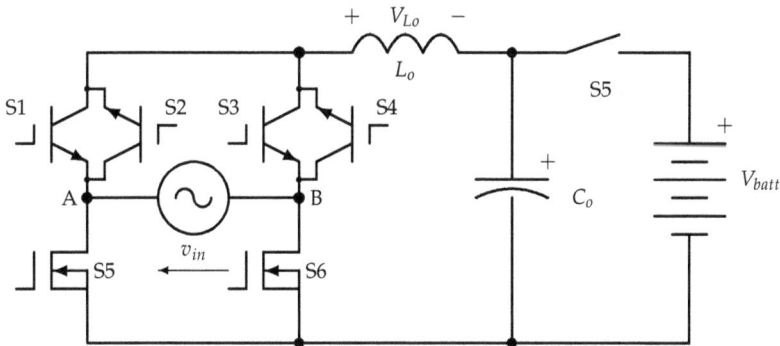

Fig. 9. Practical design of the power stage of the proposed battery charger.

to maintain the battery current profile, as well as to achieve current protection functions such as over-current protection (OCP). A current sensor will also help implement some superfast charging algorithm by constantly monitoring the battery current.

4.3 Need of paralleled capacitance
The capacitor placed in parallel with the battery is important, not only for achieving a faster response to compensate for the slow response of the battery current in response to fast current demand, but also for reducing the current ripple on the battery. Fig. 10(a) shows the battery charger in charging mode without a paralleled capacitor. Under this situation, the inductor current equals the battery charging current and so as the ripple current. This ripple current will heat up the battery due to the internal resistance. This results in shorter battery lifetime. Fig. 10(b) shows the same working condition but a capacitor is placed in parallel with the battery. Now, because the capacitor absorbs the current ripple from the inductor current, the current ripple on the battery is dramatically reduced. Note that this capacitor not only does reduce the high frequency ripple but also the low frequency ripple. But in general the high frequency ripple, especially when the inductor works in DCM, causes more losses than that of the low frequency ripple.

5. Simulation results of the proposed charger

To verify the power quality capability of the proposed battery charger, a series of simulations based on Fig. 1 have been carried out. The circuit parameters used are as follows: $L_o = 50\mu H$, $C_o = 2200\mu F$, $v_{in}=240Vrms$, $V_{batt} = 72V$ with 1Ω internal resistance and switching frequency $f_s = 100kHz$. The inductor is working in DCM. Fig. 11 shows the key waveforms of the converter when it is working in charging (buck) mode over a few line period. It can be seen that there is an input distortion due to the fact that the input voltage is less than the battery voltage for some time of the line cycle, as discussed eariler. However, when input voltage is greater than the battery voltage, the inductor L_o shapes the input current to follow the input voltage. Therefore it is possible for the proposed converter to achieve high power factor in buck operating mode. Fig. 12 shows the key waveforms of the converter when it is working in discharging (boost) mode over a half line period. It has been explained in Section 2.3 that

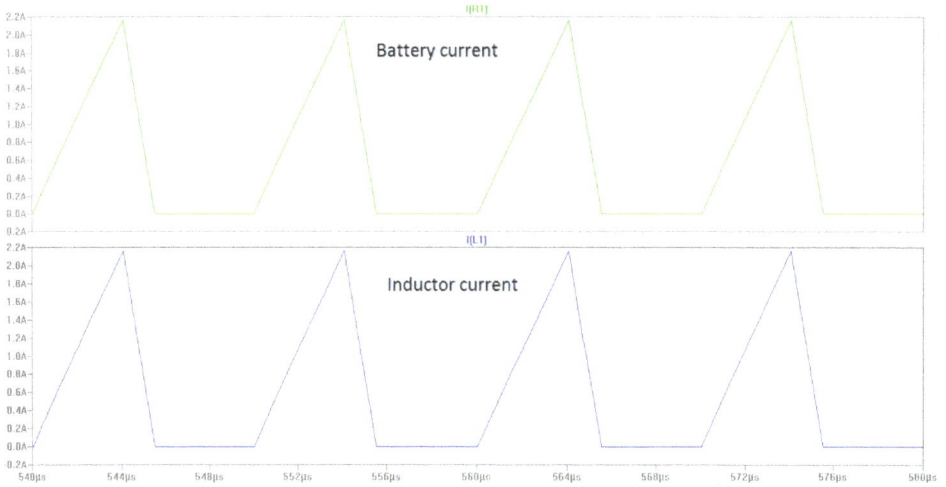

(a) Without a capacitor

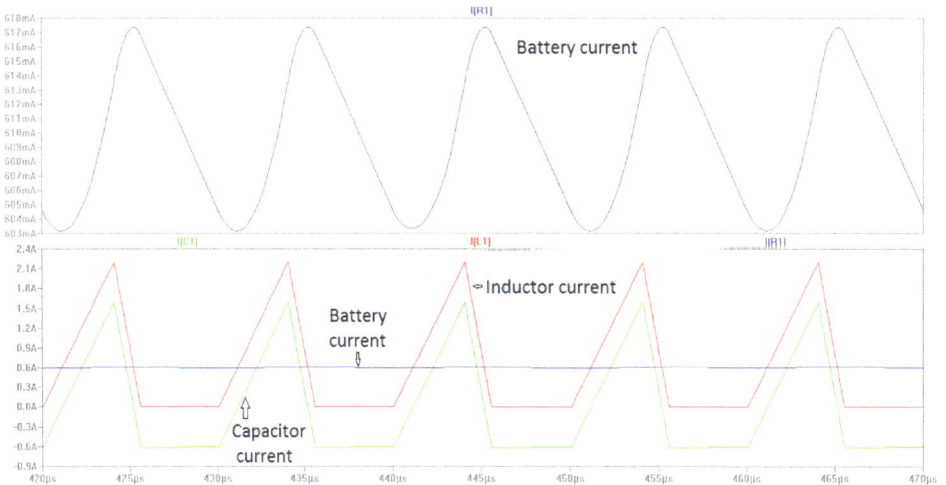

(b) With a capacitor

Fig. 10. The importance of capacitor placed in parallel with the battery.

for boost mode operation current can always flow to the ac line. However, the input voltage is lower than the battery voltage for some time during each line cycle and the boost mode requires the input ac voltage is larger than the battery voltage for proper operation, otherwise the inductor L_o cannot be reset and it will saturate the inductor during this period. Therefore to protect the inductor from saturation the converter stops operation until v_{in} is equal to and above V_{batt}. In the current setting, the input voltage angle, θ, for this equality is calculated as

$$\theta = \sin^{-1}\frac{V_{batt}}{V_m} = \sin^{-1}\frac{72}{340} = 12°$$

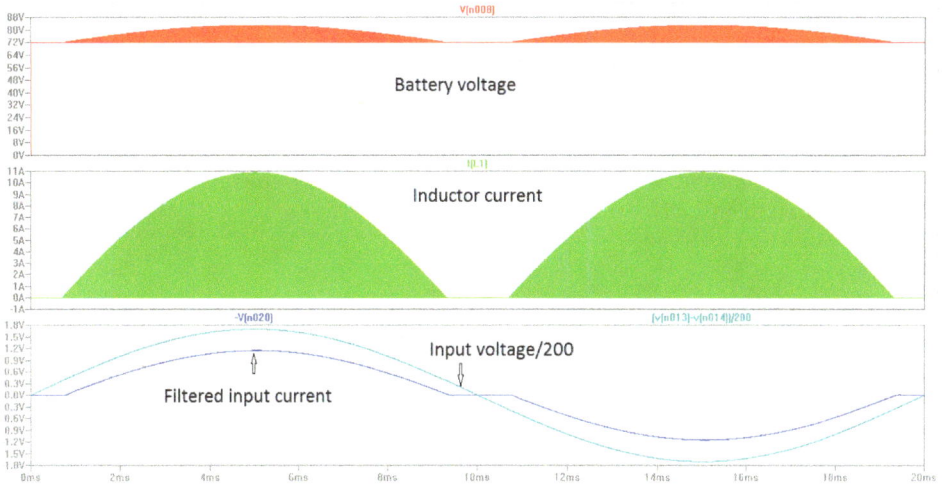

Fig. 11. Power factor correction (PFC) performed by the proposed charger during charging (buck) mode.

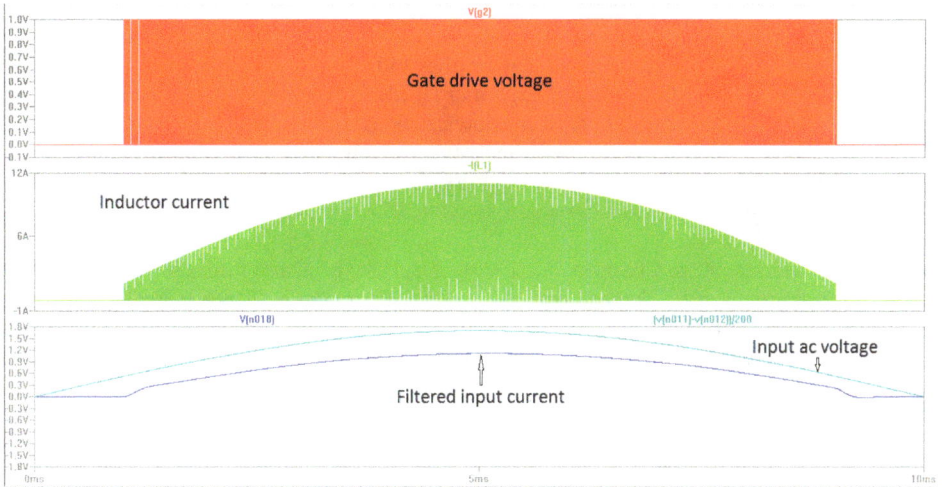

Fig. 12. The proposed charger achieves high power factor during discharging (boost) mode.

where V_m is the peak input ac voltage. The time expression of the input voltage angle is hence $12°/180° \times 10ms = 0.67ms$. The introduction of dead time of input current will certainly create current distortion but again with the good input current shaping capability during the conduction period the converter still can deliver high quality current to the ac line. In order to further improve the input current quality during the discharging mode, the converter can run in BCM or CCM and use hysteresis current control or average current mode control to track the input current to follow the input voltage. In such case, variable switching frequency will be used.

6. Conclusion

In this chapter, a simple and integrated battery charger with power quality improvement is presented. It can draw and deliver high quality current from and to the ac line by the input current shaping technique on the inductor. Circuit operation analysis and design considerations of the power converter have been discussed. A equal charge concept together with practical implementation has been explained. Simulation results have been reported to verify the theoretical analysis.

7. References

Bass, R., Harley, R., Lambert, F., Rajasekaran, V. & Pierce, J. (2001). Residential harmonic loads and ev charging, *Proceedings of Power Engineering Society Winter Meeting*, IEEE, Atlanta, USA, pp. 803–808.

C.-C., Y. & Manjrekar, M. (2005). A reconfigurable uninterruptible power supply system for multiple power quality applications, *Proceedings of IEEE Applied Power Electronics Conference and Exposition*, IEEE, Milwaukee, USA, pp. 1824–1830.

El-Haborouk,M., Darwish,M. & Mehta, P. (2000). Active power filters: A review, *IEE Proceedings, Electric Power Applications* Vol. 147(No. 5): 403–413.

Gomez, J. & Morcos, M. (2003). Impact of ev battery chargers on the power quality of distribution systems, *IEEE Transactions on Power Delivery* Vol. 18(No. 3): 975–981.

Kisacikoglu,M., Ozpineci, B. & Tolbert, L. (2010). Effects of v2g reactive power compensation on the component selection in an ev or phev bidirectional charger, *Proceedings of IEEE Energy Conversion Congress and Exposition*, IEEE, Knoxville, USA, p. 870=876.

Mi, N., Sasic, B. andMarshall, J. & Tomasiewicz, S. (2003). A novel economical single stage battery charger with power factor correction, *Proceedings of IEEE Applied Power Electronics Conference and Exposition*, IEEE, USA, pp. 760–763.

Roberts, B. (2009). Active power filters: A review, *IEEE Power and Energy Magazine* Vol. 7(No. 4): 32–41.

Singh, B., Al-Haddad, K. & Chandra, A. (1999). A review of active filters for power quality improvement, *IEEE Transactions on Industrial Electronics* Vol. 46(No. 5): 960–971.

Wu, J.-C. & Jou, H.-L. (1995). A new ups scheme provides harmonic suppression and input power factor correction, *IEEE Transactions on Industrial Electronics* Vol. 42(No. 6): 629–635.

Permissions

The contributors of this book come from diverse backgrounds, making this book a truly international effort. This book will bring forth new frontiers with its revolutionizing research information and detailed analysis of the nascent developments around the world.

We would like to thank Gregorio Romero Rey and Mᵃ Luisa Martinez Muneta, for lending their expertise to make the book truly unique. They have played a crucial role in the development of this book. Without their invaluable contribution this book wouldn't have been possible. They have made vital efforts to compile up to date information on the varied aspects of this subject to make this book a valuable addition to the collection of many professionals and students.

This book was conceptualized with the vision of imparting up-to-date information and advanced data in this field. To ensure the same, a matchless editorial board was set up. Every individual on the board went through rigorous rounds of assessment to prove their worth. After which they invested a large part of their time researching and compiling the most relevant data for our readers. Conferences and sessions were held from time to time between the editorial board and the contributing authors to present the data in the most comprehensible form. The editorial team has worked tirelessly to provide valuable and valid information to help people across the globe.

Every chapter published in this book has been scrutinized by our experts. Their significance has been extensively debated. The topics covered herein carry significant findings which will fuel the growth of the discipline. They may even be implemented as practical applications or may be referred to as a beginning point for another development. Chapters in this book were first published by InTech; hereby published with permission under the Creative Commons Attribution License or equivalent.

The editorial board has been involved in producing this book since its inception. They have spent rigorous hours researching and exploring the diverse topics which have resulted in the successful publishing of this book. They have passed on their knowledge of decades through this book. To expedite this challenging task, the publisher supported the team at every step. A small team of assistant editors was also appointed to further simplify the editing procedure and attain best results for the readers.

Our editorial team has been hand-picked from every corner of the world. Their multi-ethnicity adds dynamic inputs to the discussions which result in innovative outcomes. These outcomes are then further discussed with the researchers and contributors who give their valuable feedback and opinion regarding the same. The feedback is then collaborated with the researches and they are edited in a comprehensive manner to aid the understanding of the subject.

Apart from the editorial board, the designing team has also invested a significant amount of their time in understanding the subject and creating the most relevant covers. They scrutinized every image to scout for the most suitable representation of the subject and create an appropriate cover for the book.

The publishing team has been involved in this book since its early stages. They were actively engaged in every process, be it collecting the data, connecting with the contributors or procuring relevant information. The team has been an ardent support to the editorial, designing and production team. Their endless efforts to recruit the best for this project, has resulted in the accomplishment of this book. They are a veteran in the field of academics and their pool of knowledge is as vast as their experience in printing. Their expertise and guidance has proved useful at every step. Their uncompromising quality standards have made this book an exceptional effort. Their encouragement from time to time has been an inspiration for everyone.

The publisher and the editorial board hope that this book will prove to be a valuable piece of knowledge for researchers, students, practitioners and scholars across the globe.

List of Contributors

Marco Mauri and Luisa Frosio
Politecnico di Milano, Italy

Gabriele Marchegiani
MCMEnergyLab s.r.l, Italy

Nopporn Patcharaprakiti, Krissanapong Kirtikara and Juttrit Thongpron
King Mongkut's University of Technology Thonburi, Bangkok, Thailand
Rajamangala University of Technology Lanna, Chiang Mai, Thailand

Khanchai Tunlasakun, Dheerayut Chenvidhya, Anawach Sangswang, Veerapol Monyakul and Ballang Muenpinij
King Mongkut's University of Technology Thonburi, Bangkok, Thailand

Vasanth Reddy Bathula and Chitti Babu B.
MIC College of Technology, NIT Rourkela, India

Belkacem Mahdad
Department of Electrical Engineering, Biskra University, Algeria

Dorin Sarchiz, Mircea Dulau and Daniel Bucur
Petru Maior University, Romania

Ovidiu Georgescu
Electrica Distribution and Supply Company, Romania

Juan de Dios Sanz-Bobi, Jorge Garzón-Núñez, Roberto Loiero and Jesús Félez
Research Centre on Railway Technologies – CITEF, Universidad Politécnica de Madrid, Spain

Alexis Polycarpou
Frederick University, Cyprus

Krisda Yingkayun
Rajamangala University of Technology Lanna, Thailand

Suttichai Premrudeepreechacharn
Chiang Mai University, Thailand

Belkacem Mahdad and K. Srairi
Department of Electrical Engineering, Biskra University, Algeria

Sonia Pinto, J. Fernando Silva, Filipe Silva and Pedro Frade
DEEC; Instituto Superior Técnico, TULisbon, Cie3 – Centre for Innovation in Electrical and Energy Engineering, Portugal

C. M. Liaw and Y. C. Chang
National Tsing Hua University, National Chung Cheng University, Taiwan

Dylan Dah-Chuan Lu
School of Electrical and Information Engineering, The University of Sydney, Australia